Next-Generation Electrical Insulating Materials

Next-Generation Electrical Insulating Materials

Low-Carbon, Eco-Friendly, and Sustainable Solutions

Edited by Jun-Wei Zha

WILEY-VCH

Editor

Prof. Jun-Wei Zha
North China Electric Power University
Beijing 102206
P. R. China

Cover Image: Generated with AI using ChatGPT-4O

Typesetting: Straive, Chennai, India
Printing and Binding: CPI Group (UK) Ltd, Croydon, CR0 4YY

Library of Congress Card No.: applied for

British Library Cataloguing-in-Publication Data A catalogue record for this book is available from the British Library.

Bibliographic information published by the Deutsche Nationalbibliothek
The Deutsche Nationalbibliothek lists this publication in the Deutsche Nationalbibliografie; detailed bibliographic data are available on the Internet at <http://dnb.d-nb.de>.

Print ISBN: 978-3-527-35570-9
ePDF ISBN: 978-3-527-85406-6
ePub ISBN: 978-3-527-85405-9
oBook ISBN: 978-3-527-85407-3

C9783527355709_240226

Contents

Preface

As indispensable and critical components in the electrical and electronic engineering field, the performance and reliability of insulating materials directly dictate the safe operation, efficiency, and service life of electrical equipment. Against the global backdrop of addressing climate change and advancing the goals of the Paris Agreement, the transition toward Low-Carbon, Eco-Friendly, and Sustainable (LES) practices in this field has become the prevailing development trend for insulating materials. However, traditional insulating material systems face challenges throughout their entire lifecycle, including resource consumption, environmental pollution, and recycling difficulties, which urgently demand comprehensive solutions. The development of green insulating materials, characterized by high performance and low environmental impact, has evolved from a technical exploration to an urgent need and an inevitable choice for supporting the sustainable development of the power industry.

Currently, relevant books on the direction of LES development still concentrate on insulating gases, with lesser discussion on insulating solids and liquids. However, insulating solids and liquids are also essential. The significance of this book lies in its systematic and comprehensive description of what has been achieved in recent years in the context of the Paris Agreement for all types of electrical insulating materials. It covers a wide range of materials, including gases, solids, and liquids, with a particular focus on six key categories: plastics, resins, rubbers, paper, oils, and gases, and provides in-depth analysis of milestone breakthroughs in their transition toward greener alternatives.

This book covers a wide range of topics rather than merely introducing a single insulating material. It is an excellent guide and reference for improving the existing properties of LES insulating materials and facilitating their application, design, and development. This book can be used both as an introductory study for graduate students and as a reference for researchers and engineers. It is effective in helping students and researchers in a wide range of disciplines and fields gain insights into the new milestones and sustainable development of electrical insulating materials.

1

Overview of Development of Insulating Materials

Wenye Zhang[1] *and Jun-Wei Zha*[1,2]

[1] *University of Science and Technology Beijing, School of Chemistry and Biological Engineering, No. 30 Xueyuan Road, Haidian District, Beijing 100083, P. R. China*
[2] *North China Electric Power University, State Key Laboratory of Alternate Electrical Power System with Renewable Energy Sources, School of Electrical and Electronic Engineering, No. 2 Beinong Road, Huilongguan, Changping District, Beijing 102206, P. R. China*

1.1 Introduction

Electrical insulating materials are the foundational components that enable safe and efficient operation of power and electronic systems. By preventing unwanted current flow, insulators allow voltages to be maintained and electric fields to be controlled within equipment. Over the past two centuries, the field of insulating materials has evolved dramatically, encompassing solids such as ceramics, polymers, resins, and paper, liquids such as mineral oils, and gases such as sulfur hexafluoride (SF_6), air [1–6]. Historically, the selection of insulating materials has been a key driver for advancements in electrical technology. In the earliest days of telegraphy and power distribution (mid-19th century), engineers relied on natural materials like cotton, silk, paper, rubber, and oils – materials that were readily available but often exhibited limited durability [6–8]. As electrical systems evolved toward higher voltages and more demanding operating environments, the limitations of these early insulating materials became increasingly apparent: Many materials age, degrade, crack, or absorb moisture over time, thereby weakening their performance [9, 10]. This has driven continuous innovation in insulating materials, closely intertwined with the progress in electrical engineering itself. Each generation of insulating materials has addressed the shortcomings of its predecessors and opened up new possibilities for system design, such as enabling higher voltage transmission or reducing equipment size. Therefore, the development of insulating materials can be viewed as a synergistic evolution alongside electrical infrastructure.

In modern times, polymer materials such as plastics and resins, liquid materials such as mineral oil, synthetic esters, and silicone oil, and gas materials such as SF_6 have dominated many electrical devices due to their excellent fundamental properties, including dielectric performance, mechanical properties, chemical stability, and thermal properties [8, 11–15]. However, some new challenges that were

Next-Generation Electrical Insulating Materials: Low-Carbon, Eco-Friendly, and Sustainable Solutions, First Edition. Edited by Jun-Wei Zha.

not fully considered in earlier times have also emerged: environmental sustainability and lifecycle management of these materials have become key concerns [16–18]. In the context of urgent climate agreements, insulating materials now face new challenges and requirements: they must not only exhibit outstanding electrical performance but also remain low-carbon, eco-friendly, and sustainable throughout their entire lifecycle. Many traditional insulating materials, including cross-linked thermosetting polymers, insulating oils, and certain gas or vacuum-based insulating materials, pose recycling or environmental risks when discarded or leaked [19–21]. Additionally, as reliability requirements increase, materials capable of self-monitoring/warning of health status, self-adapting, or even self-healing damage are gaining favor [22–28]. These considerations are driving a new round of research into next-generation insulating materials, which must not only be excellent in performance but also possess low-carbon, eco-friendly, sustainable, and smart characteristics. This chapter will review this evolutionary journey, from the earliest natural insulating materials to the latest low-carbon, eco-friendly, and smart solutions. We will see how early engineers relied on natural materials such as paper, rubber, glass, and oil, and how the emergence of synthetic polymers, specialty fluids, and smart insulating materials revolutionized electrical insulation technology.

1.2 Historical Evolution of Insulating Materials

1.2.1 Early Insulating Materials in Electrical Engineering

The development of electrical insulating materials is closely related to the development of electrical technology. Early researchers and engineers experimented with any material that could separate and protect conductors. Up to about 1925, naturally occurring products such as bitumen, natural rubber, mica sheets, cotton cloth, silk, paper, wood, and ceramics were the standard insulators, as shown in Table 1.1 [32, 33]. For example, cotton and silk tape were used to wrap early wire conductors. Mica, a mineral that can be split into thin flexible sheets, was prized for its high dielectric strength and heat resistance in applications like commutators and early capacitors. Oil-impregnated paper became a widely used insulating material in insulated cables and transformer windings due to its ability to improve dielectric properties by removing air [8]. Even today, some high-voltage direct current (HVDC) cables still use impregnated paper or paper-polypropylene laminates as insulators, although they are gradually being replaced by polymer materials [34, 35]. These materials are widely used because they are simple to obtain and show high electrical impedance, but their performance under electrical and thermal stress is very limited.

One notable natural insulator was natural rubber, derived from latex, which was one of the first insulating materials used for electrical wires in the early 19th century [32]. Unvulcanized rubber was tried as early as 1810, followed by vulcanized rubber after Charles Goodyear's patent in 1844. Natural rubber softens easily at high temperatures and becomes brittle at low temperatures. The vulcanization process greatly improved the stability of rubber by adding cross-links (sulfur bonds)

Table 1.1 Early commonly used insulating materials, their properties, and applications.

Era (approx.)	Common insulating materials	Key properties and applications
1800s (early)	**Glass, porcelain:** for telegraph and early power line insulators; very durable, high dielectric constant (~6), but brittle and heavy [6]. **Cloth (cotton, silk):** wrapped on wires and coils; flexible but moisture-sensitive [9]. **Shellac, wax, and asphalt:** coatings on windings and cables; provided dielectric protection but could soften with heat [29]. **Mica:** used in motors, commutators, and early capacitors; excellent dielectric strength and temperature endurance [3].	1850s telegraph lines using glass insulator knobs on poles. First undersea cable (1851) with gutta-percha insulation.
1900~1920	**Natural rubber:** used for wire insulation and cable jackets by late 1800s; soft and prone to cracking unless vulcanized [30, 31]. **Oil-impregnated paper:** introduced for high-voltage cables/transformers; dielectric strength: up to ~10–20 kV/mm when thoroughly impregnated [8].	1910s transformers with oil-paper insulation; rubber-insulated wiring in early homes (rubber often wrapped in cloth for protection).

between polymer chains, marking one of the first instances of chemically modified insulating material to enhance performance [30, 36]. Additionally, another crucial early insulator was gutta-percha, a resinous polymer obtained from the Palaquium gutta tree. In 1843–1847, it was introduced as a wire insulating material, especially for submarine telegraph cables. Gutta-percha is a thermoplastic material that can be heated and coated onto the surface of wires, then hardened to form a tough and waterproof coating. It remained the primary insulating material for submarine cables for decades, including the first transatlantic cable in 1858. However, gutta-percha was expensive, available only from Southeast Asia. And it tended to degrade (becoming brittle) when exposed to oxygen and sunlight over time.

In overhead power lines and early electrical equipment, ceramics and glass were essential insulating materials [37]. By the 1850s, glass insulators were used on telegraph lines, and porcelain, a type of ceramic made from clay, was later found to offer superior mechanical strength and better resistance to weathering. Porcelain insulators were widely used to support high-voltage lines and busbars due to their excellent dielectric strength and weather resistance [6, 38]. These inorganic insulators are essentially permanent, as they do not undergo significant aging, but they are heavy and prone to brittleness. Thus, although polymers have replaced many insulation tasks, ceramics remain crucial in areas with high requirements for stability in extreme environments, such as substation pillar insulators and bushings.

By 1920, the electrical industry had identified a range of natural insulating materials and combinations such as paper oil systems and rubber fabric composites. However, the pursuit of improved performance including higher breakdown voltage, greater temperature resistance, and reduced aging was constrained by the inherent limitations of these natural materials. This laid the foundation for the introduction of synthetic materials.

1.2.2 Advent of Synthetic Insulating Materials

From the late 1920s to the 1940s, the field of insulating materials underwent a revolution with the emergence of synthetic polymers and resins. The first major breakthrough was the synthesis of phenolic resin by Bakelite in 1907, also known as Bakelite plastic [39]. It is regarded as the first completely synthetic plastic in human history, marking the beginning of the polymer era. In the electrical industry, Bakelite and similar phenolic resins were rapidly adopted as insulating varnishes and molded parts in the 1910s to 1930s. For example, phenolic laminates made by impregnating paper or fabric with phenolic resin and curing it were used as switchboard panels, tube sockets, and connector bases. These applications fully utilized the excellent electrical insulation properties and heat resistance of Bakelite, which could withstand temperatures of approximately 150 °C. By the 1930s, phenolic resin had largely replaced shellac and asphalt for impregnating high-voltage coils, providing more durable insulation material for transformer and motor windings [40].

Following phenolic resin, other synthetic thermosetting resins were subsequently developed. The first were urea-formaldehyde resin and melamine-formaldehyde resin (1930s), which offered superior electrical properties and lighter colors. Urea-formaldehyde resin is commonly used for appliance insulation and socket components, while melamine-formaldehyde resin, with its better moisture resistance, is used for high frequency insulation. Furthermore, in the field of rubber and elastomers, researchers have synthesized synthetic rubber that can replace natural rubber. Neoprene (polychloroprene) was invented in 1931 as one of the first oil-resistant rubber insulators. Styrene-butadiene rubber (SBR) became widely available in the late 1930s as a rubber substitute for wires and cables [32]. These synthetic elastomers could operate at higher temperatures and had improved aging characteristics compared to natural gum rubber.

More importantly, in the late 1930s, thermoplastic polymers began to be used in insulating materials. Polyvinyl chloride (PVC) was first synthesized in 1872, but it was in 1926 that Waldo Semon at B.F. Goodrich discovered how to plasticize PVC to make it flexible and processable. By 1935–1936, PVC insulating materials were being tried on wires and cables. However, early PVC had higher dielectric constant and dissipation factor, which limited it to low voltages despite improvements in compounding. Moreover, polyethylene (PE) was accidentally discovered in 1933 by researchers at Imperial Chemical Industries (Fawcett and Gibson) through the high pressure polymerization of ethylene. During World War II, PE was used in coaxial cables for radar and radio because of its excellent high frequency insulation properties, including a low dielectric constant of around 2.3 and low dielectric loss. By the late 1940s, commercial PE insulating materials began to be used in radar cables and some power cables. Early PE had a relatively low melting point (approximately 105 °C), but it exhibited extremely strong chemical inertness and high dielectric strength (approximately 25–30 kV/mm). In addition, polytetrafluoroethylene (PTFE), famously known as Teflon, was discovered accidentally by Roy Plunkett at DuPont in 1938. PTFE has unmatched chemical inertness, extremely low dielectric constant (approximately 2.0) and loss, and can withstand temperatures up to 260 °C, making it the gold standard insulating material for high-temperature and high frequency applications. Its discovery was another key moment in polymer science [41, 42].

During the early stages of synthetic material development, liquid and gas insulating materials also made innovative progress (Table 1.2). In the 1930s, polychlorinated biphenyls (PCBs) were introduced as synthetic insulating oils, with higher flash points and dielectric constants than mineral oils, and were adopted as insulating media for transformers and capacitors. PCB-based insulating oil enabled capacitors to achieve smaller sizes and higher efficiency, as its higher dielectric constant allowed for greater capacitance per unit volume. However, PCBs were later found to be highly toxic and environmentally persistent, leading to their gradual phase-out in the 1970s. Another milestone was the synthesis of SF_6 as in the early 20th century (first prepared in 1900) and its recognition as an exceptional dielectric gas. By the late 1940s, SF_6 was being studied for use in high-voltage equipment, though its widespread adoption had not yet taken hold.

This era established the pivotal role of polymers in electrical engineering. The successive development of numerous synthetic insulating materials significantly expanded the design possibilities for electrical equipment. Improvements in dielectric strength and thermal stability enabled equipment to achieve miniaturization and higher reliability. For example, in motors and transformers, enameled wires made of copper and coated with synthetic resin paint such as polyvinyl alcohol or polyester replaced traditional insulation materials like silk and shellac. This substitution allowed the operating temperature limit to increase from 90 °C to 130 °C or higher. Overall, by 1950, the foundation for modern insulation systems had been established: combinations of natural and synthetic materials were widely adopted, but there was an increasing trend toward fully synthetic solutions designed for specific requirements.

1.2.3 High-Performance Insulating Materials and Composite Systems

From the 1950s to the 1970s, polymer science experienced significant development and maturation, leading to the emergence of high-performance insulating materials capable of withstanding harsh environments such as high mechanical stress and elevated temperatures. During this period, many insulation systems that are still in use today were developed, and polymers invented in the 1930s and 1940s were gradually improved.

Table 1.2 Summarizes the new synthetic insulating materials introduced up to the middle of this century.

Material	Type	Key properties and impact
Phenolic resin	Thermoset solid	Heat resistant (150 °C) and good electrical properties; enabled molded electrical parts (switches, coil forms). First synthetic polymer insulator, replacing natural shellac and wood in many components.
PVC	Thermoplastic solid	1910s transformers with oil-paper insulation; rubber-insulated wiring in early homes (rubber often wrapped in cloth for protection).
PE	Thermoplastic solid	Excellent dielectric constant (~2.3) and high volume resistivity; dielectric strength ~25 kV/mm. Used in radar/coax cables and gradually in power cables. Low melting point limited early use to moderate temperatures.
Rubbers (neoprene, SBR)	Solid	Suitable for insulated cables in industrial environments and wire insulation in high-temperature environments (e.g., motors, vehicles, etc.), without rapid degradation.
Askarel (PCB oil)	Liquid dielectric	High dielectric constant fluid for transformers/capacitors, nonflammable.
SF_6 gas	Gas dielectric	Dielectric strength about 2–3 times that of air; chemically inert, arc-quenching; laboratory synthesis began in 1900, and practical applications began in the 1940s.
Epoxy resin	Thermoset solid	A versatile resin (reaction of epichlorohydrin and bisphenol-A, pioneered by Castan and Greenlee in the late 1930s). Commercial epoxy resins became available ~1947–1950. By 1945, epoxies were used for encapsulating electronic components and soon for laminating (e.g., in printed circuit boards and insulating structural parts). Epoxies offered excellent adhesion, mechanical strength, and electrical insulation, though they are rigid and require additives for toughness.

While polyethylene had been used in cables, its relatively low melting point meant that high operating temperatures could deform it. In the 1950s, cross-linked polyethylene (XLPE) technology, which involves forming a thermosetting network structure in polyethylene using peroxides or radiation, was introduced. XLPE combines the excellent dielectric properties of PE with significantly improved thermal stability, which allows continuous operating temperatures up to ~90–110 °C [43]. By the late 1960s and 1970s, XLPE cable insulating materials began to replace traditional oil-paper cables in medium-voltage power grids. The 1970s

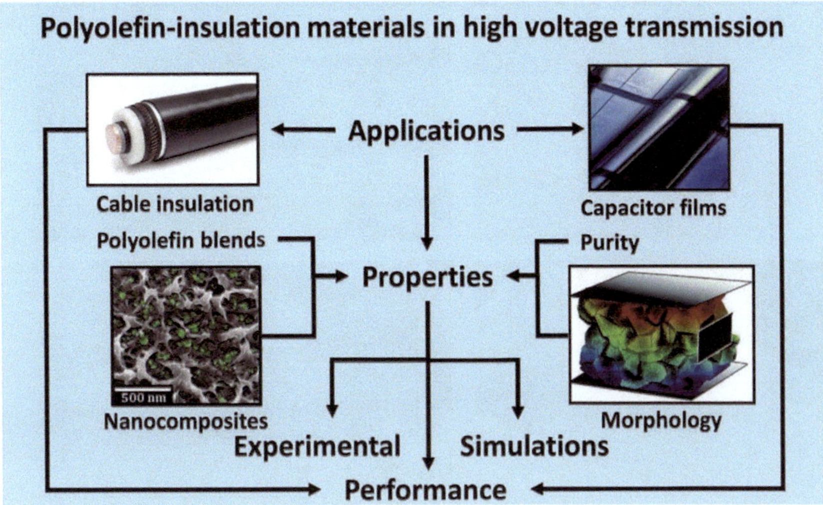

Figure 1.1 Polyolefin insulating materials for high-voltage insulation. *Source:* [12] / John Wiley & Sons / CC BY-NC 4.0.

marked the application of XLPE in medium-voltage cables, supporting voltage up to approximately 33 kV. Subsequently, improved formulations of XLPE, enhanced with additives to inhibit the formation of water trees, began to be used at high-voltage transmission levels ranging from 66 to 230 kV. Today, XLPE has become the standard insulation material for power cables due to its excellent balanced performance. In addition to PE, other types of polyolefins also play an important role in the field of electrical insulation (Figure 1.1). Isotactic polypropylene (IPP) was discovered in 1954. Although its electrical properties are similar to those of PE, with a dielectric constant of approximately 2.2 and extremely low dielectric loss, it exhibits a higher melting point of around 165 °C and can withstand higher operating temperatures [44]. In the 1960s, polypropylene (PP) began to replace paper and polystyrene as the dielectric film for capacitors, as it could be made into very thin PP films and could self-clear after dielectric breakdown [45, 46]. PP film capacitors became a common choice for AC applications in the 1970s and 1980s. However, due to the low-temperature brittleness and processing difficulties of PP, it was not initially as widely used in cables as PE. In recent years, PP has once again gained attention as an eco-friendly cable insulating material due to its ease of recycling.

Beyond PTFE, other melt-processable fluoropolymers were developed in the 1950s–1960s, such as fluorinated ethylene propylene (FEP) and ethylene tetrafluoroethylene (ETFE). These materials offer superior dielectric and thermal properties while providing more convenient processing methods such as extrusion molding. By the 1970s, FEP and ETFE were used for wire insulation in aerospace and computer cabling, where thin-walled, high-performance insulation materials were required. Notably, since its introduction around 1945, epoxy resin has rapidly become the basis for electrical insulating materials due to its high cohesive strength, strong adhesion, good flexibility, excellent thermosetting properties, and

Figure 1.2 Photos of epoxy resins used in transformer insulating bushings, curing agent, encapsulant, and electronic materials. *Source:* [48] / with permission of Elsevier.

stable chemical corrosion resistance. In the 1950s and 1960s, epoxy resin was combined with fillers such as aluminum oxide and silicon dioxide to produce molded high-voltage insulation components, including bushings, instrument transformer housings, and coil spacers for rotating machines (Figure 1.2) [47, 48]. Epoxy resin materials enabled the mass production of small insulating components with superior dielectric strength and mechanical stability. Moreover, by the 1960s, epoxy glass laminates such as G10 and FR-4 boards had become the standard material for printed circuit boards and many structural insulation boards, replacing the phenolic paper laminates previously used for high-performance applications.

A major milestone in the field of high-voltage transmission was the development of composite polymer insulators in the 1960s and 1970s. Traditional ceramic insulators were heavy and prone to brittle failure. To improve performance, researchers introduced composite insulators with a glass fiber core as the base material, covered with a polymer, typically made of silicone rubber or ethylene propylene diene monomer rubber. Silicone rubber, which debuted in the 1940s, is a high-temperature elastomer with excellent hydrophobic properties, making it suitable for external insulation applications to prevent moisture ingress and reduce leakage current [36, 49, 50]. Early applications in the 1960s were not very successful, but by the 1970s, especially the 1980s, polymer composite insulators gradually gained trust. They are lighter in weight, more resistant to damage, and can achieve excellent antipollution performance due to their hydrophobic surface, which continuously repels water. Currently, composite insulators are widely used in transmission lines and substations as an alternative to ceramic insulators (Figure 1.3).

In addition, to meet temperature requirements above 180 °C, a class of heterocyclic and aromatic polymers has emerged. Notably, DuPont commercialized polyimide

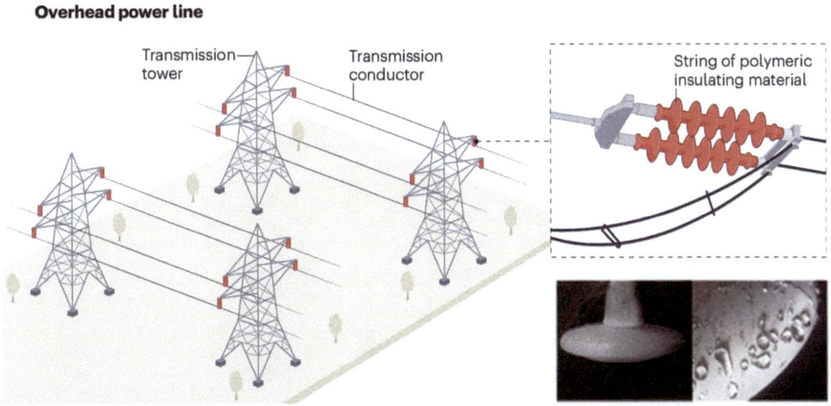

Figure 1.3 External insulation applications of silicone rubber insulating materials. *Source:* [1] / Springer Nature.

under the trademark name "Kapton®" in 1961. Kapton® polyimide film can operate within a temperature range from absolute zero (4 K) to 400 °C and only carbonizes at extremely high temperatures [13]. The industrialization of polyimide (PI) resin has a history of half a century. Its excellent high-temperature resistance, chemical stability, and mechanical properties make it an important material in high-tech fields as an engineering plastic and composite material [51]. Furthermore, its dielectric strength of 150–200 kV/mm (in film form) makes it suitable for extreme applications, such as insulation material for spacecraft wiring, high-temperature flexible circuits, and insulation material for special motors and generators, as shown in Figure 1.4. Similarly, other high-performance polymers such as polybenzimidazole (PBI), polysulfone, polyetheretherketone (PEEK, introduced in 1978), and polyphenylene sulfide (PPS) were developed in the 1960s and 1970s, providing engineers with material options to meet extreme thermal and electrical requirements.

During the 1960s, the insulation systems of large rotating machines (generators, motors, etc.) achieved technological advancements through the adoption of F-class (155 °C) and H-class (180 °C) materials. This was achieved by using mica paper tapes bonded with epoxy or PI resins on coils, and fiber glass cloth, in place of older asphaltic mica or shellac-based systems. This improvement enabled generators and motors to operate more reliably at higher temperatures for longer periods, thereby increasing their power density. For example, in the 1970s, the use of epoxy mica vacuum pressure impregnation (VPI) processing for generator stator windings eliminated air gaps and achieved significantly optimized dielectric properties.

Meanwhile, gas and liquid insulation technologies have also made significant progress. By the 1960s, SF_6 gas insulation technology had been fully adopted in the power industry. Power companies deployed SF_6 circuit breakers and gas-insulated substations (GIS), which could handle extremely high voltages (up to 500 kV) in a compact layout, something air insulation technology could not achieve. Furthermore, in terms of liquids, safer silicone oil and synthetic ester liquids replaced PCB oil as transformer coolants during the 1970s and 1980s. At the same

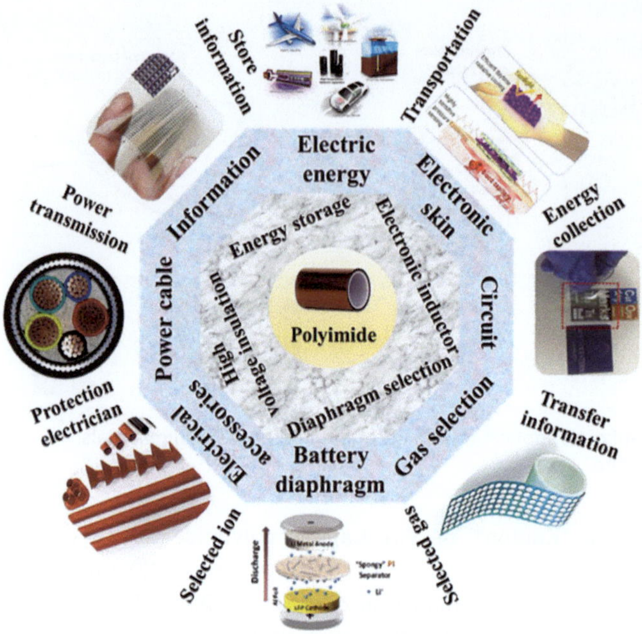

Figure 1.4 Summary diagram of PI applications. *Source:* [13] / John Wiley & Sons.

time, advancements in refining processes, including the removal of impurities and improved oxidation stability, enhanced the quality of transformer oil and contributed to extending the service life of transformers.

In summary, by the late 20th century, the electrical industry had at its disposal a rich toolkit of insulators: Solids: cross-linked polyolefins (XLPE), ethylene propylene rubber (EPR), and PVC for low voltage, epoxies and polyesters for structural parts, polyimides and silicones for high-temperature, and engineering thermoplastics for specialized uses. Liquids: mineral oil with inhibitor additives, silicone oil for high-temperature transformers, and new ester-based oils (starting to appear as fire-safe or biodegradable options). Gaseous insulating materials include air and SF_6, with SF_6 being the dominant choice in enclosed high-voltage equipment because of its reliability and space-saving benefits, despite its higher cost. The historic progress made during this period also laid a solid foundation for cutting-edge research in the present day.

1.3 Modern Insulating Materials and Current Challenges

Current electrical infrastructure relies heavily on polymer insulating materials, from underground cables and substation equipment to home wiring and electronic products, as shown in Figure 1.5. The widespread use of polymer materials is due to their excellent electrical insulation properties, mechanical flexibility, lightweight

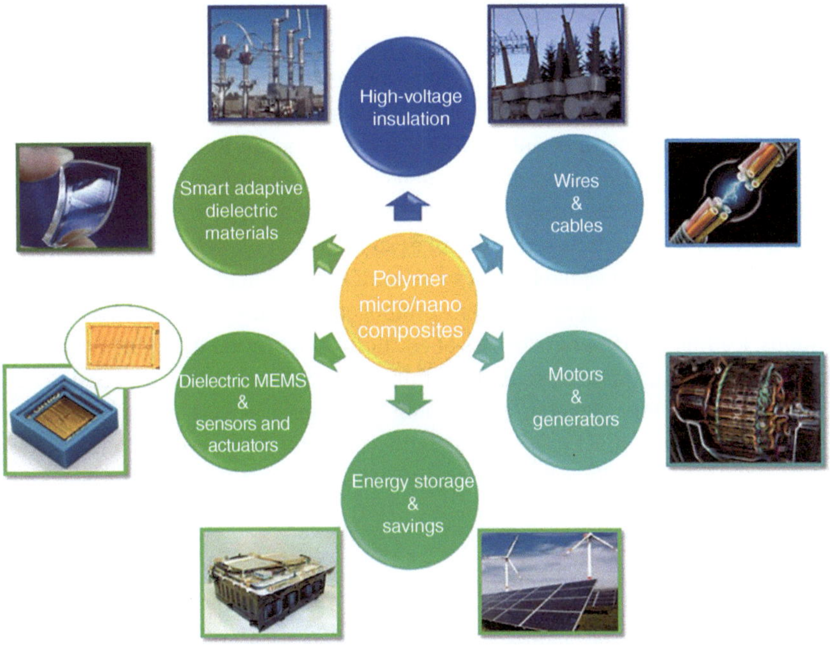

Figure 1.5 Application scenarios for modern polymer insulating materials. *Source:* [52] / MDPI / CC BY 4.0.

characteristics, and scalable manufacturing processes. The most common insulating materials currently used in practical applications include epoxy resin, XLPE, PP, EPR, PVC, PI, cellulose insulating paper, etc. In addition, liquid insulating materials such as mineral oil and natural ester fluids, as well as gas insulating materials such as SF_6, C-FN/CO_2 mixtures, and dry air, also play an important role in modern insulation systems. Despite the great success of modern insulating materials, several challenges and concerns have become apparent.

1.3.1 Aging and Degradation

The first issue is aging and performance degradation during operation. All organic insulating materials undergo aging under electrical, thermal, and environmental stress. Taking XLPE cables as an example, electrical aging can lead to the formation of electrical tree structures initiated by partial discharge, which involves microscopic sparks occurring within material defects. These microscopic damage pathways can penetrate the insulation layer and ultimately lead to insulation breakdown. Although XLPE cables have high long-term reliability, water trees may still form if trace moisture and impurities combine under electrical stress over decades [53, 54]. In inverter-powered motors or HVDC systems, due to the presence of high frequency components in the voltage waveform, polymer insulating materials may suffer from electrical erosion or space charge accumulation. This electrical aging leads to long-term degradation of material performance, occurring well before complete failure. Therefore, monitoring and diagnosing the health of insulating materials is critical for power system maintenance. Traditional condition monitoring relies on methods such as measuring loss factor, partial discharge, or chemical analysis of transformer oil. However, for solid polymer insulating materials, changes in the material may be subtle such as minor chemical changes or microporosity, making it difficult to detect aging phenomena on-site before a failure occurs [24]. This has driven the development of smart self-diagnostic materials, which will be discussed in detail in subsequent chapters. These materials can emit optical warning signals when significant internal degradation takes place. Additionally, thermal aging results from prolonged exposure to thermal-oxidative conditions, leading to brittleness or loss of elasticity. Environmental aging is caused by factors such as ultraviolet radiation, rain, ice accumulation, or radiation in space applications, and remains an ongoing challenge for insulating materials. Therefore, researchers must continue to refine material formulations, such as by adding antioxidants, stabilizers, or nano-filler reinforcements, to slow down the aging process.

1.3.2 Recycling and Disposal

As mentioned earlier, many polymer insulating materials such as XLPE, PI, EPR, and epoxy resin are thermosetting materials and cannot be recycled through melting and reprocessing. Landfilling such materials is unsustainable, while incineration, though it can recover some energy, may release carbon dioxide and potentially toxic compounds, posing significant risks to the environment and human health [9, 21]. Given the growing regulatory and economic pressures, there is an urgent need to explore recycling methods for such thermosetting materials or to design and synthesize new recyclable insulating materials. Even thermoplastic insulating materials like PVC and PE face recycling challenges when they are part of composite materials containing metals and other polymers. Additionally, separation and cleaning pose significant challenges [55, 56]. Currently, innovative solutions, including chemical recycling (depolymerization), converting polymers into fuels or other chemicals, or reprocessable cross-linked polymers (glass-state polymers), are under active research. Life cycle assessments indicate that recycling plastics can save significant amounts of energy and reduce carbon emissions. For example, recycling one ton of plastic instead of incinerating it can avoid approximately three tons of carbon dioxide emissions. The cable industry has already initiated recycling efforts, such as mechanically processing XLPE into secondary products or using crushed XLPE as a filler. However, current commercial technologies have not yet achieved fully closed-loop recycling.

1.3.3 Environmental Impact

Some traditional insulating media have significant environmental drawbacks. Certain solid insulating materials, such as PI and XLPE, are difficult to self-heal or recycle due to their permanently cross-linked networks. For instance, with a large number of early XLPE cables reaching the end of their service life and waste generated during the manufacturing process, the issue of disposal has become increasingly severe. For example, it is estimated that in Germany alone, approximately 150 000 tons of waste cable, much of which contains XLPE and PVC, are generated annually [57]. Typically, these wastes are landfilled or incinerated, causing significant environmental pollution. Mineral oils used in transformers and certain electrical equipment may leak and cause soil contamination, and their raw materials are derived from petroleum. This has prompted the transformer industry to adopt plant-derived natural esters as more biodegradable and sustainable alternatives, which is a significant advancement in the field of liquid insulation technology. Another example is SF_6, a gas widely used as an insulating and arc-quenching medium in high-voltage switchgear. However, SF_6 has an extremely high global warming potential (GWP ≈ 25 200 times that of carbon dioxide) and an atmospheric lifetime of up to 3200 years, and losses during handling make it a significant source of greenhouse gas emissions. Additionally, it is worth noting that while pure SF_6 is nontoxic, its byproducts such as SF_4, SOF_2, and S_2F_{10} are toxic and corrosive [58]. Therefore, for climate and safety reasons, alternatives such as new fluorocarbon or fluoroketone gases, or vacuum and solid insulation designs are being sought to replace SF_6 [59]. Although SF_6 gas insulation exceeds the scope of "insulation materials" in the solid/polymer sense, this example highlights the widespread industrial pursuit of eco-friendly insulation solutions to reduce its environmental footprint.

1.3.4 Performance Limits

Modern electrical equipment requirements, such as higher operating temperatures, higher electric fields (e.g., compact dry-type DC capacitors, transformers, power electronic devices, etc.), and radiation-rich environments (space and nuclear power), are pushing many insulating materials to their performance limits. Solid insulating materials such as PI, XLPE, PP, and epoxy resin now not only need to withstand thermal stresses exceeding 130 °C but must also cope with increasingly intense electric fields, radiation, and mechanical loads. This has driven the development of high-temperature polymer alloys, functionally graded materials, and nanocomposites, which incorporate fillers that suppress electric fields or create traps to enhance partial discharge resistance and inhibit space charge accumulation. Liquid insulating materials, including mineral oil, natural esters, and silicone liquids, must address limitations related to oxidative stability, moisture resistance, and thermal degradation, particularly in high-voltage and high-temperature applications. Gas insulating materials, particularly SF_6, are under pressure due to environmental concerns, leading to alternatives like C_4-FN/CO_2/N_2 mixtures, which must balance dielectric performance with environmental safety.

In summary, while the insulating materials in use today represent a pinnacle of decades of development and generally perform remarkably well, they face a dual mandate of improving basic characteristics such as high breakdown strength, low loss, and long-term reliability, while enhancing sustainability and functionality like self-healing, recyclability, and reduced environmental impact.

1.4 Toward Green and Smart Insulating Materials

To address the challenges above, the past few years have seen rapid advancements in the science of insulating materials. As shown in Figure 1.6, researchers are rethinking material design at the molecular level to support goals such as recyclability, degradability, and environmental friendliness in line with the circular economy. Meanwhile, efforts are focused on extending service life through self-healing, enabling smart functions like self-warning and adaptability, and improving overall performance such as weather resistance. This section will introduce some of these cutting-edge directions, which will be explored in detail in subsequent sections.

1.4.1 Eco-Friendly and Sustainable Insulating Materials

In recent years, a strong emphasis has been placed on low-carbon, eco-friendly, and sustainable insulating materials, which is highlighted by the Paris Agreement and global climate initiatives. The goals are to reduce the environmental impact of

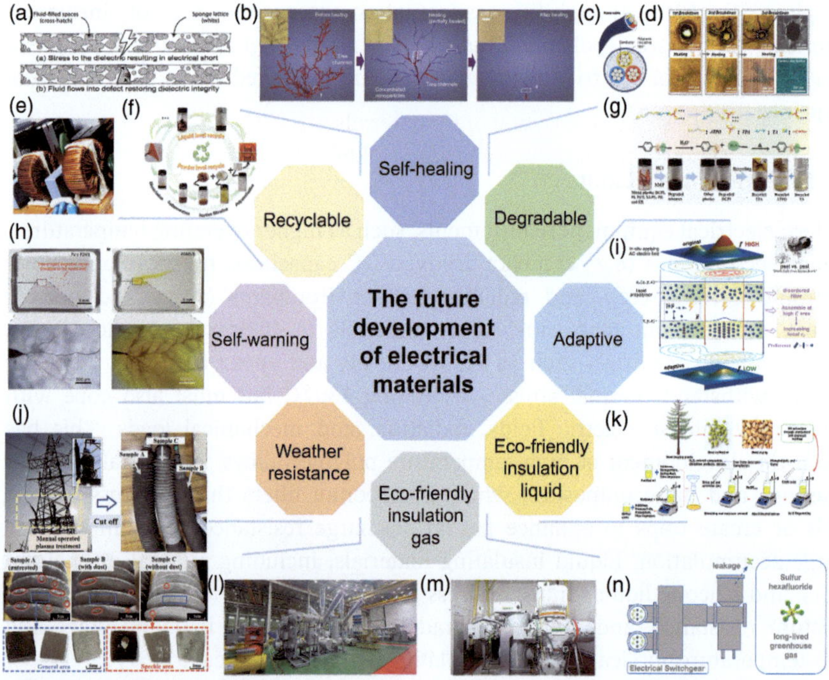

Figure 1.6 The future development of electrical materials. (a–d) Self-healing (e, f) recyclable. (g) degradable, (h) self-warding. (i) adaptive and (j) weather resistance electrical materials. (k) Eco-friendly insulation liquid. (l–n) Eco-friendly insulation gas. *Source:* (b) [23] / Springer Nature, (d) [60] / American Chemical Society, (e) [61] / American Chemical Society, (f) [61] / John Wiley & Sons, (g) Baoquan Wan et al., 2023 / With permission of John Wiley & Sons, (i) [62] / John Wiley & Sons / CC BY 4.0, (j) [49] / with permission of Elsevier, (k and M) [63] /American Chemical Society and (n) [63] / with permission of American Chemical Society.

insulating materials across their life cycle, from production to end-of-life disposal, and to ensure safety and reliability under green energy transitions.

Traditional thermoset materials such as XLPE, epoxy, and PI are not recyclable because, once cured, they cannot be remelted. This results in significant disposal challenges, with waste materials like scrap cables and electronic components contributing to the growing problem of electronic waste [11, 13, 64]. To address this issue, one revolutionary idea is to construct a polymer network that functions like a thermosetting material during operation, offering high mechanical strength and good heat resistance, but can be reverted to a dynamic state when recycling is required. Such polymers are typically referred to as covalent adaptive networks (CANs) or glass-like polymers (vitrimers) [65–67]. This concept was first proposed in the field of general polymer science in the 2010s and has begun to be gradually applied to the field of insulating materials. For example, Wan et al. [26] designed a dynamic covalent adaptable polyimide (DCPI) composite dielectric film to address the limitations of non-recyclability and limited durability of traditional PI materials after curing. By introducing dynamic imide bonds into the PI molecular structure, the DCPI maintains high insulation and mechanical strength while gaining degradability and reprocessability. As shown in Figure 1.7, the film can be completely degraded in acidic solutions at room temperature, allowing the recovery of high purity monomers for resynthesizing PI films. Notably, the regenerated films exhibit almost no loss in mechanical and electrical performance, offering a viable solution for the closed-loop recycling of PI materials.

There is a renewed interest in using renewable and degradable materials for insulation, especially where ultra-long life is not required. For instance, natural ester fluids derived from vegetable oils such as soybean or rapeseed have emerged as alternatives to mineral oil in transformer applications. These fluids are biodegradable and have higher fire points, which enhances fire safety. By 2020, natural ester insulated transformers had been widely used in distribution and even transmission

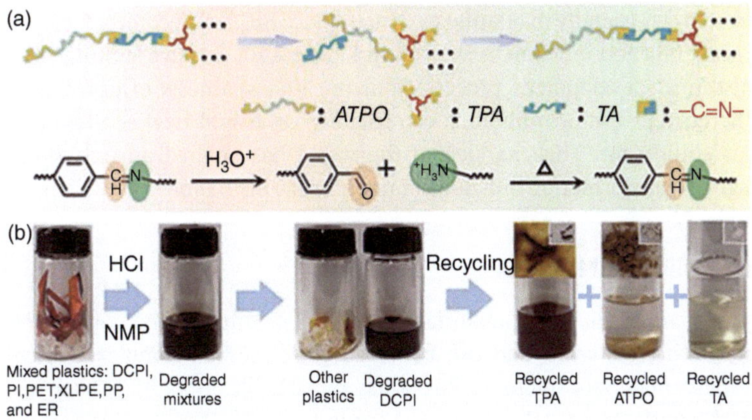

Figure 1.7 (a) Schematic and recycle process of DCPI films. (b) Depolymerization and recycling process of DCPI from the mixed plastic wastes containing PI, PET, XLPE, PP, and ER. *Source:* [26] / John Wiley & Sons.

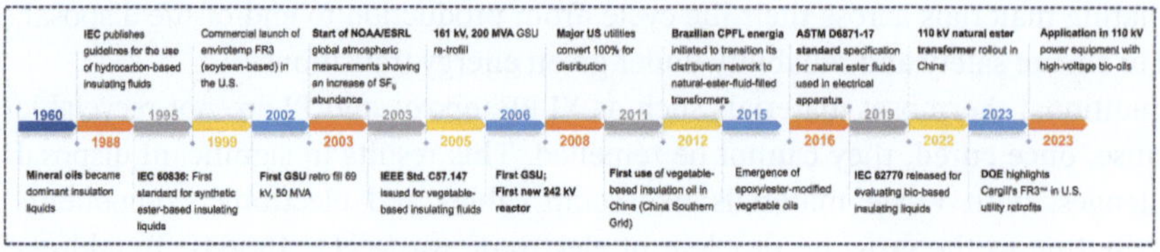

Figure 1.8 The historical development of insulating oil (from mineral oil to eco-friendly and sustainable insulating oil). *Source:* Adapted from [4, 69].

levels, effectively reducing the risk of soil contamination in the event of leakage. In addition, we have summarized the historical development of insulating liquids, from the initial use of mineral insulating oil (1960) to the first synthetic ester-based insulating liquid (1995), and finally to the current use of eco-friendly bio-based insulating oil in electrical equipment (Figure 1.8). This provides a brief overview of how insulating liquids have evolved from initial performance requirements to current environmentally driven green insulation. In the field of solid insulation, researchers are exploring the use of cellulose nanofibers, starch-based polymers, and other biopolymers as dielectric materials. Wang et al. [69] prepared a biodegradable bio-based fluorinated epoxy resin by introducing Schiff base bonds and fluorine substituents into traditional epoxy resins, achieving excellent mechanical and dielectric properties, as well as the ability to be recycled and acid-catalyzed degradation (Figure 1.9).

As the global energy transition towards low-carbon and sustainable development, power systems face the major challenge of reducing greenhouse gas emissions. In the field of gas insulation, traditional SF_6 gas is widely used in high-voltage electrical equipment, such as gas-insulated switchgear and gas-insulated transformers, due to its high insulation strength and excellent arc-quenching performance [18, 70, 71]. However, SF_6 has a GWP as high as 25 200 and an atmospheric lifetime of up to 3200 years, making it one of the primary restricted industrial greenhouse gases. Therefore, the synthesis and search for high-performance, high-safety, and long-term reliable eco-friendly insulating gases to replace SF_6 has become a focus of attention in academia and industry. As shown in Figure 1.10, we have summarized the entire historical development process from the first synthesis of gas media in 1889 to the current use of numerous eco-friendly gases and new eco-friendly insulating gas equipment. This marks that the world has entered an important historical stage of transformation and development in gas insulation.

1.4.2 Smart Insulating Materials

As a new type of material, smart materials have broken through the concept of traditional insulating materials and can better adapt to various complex service conditions. These materials often exhibit some remarkable properties, such as self-warning of health status, self-adaptive changes to the working environment, and self-healing after mechanical and electrical damages.

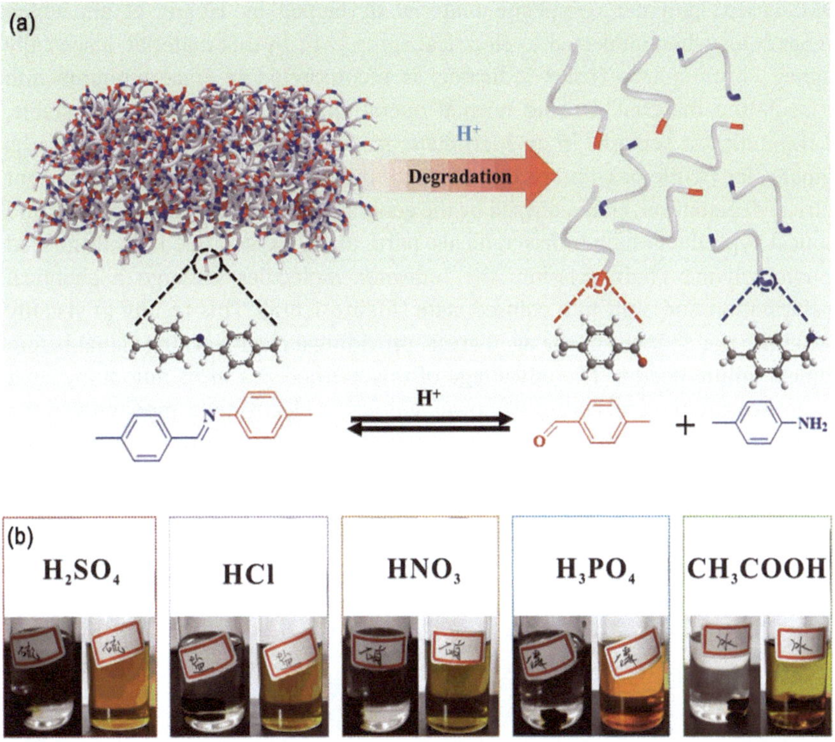

Figure 1.9 (a) Degradation mechanism of Schiff base epoxy thermosets. (b) Digital photographs of SA-BTB-EP/DDM and degradation rates at different acids. *Source:* [69] / Springer Nature.

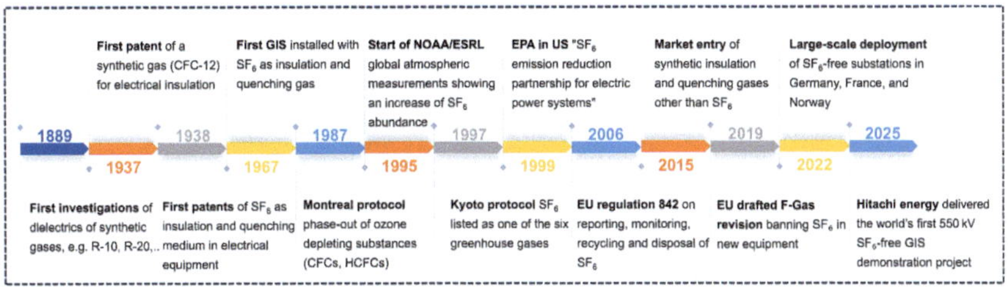

Figure 1.10 Timeline and key events in the development of insulating gases (from SF_6 to eco-friendly gases).

Electrical aging is a major challenge faced by most electrical equipment, and the emergence of smart insulating materials has provided an innovative solution to this issue. Electrical aging often does not significantly change bulk electrical measurements until near failure. A visual or optical early warning could allow operators to replace or repair insulating materials well before failure, improving system reliability and safety. A typical example is the polydimethylsiloxane-based

(PDMS-based) polymer composite material developed by Huang et al., which changes color when subjected to electrical aging [24]. In this material, a specially designed chromophore (color indicator) is incorporated in trace amounts into the insulating material. During normal operation, the material remains stable, and the indicator remains in an inert state, so the insulating material retains its original color (white or colorless). However, if the polymer experiences significant electrical degradation, characterized by the generation of high-energy electrons and chemical byproducts such as free radicals, particularly oxygen free radicals formed through polymer chain scission, the indicator molecules undergo a chemical transformation and shift to a colored state (Figure 1.11a). This results in visually detectable color changes in the aged areas, providing a clear warning signal before complete failure occurs. The advantage of this method lies in its autonomy, as it does not require the embedding of power sources or electronic components in the material. The chemical reaction of the material itself provides the signals, which are continuous and distributed.

During operation, insulating materials inevitably face unfavorable conditions such as voltage fluctuations, sudden temperature changes, and contamination.

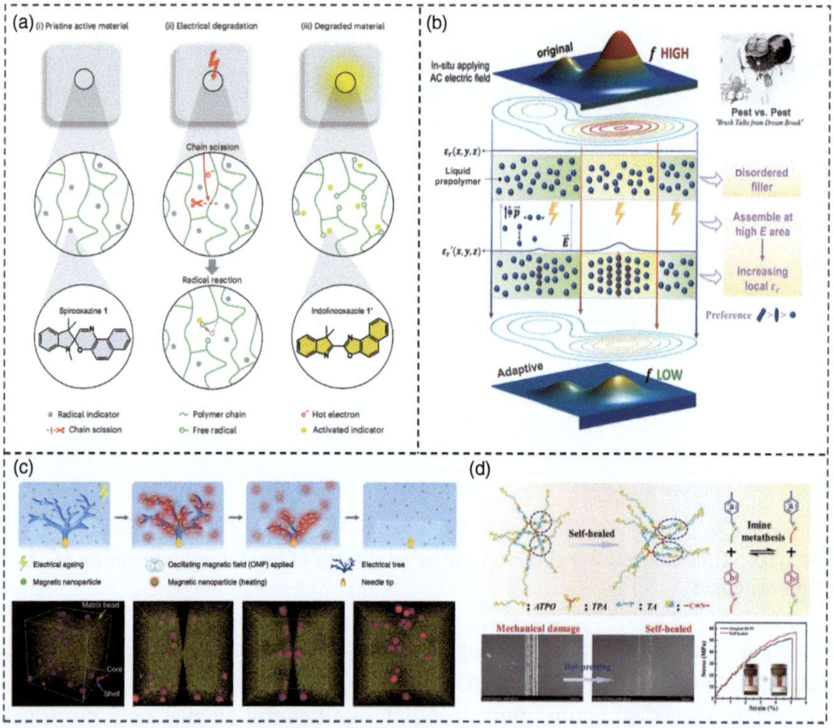

Figure 1.11 (a) Electrical-degradation-indicating materials and visualization of the free radical induced chromogenic mechanism. *Source:* [24] / Springer Nature. (b) 3D schematic of the permittivity-electric field evolution. *Source:* [62] / John Wiley & Sons / CC BY 4.0. (c) Schematic diagram of electric trees self-healing induced by functionalized nanoparticles. *Source:* [23] / Springer Nature. (d) Schematic and mechanism of self-healable process of DCPI films based on dynamic imine bonds. *Source:* [26] / John Wiley & Sons.

Therefore, the development of new self-adaptive insulating materials is particularly important. These materials can sense and respond to local electric field variations, actively adjusting their permittivity or conductivity to alleviate electric field distortions and enhance breakdown strength and system stability [25]. Moreover, self-adaptive insulating materials enable demand-driven design, which reduces redundancy and improves system compactness and cost-effectiveness. They also signify the transition from the "passive insulation" mechanism of traditional insulating materials to the "active regulation" mechanism of adaptive insulating materials, offering unique advantages under complex electrical, thermal, and mechanical stresses. As shown in Figure 1.11b, Shen et al. [62] proposed a fabrication method for adaptive dielectric gradient insulators based on in situ alternating current field-induced self-assembly. High aspect ratio fillers are driven to align in chain-like structures in high field regions, enabling spatial tuning of permittivity and mitigating field concentration. This method integrates electric field responsiveness, self-organized structuring, and structure freezing, demonstrating smart environmental adaptability and providing a new pathway for adaptive high-voltage insulation design.

Additionally, electrical insulation failures are usually caused by microscale defects that grow under the influence of electrical, thermal, and mechanical stresses. If these minor damages could heal themselves, the service life, and reliability of insulating materials could be greatly enhanced. This has inspired researchers to study self-healing insulating materials, i.e. materials that can repair electrical or mechanical damages (or at least mitigate its effects) without human intervention. Research on self-healing insulating materials currently focuses on two types: extrinsic and intrinsic strategies. Among these, the main method for extrinsic self-healing strategies is the microcapsule method, embedding tiny capsules of a liquid healing agent in the solid insulation. When electrical trees or cracks grow, they rupture nearby capsules, releasing monomers or liquid to fill the gaps and polymerize, thereby repairing the insulating material. As shown in Figure 1.11c, by introducing superparamagnetic γ-Fe_2O_3 nanoparticles into the electrical cable insulating material PP, effective self-healing of electrical damage has been achieved [23]. Driven by conformational entropy, surface-modified γ-Fe_2O_3 nanoparticles can detect defects caused by electrical treeing. Simultaneously, these nanoparticles autonomously migrate to the electrical treeing channel region and utilize an oscillating magnetic field to perform localized heating of the damaged area. This process increases the local temperature of the defect region by 24 K compared to the surrounding polymer matrix, thereby triggering targeted melting and damage repair of the PP. Another strategy is to enable intrinsic self-healing

capabilities in polymer networks after damage through molecular structure design. This can be achieved by incorporating dynamic covalent bonds such as disulfide bonds, Diels-Alder bonds, borate ester bonds, imide bonds, and hydrazone bonds, as well as reversible noncovalent interactions including van der Waals forces, hydrogen bonds, ionic bonds, host-guest interactions, and coordination bonds into the network structure [67, 72–77]. The reversible rearrangement of dynamic bonds is used to achieve self-healing after damage, which effectively extends the service life of the material. As shown in Figure 1.11d, the dynamic reversible properties of imine bonds were used to achieve self-healing of PI after mechanical damage, and the self-healed PI basically maintained its original performance level [26].

1.5 Summary and Outlook

The development of insulating materials has been driven by the evolving demands of electrical technology. In recent years, the urgent need for low-carbon, eco-friendly, and sustainable development has also become a significant driving force behind this progress. In this chapter, we first review the evolution of insulating materials, tracing the development from simple materials such as glass, rubber, oil, and paper to advanced engineering materials including cross-linked polymers, SF_6 gas, and composite materials. These advancements have enabled us to rely primarily on polymer dielectric materials for electrical infrastructure, which can maintain excellent properties for decades even under harsh conditions. However, modern materials not only require outstanding performance but must also strike a balance between high-performance, smart, low-carbon, eco-friendly, and sustainable characteristics. This also represents the new set of challenges facing insulating materials today: addressing environmental concerns such as how to recycle, degrade, or safely dispose of large volumes of polymer insulating materials; ensuring long-term reliability by managing performance degradation during service life and enabling self-healing or self-warning after damage; and achieving performance breakthroughs, including the ability to withstand higher operating temperatures and electric fields.

In summary, with the comprehensive implementation of the green transformation of electrical insulation, the research and development of insulation materials is entering a new era centered on low-carbon, eco-friendly, sustainable, and smart. Meanwhile, the research and development of insulation materials has gradually become a multidisciplinary challenge involving electrical engineering, materials science, and chemistry. The accumulated experience over the past century regarding the relationship between material structure and performance is being utilized by researchers to design new types of electrical insulating materials that combine excellent basic properties with multifunctionality. Subsequent chapters will delve deeper into insulating materials from the perspectives of different insulation types and the latest research findings.

References

1 Wang, T.-Y., Mao, J., Zhang, B. et al. (2024). Polymeric insulating materials characteristics for high-voltage applications. *Nature Reviews Electrical Engineering* 1 (8): 516–528.

2 Xu, S., Chen, L., Gong, M. et al. (2017). Characterization and engineering application of a novel ceramic composite insulation material. *Composites Part B: Engineering* 111: 143–147.

3 Sun, W.B., Han, Z.M., Yue, X. et al. (2023). Nacre-inspired bacterial cellulose/mica nanopaper with excellent mechanical and electrical insulating properties by biosynthesis. *Advanced Materials* 35 (24): 2300241.

4 Fofana, I. (2013). 50 years in the development of insulating liquids. *IEEE Electrical Insulation Magazine* 29 (5): 13–25.

5 Wanatasanappan, V.V., Rezman, M., and Abdullah, M.Z. (2022). Thermophysical properties of vegetable oil-based hybrid nanofluids containing Al_2O_3–TiO_2 nanoparticles as insulation oil for power transformers. *Nanomaterials* 12 (20): 3621.

6 Francis, M. (1944). Porcelain and other ceramic insulating materials. *Nature* 154: 474.

7 Kuffel, J. and Kuffel, P. (2000). *High Voltage Engineering Fundamentals.* Elsevier.

8 Wei, W., Chen, H., Zha, J., and Zhang, Y. (2023). Research progress on low dielectric constant modification of cellulose insulating paper for power transformers. *Frontiers of Chemical Science and Engineering* 17 (8): 991–1009.

9 Hirschler, M.M. (2000). *Electrical Insulating Materials: International Issues.* ASTM International.

10 Thiviyanathan, V.A., Ker, P.J., Leong, Y.S. et al. (2022). Power transformer insulation system: a review on the reactions, fault detection, challenges and future prospects. *Alexandria Engineering Journal* 61 (10): 7697–7713.

11 D'Auria, S., Pourrahimi, A.M., Favero, A. et al. (2023). Polyethylene based ionomers as high voltage insulation materials. *Advanced Functional Materials* 33 (36): 2301878.

12 Bjurström, A., Edin, H., Hillborg, H. et al. (2024). A review of polyolefin-insulation materials in high voltage transmission; from electronic structures to final products. *Advanced Materials* 36 (52): 2401464.

13 Wan, B., Dong, X., Yang, X. et al. (2023). Rising of dynamic polyimide materials: a versatile dielectric for electrical and electronic applications. *Advanced Materials* 35 (39): 2301185.

14 Zhang, D., Xu, H., Yang, D. et al. (2024). Study on the aging characteristics of pet for SF_6 gas-insulated transformer. *AIP Advances* 14 (7): 075019.

15 Lyutikova, M., Korobeynikov, S., Rao, U.M., and Fofana, I. (2022). Mixed insulating liquids with mineral oil for high voltage transformer applications: a review. *IEEE Transactions on Dielectrics and Electrical Insulation* 29: 454–461.

16 Wu, Y., Hu, Y., Lin, H., and Zhang, X. (2024). An anhydride-cured degradable epoxy insulating material exhibiting recyclability, reusability, and excellent electrical performance. *Green Chemistry* 26 (4): 2258–2268.

17 Oparanti, S.O., Fofana, I., Zarrougui, R. et al. (2024). Improving some physicochemical characteristics of environmentally friendly insulating liquids for enhanced sustainability in subpolar transformer applications. *Sustainable Materials and Technologies* 41: e00996.

18 Xiao, S., Shi, S., Li, Y. et al. (2021). Review of decomposition characteristics of eco-friendly gas insulating medium for high-voltage gas-insulated equipment. *Journal of Physics D: Applied Physics* 54 (37): 373002.

19 Pan, B., Wang, G., Shi, H. et al. (2020). Green gas for grid as an eco-friendly alternative insulation gas to SF_6: a review. *Applied Sciences* 10 (7): 2526.

20 Pourrahimi, A.M., Mauri, M., D'Auria, S. et al. (2024). Alternative concepts for extruded power cable insulation: from thermosets to thermoplastics. *Advanced Materials* 36 (52): 2313508.

21 Bryce, T. and Nicholls, B.P.F. (2024). Closing the loop on thermoset plastic recycling. *Science* 384 (6692): 156–157.

22 Zhou, Y. and He, J. (2024). Development of environmentally friendly high-capacity power cables. *Nature Reviews Electrical Engineering* 1 (9): 565–566.

23 Yang, Y., He, J., Li, Q. et al. (2018). Self-healing of electrical damage in polymers using superparamagnetic nanoparticles. *Nature Nanotechnology* 14 (2): 151–155.

24 Huang, X., Zhang, S., Zhang, P. et al. (2023). Autonomous indication of electrical degradation in polymers. *Nature Materials* 23 (2): 237–243.

25 Yang, X., Hu, J., Wang, S. et al. (2022). A dielectric polymer/metal oxide nanowire composite for self-adaptive charge release. *Nano Letters* 22 (13): 5167–5174.

26 Wan, B., Xiao, M., Dong, X. et al. (2023). Dynamic covalent adaptable polyimide hybrid dielectric films with superior recyclability. *Advanced Materials* 36 (52): 2304175.

27 Ab Ghani, S., Muhamad, N.A., Noorden, Z.A. et al. (2018). Methods for improving the workability of natural ester insulating oils in power transformer applications: a review. *Electric Power Systems Research* 163: 655–667.

28 Kharal, H.S., Kamran, M., Ullah, R. et al. (2019). Environment-friendly and efficient gaseous insulator as a potential alternative to SF_6. *Processes* 7 (10): 740.

29 Pan, X.F., Bao, Z., Xu, W. et al. (2022). Recyclable nacre-like aramid-mica nanopapers with enhanced mechanical and electrical insulating properties. *Advanced Functional Materials* 33 (9): 2210901.

30 Riba, J.-R., González, N., Canals, T., and Cantero, R. (2019). Identification of natural rubber samples for high-voltage insulation applications. *Computers & Chemical Engineering* 124: 197–205.

31 Gonzalez, N., Riba, J.-R., dels Angels Custal, M., and Armelin, E. (2017). Improvement of insulation effectiveness of natural rubber by adding hydroxyl-functionalized barium titanate nanoparticles. *IEEE Transactions on Dielectrics and Electrical Insulation* 24 (5): 2881–2889.

32 Zuidema, C., Kegerise, W., Fleming, R. et al. (2011). A short history of rubber cables. *IEEE Electrical Insulation Magazine* 27 (4): 45–50.

33 Shen, J., Yang, Z., Lian, J. et al. (2023). An all-natural, bioinspired, biodegradable electrical insulating composite based on lignocellulose and mica tailings. *International Journal of Biological Macromolecules* 253: 127222.

34 Yang, X., Zhao, H., Wang, X. et al. (2024). Modulated electrical performance of cross-linked polyethylene by grafted charge-attracting molecules for high voltage direct current cable insulation. *Materials & Design* 240: 112884.

35 Li, G., Wang, Z., Lan, R. et al. (2023). The lifetime prediction and insulation failure mechanism of XLPE for high-voltage cable. *IEEE Transactions on Dielectrics and Electrical Insulation* 30 (2): 761–768.

36 Nazir, M.T., Khalid, A., Kabir, I. et al. (2021). Flame retardancy and excellent electrical insulation performance of RTV silicone rubber. *Polymers* 13 (17): 2854.

37 Yang, H., Jiang, Y., Liu, H. et al. (2018). Mechanical, thermal and fire performance of an inorganic-organic insulation material composed of hollow glass microspheres and phenolic resin. *Journal of Colloid and Interface Science* 530: 163–170.

38 Grigoriev, S.N., Volosova, M.A., and Okunkova, A.A. (2023). Advances in electrical discharge machining of insulating ceramics. *Materials* 16 (17): 5959.

39 Baekeland, L. (1993, 1993). *The Bakelizer: National Museum of American History, Smithsonian Institution: A National Historic Chemical Landmark, November 9, 1993*, 1–4. American Chemical Society.

40 Yu, Z.-L., Gao, Y.-C., Qin, B. et al. (2024). Revitalizing traditional phenolic resin toward a versatile platform for advanced materials. *Accounts of Materials Research* 5 (2): 146–159.

41 Dhanumalayan, E. and Joshi, G.M. (2018). Performance properties and applications of polytetrafluoroethylene (PTFE)—a review. *Advanced Composites and Hybrid Materials* 1 (2): 247–268.

42 Yuan, L., Zheng, X., Zhu, W. et al. (2024). Study on the electrical insulation properties of modified PTFE at high temperatures. *Polymers* 16 (3): 316.

43 Wang, J.-Q., Li, W.-K., Zhang, W.-Y. et al. (2024). Aging and life control of cross-linked polyethylene as cable insulation material. *Acta Physica Sinica* 73 (7): 078801.

44 Yu, B.-Q., Xia, B., Yang, X.-Y. et al. (2023). Electric field regulation of polypropylene insulation for high voltage DC cables. *Acta Physica Sinica* 72 (6): 068402.

45 Liu, X.-J., Zheng, M.-S., Chen, G. et al. (2022). High-temperature polyimide dielectric materials for energy storage: theory, design, preparation and properties. *Energy & Environmental Science* 15 (1): 56–81.

46 Li, H., Zhou, Y., Liu, Y. et al. (2021). Dielectric polymers for high-temperature capacitive energy storage. *Chemical Society Reviews* 50 (11): 6369–6400.

47 Jin, F.-L., Li, X., and Park, S.-J. (2015). Synthesis and application of epoxy resins: a review. *Journal of Industrial and Engineering Chemistry* 29: 1–11.

48 Matykiewicz, D. and Skórczewska, K. (2022). Characteristics and application of eugenol in the production of epoxy and thermosetting resin composites: a review. *Materials* 15 (14): 4824.

49 Wang, Q., Liu, S., Liu, S. et al. (2024). Super-hydrophobic silicone rubber for outdoor electrical insulation. *Nano Today* 58: 102406.

50 Yang, X., Huang, W., Dong, H., and Zha, J.W. (2025). Smart polydimethylsiloxane materials: versatility for electrical and electronic devices applications. *Advanced Materials* 37 (17): e2500472.

51 Dong, X., Wan, B., and Zha, J.-W. (2024). Versatile landscape of low-K polyimide: theories, synthesis, synergistic properties, and industrial integration. *Chemical Reviews* 124 (12): 7674–7711.

52 Pleşa, I., Noţingher, P., Schlögl, S. et al. (2016). Properties of polymer composites used in high-voltage applications. *Polymers* 8 (5): 173.

53 Roy, S.S., Paramane, A., Singh, J. et al. (2023). Characterization and identification of electrical tree growth stages inside high-voltage cable insulation. *IEEE Transactions on Instrumentation and Measurement* 72: 1–9.

54 Zheng, X.-X., Pan, Y.-C., and Sun, W.-F. (2022). Water-tree characteristics and its mechanical mechanism of crosslinked polyethylene grafted with polar-group molecules. *International Journal of Molecular Sciences* 23 (16): 9450.

55 Lewandowski, K. and Skórczewska, K. (2022). A brief review of poly(vinyl chloride) (PVC) recycling. *Polymers* 14 (15): 3035.

56 Tian, Y., Han, M., Gu, D. et al. (2024). PVC dechlorination for facilitating plastic chemical recycling: a systematic literature review of technical advances, modeling and assessment. *Sustainability* 16 (19): 8331.

57 Blinová, L. and Godovčin, P. (2021). Importance of recycling the waste-cables containing copper and PVC. *Research Papers Faculty of Materials Science and Technology Slovak University of Technology* 29 (48): 1–21.

58 Yang, Y., Gao, K., Ding, L. et al. (2021). Review of the decomposition characteristics of eco-friendly insulation gas. *High Voltage* 6 (5): 733–749.

59 Xiao, S., Zhang, X., Tang, J., and Liu, S. (2018). A review on SF_6 substitute gases and research status of CF_3I gases. *Energy Reports* 4: 486–496.

60 Li, Z., Wen, Y., Song, Z. et al. (2023). Dynamic cross-linked polyethylene networks with high energy storage and electrical damage self-healability. *ACS Macro Letters* 12 (10): 1409–1415.

61 Wan, B., Yang, X., Dong, X. et al. (2022). Dynamic sustainable polyimide film combining hardness with softness via a "mimosa-like" bionic strategy. *Advanced Materials* 35 (2): 2207451.

62 Shen, Z., Jia, Z., Hao, Y. et al. (2022). A fabrication method for adaptive dielectric gradient insulating components. *High Voltage* 8 (1): 59–69.

63 Rabie, M. and Franck, C.M. (2018). Assessment of eco-friendly gases for electrical insulation to replace the most potent industrial greenhouse gas SF_6. *Environmental Science & Technology* 52 (2): 369–380.

64 Wu, W., Feng, H., Xie, L. et al. (2024). Reprocessable and ultratough epoxy thermosetting plastic. *Nature Sustainability* 7 (6): 804–811.

65 Yokoyama, K. and Guan, Z. (2024). A vitrimer acts as a compatibilizer for polyethylene and polypropylene blends. *Angewandte Chemie International Edition* 63 (20): e202317264.

66 Ng, K.W.J., Lim, J.S.K., Gupta, N. et al. (2023). A facile alternative strategy of upcycling mixed plastic waste into vitrimers. *Communications Chemistry* 6 (1): 158.

67 Zheng, N., Xu, Y., Zhao, Q., and Xie, T. (2021). Dynamic covalent polymer networks: a molecular platform for designing functions beyond chemical recycling and self-healing. *Chemical Reviews* 121 (3): 1716–1745.

68 Rafiq, M., Shafique, M., Azam, A., and Ateeq, M. (2020). The impacts of nanotechnology on the improvement of liquid insulation of transformers: emerging trends and challenges. *Journal of Molecular Liquids* 302: 112482.

69 Wang, Z., Hou, G., Xie, J. et al. (2023). Degradable bio-based fluorinated epoxy resin with excellent flame-retardant, dielectric, hydrophobic, and mechanical properties. *Advanced Composites and Hybrid Materials* 6 (4): 128.

70 Yao, Y., Yang, Z., Zhang, B. et al. (2024). Discovery of a novel binary azeotrope with positive synergistic insulation strength as eco-friendly SF_6-alternative. *Journal of Physics D: Applied Physics* 58 (3): 035503.

71 Wang, Y., Huang, D., Liu, J. et al. (2019). Alternative environmentally friendly insulating gases for SF_6. *Processes* 7 (4): 216.

72 Lei, Z., Chen, H., Huang, S. et al. (2024). New advances in covalent network polymers via dynamic covalent chemistry. *Chemical Reviews* 124 (12): 7829–7906.

73 Maes, S., Badi, N., Winne, J.M., and Du Prez, F.E. (2025). Taking dynamic covalent chemistry out of the lab and into reprocessable industrial thermosets. *Nature Reviews Chemistry* 9 (3): 144–158.

74 Li, B., Cao, P.-F., Saito, T., and Sokolov, A.P. (2022). Intrinsically self-healing polymers: from mechanistic insight to current challenges. *Chemical Reviews* 123 (2): 701–735.

75 Ratwani, C.R., Kamali, A.R., and Abdelkader, A.M. (2023). Self-healing by Diels-Alder cycloaddition in advanced functional polymers: a review. *Progress in Materials Science* 131: 101001.

76 Ren, Y. and Dong, X. (2024). Dynamic polymeric materials via hydrogen-bond cross-linking: effect of multiple network topologies. *Progress in Polymer Science* 158: 101890.

77 Yang, Y., Dang, Z.M., Li, Q., and He, J. (2020). Self-healing of electrical damage in polymers. *Advanced Science* 7 (21): 2002131.

2

Key Issues of Conventional Insulating Materials

Yang Zhang[1] *and Jun-Wei Zha*[1,2]

[1] *University of Science and Technology Beijing, School of Chemistry and Biological Engineering, No. 30 Xueyuan Road, Haidian District, Beijing 100083, P. R. China*
[2] *North China Electric Power University, State Key Laboratory of Alternate Electrical Power System with Renewable Energy Sources, School of Electrical and Electronic Engineering, No. 2 Beinong Road, Huilongguan, Changping District, Beijing 102206, P. R. China*

2.1 Introduction

As the global energy system evolves rapidly, electrical insulation materials integral to power equipment carry the crucial responsibility of safeguarding energy security and stability [1]. However, in the context of dual-carbon targets and the transition to a circular economy, traditional insulation materials have increasingly exposed environmental and health issues across their entire life cycle, from manufacture and service to disposal, demanding systematic innovation.

As shown in Figure 2.1, at the manufacturing stage, the preparation of conventional insulation materials is beset by high energy consumption and severe pollution. Solid dielectrics exemplified by epoxy resins and cross-linked polyethylene (XLPE) rely on petroleum-based synthesis; their feedstock extraction and polymerization consume vast amounts of fossil energy and release volatile organic compounds (VOCs) and carcinogenic byproducts, endangering workers' health and the surrounding ecosystem [2]. The purification of liquid dielectrics, such as mineral oil or synthetic esters, often involves organic solvents, and improper handling can lead to water contamination [3]. Meanwhile, the production of gaseous insulators like sulfur hexafluoride (SF_6) depends on high-temperature, high-pressure processes that generate substantial carbon emissions and unregulated fluorinated off-gases, further exacerbating climate change and ozone depletion [4, 5].

During the service phase, environmental risks associated with all three insulation classes continue to accumulate. Solid dielectrics with their irreversibly cross-linked networks are subjected to coupled thermal, electrical, and mechanical stresses in operation [6], unavoidably developing micro-damage that degrades performance and shortens service life. Liquid insulating media are prone to leakage due to aging or seal failure, allowing contaminants to infiltrate soil and groundwater and posing long-term ecological threats [7]. Gaseous insulators, by virtue of molecular

Next-Generation Electrical Insulating Materials: Low-Carbon, Eco-Friendly, and Sustainable Solutions, First Edition. Edited by Jun-Wei Zha.

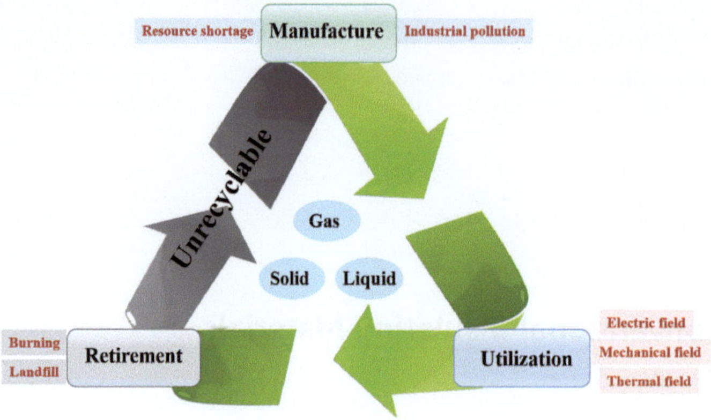

Figure 2.1 Critical issues under the full life cycle of conventional insulation materials.

diffusion, inevitably escape containment and, under combined electrical thermal stress, decompose, further undermining their dielectric integrity [8].

During the disposal stage, all three insulation material classes face sustainability challenges [9]. Thermoset plastics, owing to their excessive chemical stability, resist safe landfill or incineration; some degradation products are bio-toxic and may leach into the food chain via rainwater [10]. End-of-life liquid insulating oils lack mature treatment technologies. If residual toxic compounds, such as polycyclic aromatic hydrocarbons (PAHs), are not completely degraded prior to discharge, they can severely disrupt ecological balance. Gaseous dielectrics suffer from low recovery efficiencies, resulting in the direct release of residual fluids whose ultra-high global warming potentials (GWPs) impose far-reaching climate impacts [11, 12].

Even more, traditional insulation materials stand in stark conflict with global green-governance trends. At present, the international community is driving industrial-chain transformation through policies, regulations, and technological innovation, for example, restricting the use of fluorinated gases and tightening hazardous waste treatment standards. This compels the insulation materials sector to break free from its conventional paradigm and adopt a closed-loop model of "low-carbon design, low-toxicity application, regenerative recycling," thereby achieving source-level reductions in environmental burdens and maximizing resource value. Accordingly, this chapter will systematically dissect the life-cycle challenges and environmental loads of solid, liquid, and gaseous insulators; reveal their deep impacts on ecosystems; and, grounded in circular economy principles, propose a theoretical framework for sustainable solutions.

2.2 Key Issues in Conventional Solid Insulation

Traditional solid insulation materials such as polyvinyl chloride (PVC), polyethylene (PE), XLPE, epoxy resin, and polyimide are widely used due to their excellent dielectric performance and processing adaptability. However, their environmental footprint spans the entire life cycle from production and service to end-of-life disposal, posing environmental risks at each stage [10].

2.2.1 Key Issues in the Production Phase of Conventional Solid Insulating Materials

The manufacturing of traditional solid insulation materials faces multifaceted environmental and technical challenges. Their synthesis relies heavily on petroleum-derived feedstocks such as bisphenol A (BPA) and phenol. It involves high-temperature condensation and cross-linking steps that intensify fossil energy consumption and carbon emissions. During polymerization, VOCs (e.g. formaldehyde, benzene derivatives) and unreacted monomers (e.g. epichlorohydrin) are released, posing potent irritant and carcinogenic risks to both ecosystems and human health [13]. Although the irreversible cross-linked network of thermosets confers excellent dielectric performance, it precludes effective physical or chemical recycling after service, generating persistent industrial solid waste. Furthermore, the widespread use of halogenated flame retardants such as brominated epoxy resins complicates the removal of halogenated organics in wastewater. Owing to inadequate end-of-pipe treatment technologies, many manufacturers inadvertently emit toxic substances through unregulated discharges.

Approximately 90% of conventional epoxy resins rely on BPA-based epoxy prepolymers, such as diglycidyl ether of bisphenol A (DGEBA), during synthesis. BPA, however, is entirely derived from fossil resources. It is estimated that resin production consumes nearly 300 million tons of fossil-based raw materials annually, intensifying concerns over resource depletion. The nonrenewable nature of fossil feedstocks, characterized by prolonged regeneration cycles and limited reserves, poses significant challenges for material supply sustainability [14].

More critically, BPA has been identified as a reproductive toxicant. Its presence throughout production and application stages raises significant health concerns, particularly for pregnant women, infants, and children, including potential impairments to immune and reproductive systems and disruptions in neurological development [14]. In parallel, polyurethanes (PUs), the sixth most widely used polymer family, account for approximately 8% of the global polymer market [15]. Conventional PUs are synthesized through the polyaddition reaction between polyols and isocyanates. Presently, about 75% of commercial polyols are petroleum-derived polyether polyols, and the remaining 25% are polyester based, also originating from fossil resources. Common isocyanate monomers, such as methylene diphenyl diisocyanate (MDI) and toluene diisocyanate (TDI), are likewise fossil based. This heavy dependency on fossil feedstocks creates a dual challenge: on one hand, the continued consumption of petroleum exacerbates greenhouse gas emissions, contributing to global climate change; on the other, the depletion of fossil reserves threatens long-term industrial viability. Moreover, fluctuating oil prices and the emission of VOCs further underline the environmental and economic unsustainability of current production paradigms.

Hence, there is an urgent need to develop renewable, bio-based alternatives to facilitate a greener and more sustainable transformation of the polymer industry [16]. The conventional "high carbon, high toxicity" manufacturing paradigm for solid insulation materials is increasingly misaligned with the global shift toward green manufacturing. Urgent advancements in raw material substitution and process optimization are essential to enable sustainable industrial upgrading.

2.2.2 Key Issues in the Service Phase of Conventional Solid Insulating Materials

During long-term service, solid insulation materials are subjected to the coupled effects of thermal, electrical, and mechanical fields. In practice, inevitable defects introduced during manufacturing can cause local electric field intensification, leading to partial discharge and the subsequent initiation of electrical treeing phenomena [17]. As shown in Figure 2.2, in the initiation phase, electrical trees typically exhibit decelerated propagation characterized by intermittent, stepwise growth bursts. This is followed by a relatively quiescent or dormant growth stage, often accompanied by tip splitting. The formation of branched channels reduces the local electric field at the tree tip and may temporarily halt tree progression. For sustained propagation, accelerated growth resumes as the leading branches approach the counter electrode. Ultimately, in the runaway stage, the growth transitions into catastrophic electrical breakdown, resulting in the formation of through-thickness channels ranging from micrometers to millimeters in diameter [18].

In high-field regions, solid insulation materials are prone to collision ionization and dielectric loss-induced heating. Joule heat and polarization relaxation heat synergistically accumulate within the porous, dendritic tree structures, forming

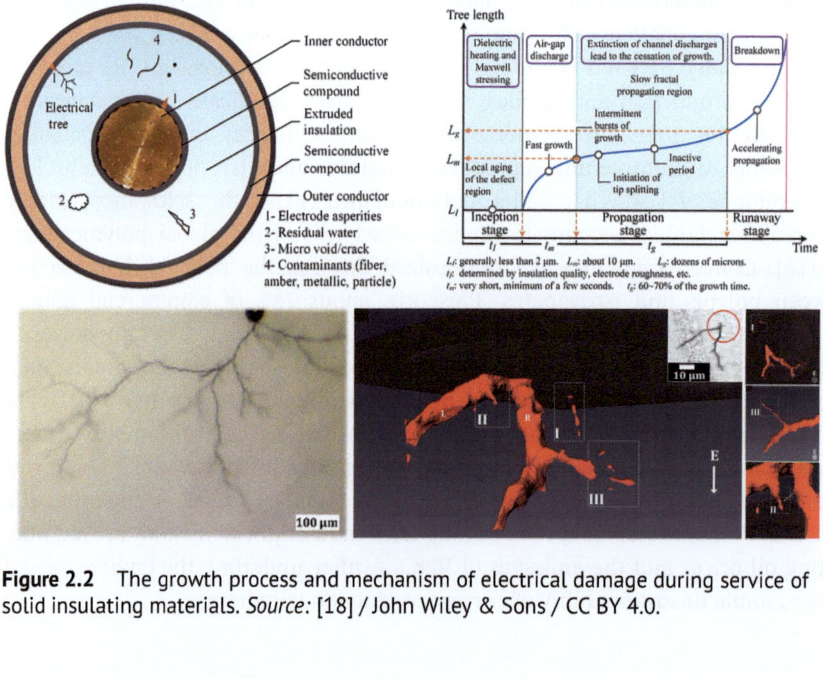

Figure 2.2 The growth process and mechanism of electrical damage during service of solid insulating materials. *Source:* [18] / John Wiley & Sons / CC BY 4.0.

Figure 2.3 Forms of damage to solid insulating materials during service: (a) electrical damage; (b) fire from heat accumulation; (c) stress cracking after aging.

localized heat cores and resulting in steep thermal flux density gradients within the material. Such nonuniform thermal distribution induces differential segmental relaxation rates of polymer chains in the matrix, leading to interfacial thermal stress concentration and degradation of the thermal stability of binding agents. Under the combined influence of thermal expansion and material anisotropy, micro-slippage and dislocation multiplication occur at the polymer filler interfaces (Figure 2.3). These processes ultimately trigger a coupled deterioration of dielectric and mechanical properties: the electrical tree network evolves into penetrating cracks; molecular chain scission and chemical bond degradation drastically reduce the material's ultimate tensile strength; and multiscale, synergistic damage mechanisms critically undermine the operational lifetime and reliability of the insulating medium.

2.2.3 Key Issues in the Disposal Phase of Conventional Solid Insulating Materials

Moreover, the intrinsic chemical inertness of polymer chains in solid insulation materials results in extremely low natural degradation rates. Certain cross-linked polymers may persist in the environment for centuries, gradually fragmenting into microplastics that exert long-term detrimental effects on soil ecosystems. At the end-of-life, these materials are typically disposed of via landfilling or incineration. Landfilling not only consumes valuable land resources but also poses environmental risks. Slow hydrolysis can release chlorinated hydrocarbons, such as phthalate plasticizers from PVC, which migrate through groundwater and cause diffuse source pollution. Incineration, on the other hand, risks the release of persistent organic pollutants (POPs) such as dioxins and furans, which are bioaccumulative and carcinogenic, and may disperse through the atmosphere, contributing to transboundary pollution.

Due to their low biodegradability under natural conditions, solid insulation materials undergo surface-level physical disintegration and chemical oxidation, generating microplastic particles (<5 mm). These particles can be transported across environmental media, entering soil and aquatic systems via surface runoff or being carried airborne through wind, leading to widespread distribution [19]. In terrestrial environments, microplastics disrupt soil aggregate structures, suppress microbial activity, and impair nutrient cycling. In aquatic ecosystems, biofilms forming on the surface of microplastics selectively adsorb heavy metals and hydrophobic organic contaminants, turning the particles into carriers of complex pollutants.

These contaminants can be internalized by plankton and accumulate through the food chain, posing increasing toxicological risks and serious threats to public health and environmental safety.

Designing appropriate recovery and recycling strategies for solid insulation materials remains a pressing challenge. Conventional thermosetting resins (e.g., phenolic resins, unsaturated polyesters) possess highly cross-linked three-dimensional networks that are difficult to depolymerize through either physical or chemical means. As a result, mechanically recycled thermosets are typically downcycled into low-value fillers. While thermoplastic materials (e.g., PE, PVC) can be remelted and reprocessed, repeated thermal cycles reduce molecular weight and crystallinity, significantly compromising their dielectric strength. The breakdown field strength of recycled thermoplastics is commonly 20–40% lower than that of virgin materials, failing to meet the electrical insulation requirements for reuse in high-voltage equipment.

Therefore, the development of next-generation recyclable solid insulation materials with minimal or no performance loss has emerged as a critical research frontier. Balancing recyclability, material integrity, and insulation performance is essential to achieve sustainable circularity in electrical insulation systems.

2.3 Key Issues in Conventional Liquid Insulation

Liquid insulating materials (insulating oils) are primarily employed as core dielectric media in equipment such as transformers and capacitors. Composed mainly of mineral oils and synthetic esters, these materials pose environmental hazards not only during operation due to potential leakage but more critically at the end-of-life stage, where pollutant release is characterized by long-term persistence and composite toxicity. Throughout their life cycle, insulating oils harbor considerable ecotoxicological risks, making their management a central concern in the sustainable development of electrical insulation systems [20].

2.3.1 Key Issues in the Production Phase of Conventional Liquid Insulation

Mineral oil, as a representative traditional insulating fluid, exhibits multiple environmental and process-related drawbacks throughout its production cycle. As a petroleum-derived product, its feedstock extraction is inherently dependent on fossil fuel mining, exacerbating the depletion of nonrenewable resources. The manufacturing process involves complex physical and chemical treatment steps, including vacuum distillation, solvent refining, and conventional acid washing, resulting in high energy consumption and the release of pollutants such as sulfur-containing wastewater and acidic gaseous emissions. To enhance oxidative stability, a controlled aromatic hydrocarbon content must be maintained during refining. However, residual sulfur compounds in fossil-based feedstocks can be converted into corrosive species during operation, posing threats to equipment integrity. Moreover, the high-temperature refining steps unavoidably lead to the formation of PAHs, persistent toxic substances. Due to their hydrophobic nature, PAHs tend to accumulate in the environment and enter biological systems through bioaccumulation, potentially inducing long-term genetic toxicity and increasing the risk of heritable diseases.

2.3.2 Key Issues in the Service Phase of Conventional Liquid Insulation

The environmental burden of liquid insulating materials is strongly inversely correlated with their effective service life [21]. Among various degradation mechanisms, thermo-oxidative aging is a dominant factor responsible for performance deterioration. Elevated temperatures accelerate the scission and recombination of hydrocarbon chains, leading to the formation of polar degradation products such as carboxylic acids and ketones. These byproducts contribute to internal corrosion and the generation of conductive particulates within electrical equipment, ultimately resulting in premature oil failure [22] (Figure 2.4).

As reported by Liao et al., under sustained thermal, electrical, and mechanical field coupling, the alkane components in insulating oils undergo a progressive oxidation process involving a three-stage mechanism: chain initiation, chain propagation, and chain termination. During initiation, various alkyl free radicals are generated; in the propagation stage, alcohols and water are produced; and in the termination stage, acidic species such as carboxylic acids dominate [24]. The accumulation of alcohols, acids, and moisture not only alters the dielectric properties of the oil but also serves as catalytic agents, further accelerating oxidation reactions and shortening the effective service life of the insulating fluid [22].

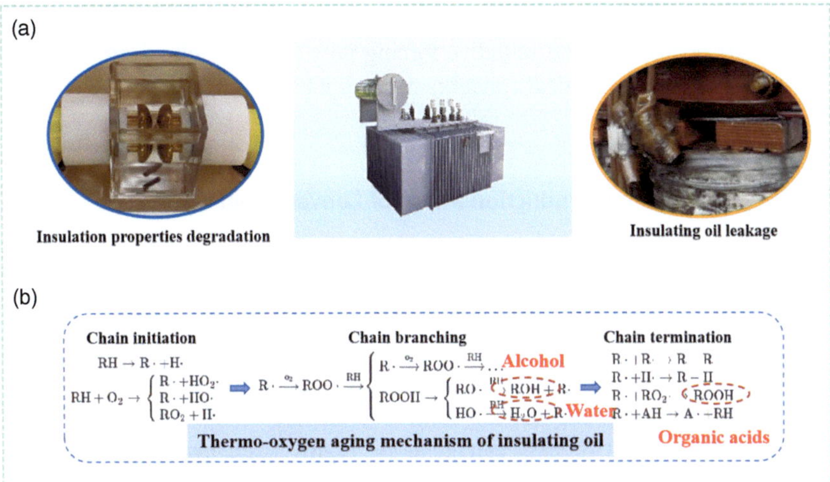

Figure 2.4 (a) Key issues during the service life of conventional insulating oils. *Source:* [23] / American Chemical Society. (b) Thermo-oxygen aging mechanism of insulating oil. *Source:* Adapted from [24].

Additionally, electrical treeing under high electric field stress and molecular chain scission induced by mechanical vibrations exacerbate dielectric breakdown. These combined effects accelerate the decline in insulation strength, increase the frequency of equipment maintenance or replacement, and ultimately lead to a nonlinear increase in the volume of discarded insulating oil [25].

2.3.3 Key Issues in the Disposal Phase of Conventional Liquid Insulation

Conventional disposal and regeneration methods for aged insulating oils primarily rely on filtration and adsorption techniques. However, these approaches exhibit limited efficacy in impurity removal and are largely ineffective in restoring the oil's original dielectric performance [26]. Residual polar contaminants such as water, acids, alcohols, and furans not only compromise the insulation properties but also catalyze further thermo-oxidative degradation of the oil. Although advanced methods like adsorption–desorption cycles can partially separate pollutants, they often rely on activated carbon, whose thermal regeneration may lead to the release of VOCs, generating secondary gaseous pollution. These limitations highlight the urgent need for the development of next-generation electrical insulating oils, which remains both a research priority and a significant technical challenge.

2.4 Key Issues in Conventional Gas Insulation

SF_6 is widely employed in gas-insulated switchgear (GIS) due to its exceptional dielectric strength and arc-quenching capability. However, SF_6 is also an extremely potent greenhouse gas, with a GWP 23 500 times that of CO_2 (1 kg $SF_6 \approx 23.5$ metric tons CO_2 equivalent) [27], and an atmospheric lifetime exceeding 3200 years [28]. Its persistence and radiative forcing pose a severe long-term threat to the environment, raising critical concerns regarding its continued use in electrical insulation systems.

2.4.1 Key Issues in the Production Phase of Conventional Gas Insulation

The industrial-scale production of conventional insulating gases such as SF_6 faces significant toxicological and environmental challenges, primarily stemming from the hazardous nature of the high-temperature fluorination processes involved. Fluorine gas (F_2), a key reactant, is highly corrosive and acutely toxic, necessitating stringent leak-prevention engineering controls to ensure occupational safety [29]. Moreover, the synthesis process can generate highly toxic byproducts such as S_2F_{10}; improper control during the distillation and purification stages can elevate the risk of occupational exposure and persistent environmental contamination. More critically, SF_6 itself is classified as a super greenhouse gas with a GWP of 23 500 and an atmospheric lifetime exceeding 3200 years [30]. Fugitive emissions during production and application phases substantially exacerbate climate change pressures. In addition, coproduced perfluorinated compounds (PFCs) exhibit extreme thermal and chemical stability, leading to long-term accumulation in ecosystems and the emergence of complex environmental toxicological effects.

As such, the entire production chain of SF_6 is intrinsically associated with high environmental loads and toxic hazards, underscoring the urgent need to develop environmentally benign synthesis routes or next-generation eco-friendly insulating gases to replace SF_6 in electrical equipment.

2.4.2 Key Issues in the Service Phase of Conventional Gas Insulation

Due to inevitable manufacturing and operational defects in GIS, localized discharges, overheating, or intense sparking and arcing may occur within the equipment. These phenomena induce the decomposition of the SF_6 insulating medium, generating a variety of low-fluorinated sulfur species and fluorine radicals, such as SF_5, SF_4, SF_3, SF_2, SF, and fluorine (F) [4].

Furthermore, owing to the inherent sealing limitations during GIS assembly, trace amounts of moisture (H_2O) and oxygen (O_2) are almost always present as impurities. The reactive low-fluorinated sulfur species readily interact with these impurities, forming a wide array of secondary decomposition products, including SO_2, SOF_2, SO_2F_2, H_2S, and HF [31]. These reactions disrupt the equilibrium between SF_6 decomposition and recombination, triggering cascading degradation pathways that significantly compromise the insulating effectiveness and longevity of the gas medium.

Under arcing conditions, SF_6 further decomposes into highly toxic gases such as SF_4 and SOF_2, posing direct respiratory hazards to personnel. Moreover, decomposition products may react with residual moisture to form hydrofluoric acid (HF), a corrosive substance that attacks metal components and accelerates insulation aging. This leads to a "leakage–decomposition–re-leakage" feedback loop, compounding both safety and reliability concerns in GIS-based systems (Figure 2.5).

In-service GIS systems are inherently prone to SF_6 leakage over time. According to international standards such as IEC 62271-203 and IEEE C37.122, the maximum allowable annual leakage rate per gas compartment is 0.5%. Over a typical 40-year service lifetime, this could cumulatively result in up to 20% of the total SF_6 inventory being released into the atmosphere [32]. Given SF_6's extremely high GWP and long atmospheric residence time, such leakage levels pose a substantial long-term threat to the atmospheric environment, significantly undermining the environmental sustainability of GIS-based insulation systems.

2.4.3 Key Issues in the Disposal Stage of Conventional Gas Insulation Materials

Current recovery strategies for decommissioned SF_6 primarily involve physical adsorption, cryogenic condensation, membrane separation, and chemical catalytic decomposition techniques. Physical adsorption utilizes porous materials such as activated carbon to concentrate low-concentration SF_6, but its efficiency is inherently limited by the adsorption capacity of the medium. Cryogenic

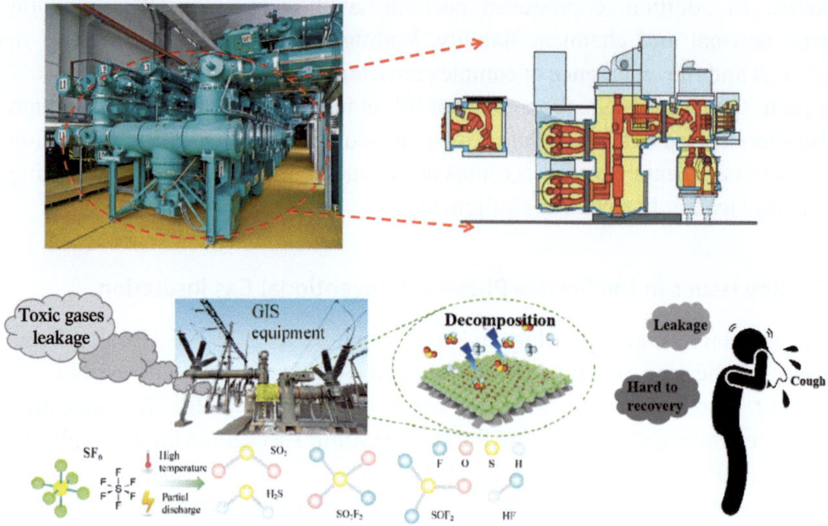

Figure 2.5 GIS devices and internal insulation gas distribution; key issues faced during the service life of insulation gases. *Source:* [32] / with permission of Elsevier.

condensation, which liquefies SF_6 (boiling point: −64 °C) using liquid nitrogen, can achieve high-purity recovery, yet suffers from high energy consumption and operational costs [33]. Membrane separation leverages differences in gas molecular size and diffusivity to enable selective permeation, with separation efficiencies exceeding 80% under optimal conditions [34]. However, performance is highly sensitive to mixed contaminants such as moisture and particulates. Chemical catalytic decomposition involves breaking down SF_6 into simpler molecules under high temperature or in the presence of catalysts. Nevertheless, this method generates toxic byproducts, including HF and SOF_2, which require additional posttreatment to mitigate environmental and health risks. Key technological bottlenecks include the efficient enrichment of low-concentration mixed gases (with typical recovery rates below 70%), synergistic interference from coexisting pollutants (e.g. SOF_2, CF_4) on separation systems, and the high energy footprint of most recovery processes [35].

As a result, the safe and effective handling of decommissioned SF_6, along with the rational design of next-generation eco-friendly insulating gases, represents one of the most critical and challenging frontiers in the development of sustainable gas insulation technologies.

2.5 Eco-Cycling Requirements for Conventional Insulation Materials

Current electrical insulation systems spanning solid, liquid, and gaseous states face critical challenges in aligning with circular economy principles. Solid insulation materials, such as epoxy resins and other thermosetting polymers, exhibit inherent difficulty in degradation due to their irreversible cross-linked networks; moreover, thermal decomposition often generates toxic byproducts such as dioxins. Liquid insulating media, including mineral oils and synthetic esters, pose significant environmental risks in the event of leakage, as they contain PAHs and heavy metal contaminants. The ecological remediation costs can be three to five times higher than conventional landfill treatment. For gaseous insulators, SF_6, a potent greenhouse gas with a GWP 23 500 times that of CO_2, remains difficult to recycle efficiently due to sub-60% recovery rates at low concentrations and high energy consumption, thus hindering resource-closed-loop utilization. Meanwhile, international policy frameworks are accelerating the transition toward circularity. The European Union's *Circular Economy Action Plan* mandates that by 2030, over 70% of insulation materials should originate from renewable sources. In China, the "dual carbon" strategy emphasizes full life-cycle carbon tracking for electrical equipment. Additionally, the revised *Basel Convention* imposes stricter controls on the transboundary movement of halogenated insulation waste, compelling the industry to overcome existing technological bottlenecks. In response, research is converging on three key directions: (1) Development of bio-based and degradable materials, such as enzyme-responsive epoxies and polylactic acid (PLA)-based insulating oils, enabling controllable degradation and carbon footprint reduction through molecular-level design. (2) Construction of dynamic covalent bond networks, allowing conventional thermosets to undergo depolymerization and reconfiguration under thermal or photonic stimuli, thus supporting modular disassembly and material regeneration. (3) Innovation in closed-loop processing for gaseous media, involving techniques such as selective adsorption and plasma-catalyzed decomposition of SF_6 into reusable sulfur fluoride feedstocks, and concurrent development of eco-friendly insulating gases to close the material loop. Looking ahead, integrating materials genome engineering and multiscale simulation will be vital to optimizing the trade-off between performance and recyclability, fostering a transition from linear consumption models to regenerative material paradigms. These advancements will lay the low-carbon technological foundation for next-generation power systems.

References

1 Huang, X., Han, L., Yang, X. et al. (2022). Smart dielectric materials for next-generation electrical insulation. *iEnergy* 1 (1): 19–49.

2 Wang, C., Goldman, T.M., Worrell, B.T. et al. (2018). Recyclable and repolymerizable thiol–X photopolymers. *Materials Horizons* 5 (6): 1042–1046.

3 Li, C., Chong, Y., and Chen, P. (2024). Comparison on the pyrolysis characteristics of PDMS, PIB, and NTO based on ReaxFF. *IEEE Transactions on Dielectrics and Electrical Insulation* 31 (1): 12–21.

4 Fang, X., Hu, X., Janssens-Maenhout, G. et al. (2013). Sulfur hexafluoride (SF_6) emission estimates for China: an inventory for 1990–2010 and a projection to 2020. *Environmental Science & Technology* 47 (8): 3848–3855.

5 Xiao, S., Shi, S., Li, Y. et al. (2021). Review of decomposition characteristics of eco-friendly gas insulating medium for high-voltage gas-insulated equipment. *Journal of Physics D: Applied Physics* 54 (37).

6 Tsai, H.-Y., Fujita, T., Wang, S., and Naito, M. (2021). Environmentally friendly recycling system for epoxy resin with dynamic covalent bonding. *Science and Technology of Advanced Materials* 22 (1): 532–542.

7 Wan, B., Dong, X., Yang, X. et al. (2022). High strength, stable and self-healing copolyimide for defects induced by mechanical and electrical damages. *Journal of Materials Chemistry C* 10 (31): 11307–11315.

8 Liu, Y., Zhao, J., Peng, Y. et al. (2020). Comparative study on the properties of epoxy derived from aromatic and heteroaromatic compounds: the role of hydrogen bonding. *Industrial & Engineering Chemistry Research* 59 (5): 1914–1924.

9 Montarnal, D., Capelot, M., Tournilhac, F., and Leibler, L. (2011). Silica-like malleable materials from permanent organic networks. *Science* 334 (18): 965–968.

10 Unger, C., Schmalz, H., Lipp, J. et al. (2024). A closed-loop recyclable low-density polyethylene. *Advanced Science* 11 (13).

11 Xiao, S., Xia, H., Li, Z. et al. (2025). Enhancing corrosion resistance of copper for C_4F_7N based eco-friendly gas-insulated equipment. *High Voltage* 10: 1125–1134.

12 Wang, T., Hu, X., Yang, H. et al. (2025). Design of Pt-C_3N_2 composites for efficient real-time monitoring of fluorocarbon insulation gas decomposition and its application in GIS equipment. *Surfaces and Interfaces* 68: 106721.

13 Cantoni, B., Riguzzi, A.C., Turolla, A., and Antonelli, M. (2021). Bisphenol a leaching from epoxy resins in the drinking water distribution networks as human health risk determinant. *Science of the Total Environment.* 783: 146908.

14 Jiang, Y., Li, J., Li, D. et al. (2024). Bio-based hyperbranched epoxy resins: synthesis and recycling. *Chemical Society Reviews* 53 (2): 624–655.

15 Skoczinski, P., Krause, L., Raschka, A. et al. (2021). Current status and future development of plastics: solutions for a circular economy and limitations of environmental degradation. In: *Enzymatic Plastic Degradation. Methods in Enzymology* (ed. P. Skoczinski), 1–26. Elsevier.

16 Tran, M.H. and Lee, E.Y. (2023). Production of polyols and polyurethane from biomass: a review. *Environmental Chemistry Letters* 21 (4): 2199–2223.

17 Nedjar, M. and Beroual, A. (2008). Electrical aging of polyurethane under AC voltage. *Journal of Applied Polymer Science* 109 (2): 789–794.

18 Yang, Y., Dang, Z.M., Li, Q., and He, J. (2020). Self-healing of electrical damage in polymers. *Advanced Science* 7 (21): 2002131.

19 Jin, M., Sun, M., Liu, J. et al. (2024). Influence of operating parameters on the yield of micro-plastics from plastics incineration. *Science of the Total Environment* 912: 169347.

20 Qin, J., Peng, X., Qiu, Q., and Tang, C. (2022). A new type of nano APTES-hBN modified palm oil as natural ester insulating oil with upgraded thermal aging characteristics. *Renewable Energy* 200: 743–750.

21 Oliveira, M.M., Medeiros, L.H., Bacin, P.O. et al. (2025). Analytical and numerical evaluation of an oil-water heat exchanger applied on power transformer considering the mineral/vegetable insulating oil replacement. *Case Studies in Thermal Engineering* 73: 106524.

22 Yao, Y. (2016). A triarchic theory of granular computing. *Granular Computing* 1 (2): 145–157.

23 Wu, Y., Zhang, J., Chen, B. et al. (2024). Two-element mixed natural ester insulation oil doped with Kh550-TiO_2 nanoparticles for enhanced antiaging properties. *ACS Applied Nano Materials* 7 (5): 4721–4730.

24 Liao, R., Lin, Y., Guo, P. et al. (2015). Thermal aging effects on the moisture equilibrium curves of mineral and mixed oil-paper insulation systems. *IEEE Transactions on Dielectrics and Electrical Insulation* 22: 842–850.

25 Siddique, A., Yaqoob, M., Aslam, W. et al. (2024). A systematic review on promising development of cost-effective, biodegradable, and environment friendly vegetable based nanofluids as a future resource for green transformer insulation oil. *Journal of Molecular Liquids* 403: 124836.

26 Wang, S., Dong, L., Peng, R., and Tang, H. (2025). Green synthesis of trifluoromethanesulfonyl fluoride as an eco-friendly alternative to SF6 gas insulation and analysis of its acute inhalation toxicity. *Molecules* 30 (10): 2241.

27 Ye, F., Xiao, F., Zhan, A. et al. (2025). QSAR-based prediction of acute inhalation toxicity and SHAP interpretability analysis of fluorocarbon environmental-friendly insulating gases. *Environmental Research* 285: 122340.

28 Wang, G., Kim, W.-H., Kil, G.-S. et al. (2019). Green gas for a grid as an eco-friendly alternative insulation gas to SF_6: from the perspective of partial discharge under AC. *Applied Sciences* 9 (4): 651.

29 Gao, W., Posada, L.F., Shiravand, V. et al. (2024). High-throughput compatibility screening of materials for SF_6-alternative insulation. *Environmental Science & Technology* 58 (30): 13296–13306.

30 Shi, Y., Wang, J., Li, Y. et al. (2025). Unveiling the ambient CO_2-mediated K_2CO_3 activation pathway in KOH post-treated activated carbon for exceptional SF_6 adsorption. *Chemical Engineering Journal* 520: 166100.

31 Zhang, H., Huang, Z., Jiang, J. et al. (2024). DFT study of Cu_n (n = 1–3) cluster-modified WSe_2 monolayers for the detection of SF_6 decomposition gases (H_2S, SO_2, SOF_2 and SO_2F_2). *Surfaces and Interfaces* 52: 104904.

32 Koch, H. (2014). Introduction. In: *Gas Insulated Substations* (ed. H.J. Koch), 1–19. Wiley.

33 Cao, C., Zhao, Y., Dai, Y. et al. (2025). ZIF-8 membranes for highly efficient SF_6 recovery. *Separation and Purification Technology* 369: 133005.

34 Wu, W., Cao, C., Zhao, Y. et al. (2025). Mixed matrix membranes embedded stiffer ZIF-8 for highly efficient SF_6 recovery. *Journal of Membrane Science* 734: 124399.

35 Gang, W., Dongyi, W., Xiaoyu, W. et al. (2025). Study on the adsorption mechanism of typical SF_6 decomposition products SOF_2 and SO_2F_2 on the Γ-Al_2O_3(110) surface: a first-principles study. *Separation and Purification Technology* 361.

3

Characteristics and Types of Green Insulating Materials

Baoquan Wan[1] *and Jun-Wei Zha*[2]

[1] *Tsinghua University, State Key Laboratory of Power System Operation and Control, Department of Electrical Engineering, Baoquan Wan: No. 30 Shuangqing Road, Haidian District Beijing 100084, P. R. China*
[2] *North China Electric Power University, State Key Laboratory of Alternate Electrical Power System with Renewable Energy Sources, School of Electrical and Electronic Engineering, Jun-Wei Zha: No. 2 Beinong Road, Huilongguan, Changping District, Beijing 102206, P. R. China*

3.1 Introduction

Electrical insulating materials are vital for the safe and efficient operation of power equipment and electronic devices [1–3]. Traditionally, the most widely used insulators, such as oil-impregnated paper, mineral oil (MO), porcelain, cross-linked polyethylene (XLPE), and sulphur hexafluoride (SF_6) gas, have been selected primarily for their excellent dielectric and thermal properties [4, 5]. However, many traditional materials pose environmental and sustainability challenges, including high carbon footprints, toxicity, difficult disposal, and greenhouse gas emissions. In response, green insulating materials have emerged, aiming to maintain high electrical properties while reducing environmental impact. Green insulating materials can be defined as insulating solids, liquids, or gases that are low-carbon, eco-friendly, and sustainable. They are typically derived from renewable resources, have lower greenhouse gas potential, and are nontoxic, recyclable, self-healable, or biodegradable, all while complying with modern health and environmental standards.

Over the last decade, intensifying environmental and safety requirements have driven the electrical industry to seek eco-friendly insulation alternatives [6–10]. Polychlorinated biphenyls (PBs), once used in transformers for their superb dielectric and fire-resistant properties, were completely banned in the late 1970s due to their extreme toxicity and persistence in the environment. Similarly, SF_6, a man-made gas with excellent insulating ability, has come under scrutiny as the most potent greenhouse gas known. At the same time, concerns about plastic waste and disposal have highlighted the shortcomings of thermosetting polymer insulation materials such as XLPE, which cannot be remelted or recycled and are typically incinerated or landfilled. This trend has catalyzed the development

Next-Generation Electrical Insulating Materials: Low-Carbon, Eco-Friendly, and Sustainable Solutions, First Edition. Edited by Jun-Wei Zha.

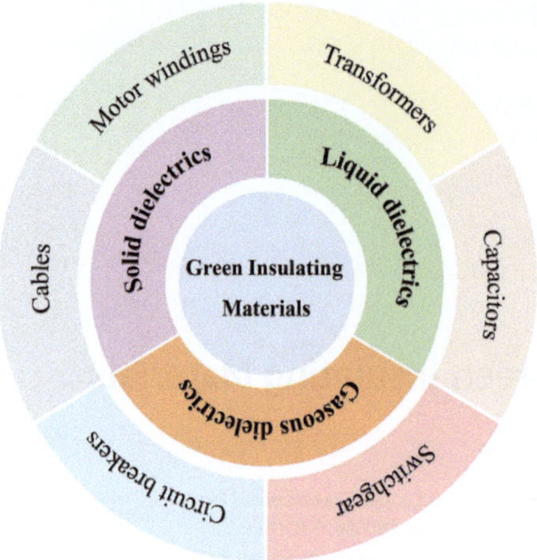

Figure 3.1 Types and applications of green insulating materials.

of next-generation insulating materials that can meet stringent performance requirements while minimizing environmental impact.

Green insulating materials include all three types of insulators: solid, liquid, and gas, each of which is suitable for specific applications, as shown in Figure 3.1. In general, solid dielectrics (plastics, rubber, composites, ceramics) provide insulation in cables, motor windings, printed circuit boards (PCBs), and electrical insulation. Liquid dielectrics (oils and fluids) are extensively used in high-voltage transformers, capacitors, and as impregnating media for paper/polymeric insulation. Gaseous dielectrics fill equipment like switchgear, circuit breakers, and gas-insulated lines, providing insulation and arc-quenching ability. For each type of insulating material, researchers are developing more eco-friendly alternatives, such as recyclable, self-healable, or biodegradable polymers for solid insulation, natural ester oils extracted from organic matter to replace mineral oils, and gases or vacuum systems with zero or extremely low global warming potential (GWP) to replace SF_6. These new materials are designed to match or exceed the electrical insulation properties of traditional materials, while also offering improvements in heat resistance, mechanical properties, ageing stability, safety, and disposal or recyclability.

This chapter provides a comprehensive review of green insulating materials in each major application category: power cables (solid dielectrics), transformers (liquid and solid insulations), high-voltage switchgear (gaseous or vacuum insulation), and electronic insulation (boards, wires, encapsulants). For each category, we discuss the key performance characteristics of emerging eco-friendly materials compared to traditional insulators, including electrical (dielectric strength, dielectric constant, dielectric loss, breakdown behavior), thermal, mechanical, and ageing properties. We also highlight relevant industry standards, safety considerations, and environmental metrics (such as flammability classes, biodegradability, and recyclability). Finally, the chapter reflects on how the material characteristics influence their adoption in practice, and the outlook for sustainable insulation solutions in the electrical industry.

3.2 Green Solid Insulating Materials for Power Cables and Accessories

Power cables rely on solid polymeric insulation to withstand high voltages over long service lifetimes [11–13]. For decades, XLPE has been the dominant insulation for medium and high voltage cables due to its excellent dielectric properties, high breakdown strength, and good thermal-mechanical balance [14, 15]. XLPE is a thermosetting plastic formed by chemically cross-linking polyethylene. This process creates a stable, heat-resistant network, but also makes the material nonmeltable and nonrecyclable. At the end of their service life, XLPE cables cannot be reshaped and are difficult to separate from metal conductors. Therefore, XLPE waste is typically disposed of through incineration or landfill, which raises environmental concerns. Moreover, the cross-linking step in manufacturing consumes additional energy and additives (peroxides), contributing to the material's carbon footprint. Polyvinyl chloride (PVC) is used in low-voltage wiring and cable sheathing. However, PVC contains chlorine and toxic plasticizers, and when burned, it produces corrosive halogen gases [16]. Stricter fire safety codes now discourage PVC in indoor installations in favor of halogen-free alternatives. Ethylene-propylene rubber (EPR) and related elastomers have seen use in medium-voltage cables, but they too are typically cross-linked and not easily recyclable [17, 18]. Traditional cable polymers pose challenges to sustainability. They introduce harmful substances and are mostly thermosetting or halogenated materials, which hinder recycling and can release toxic fumes in fires.

Green solid insulation for cables focuses on developing recyclable, halogen-free polymer systems that maintain high electrical properties. A leading candidate is polypropylene (PP), a thermoplastic polyolefin that, unlike XLPE, does not require cross-linking and can be repeatedly melted and reprocessed [19, 20]. Neat PP has desirable traits, low dielectric loss, and moderate permittivity similar to polyethylene, and the ability to be recycled at end-of-life. PP can be melted and reused, enabling full recyclability, whereas XLPE's thermoset network means XLPE cable scrap must be incinerated or landfilled. This recyclability directly translates to reduced waste and lower lifecycle environmental burden for PP-based cable insulation. Furthermore, producing cable insulation via the melt-processing of PP (without chemical cross-linkers) lowers energy consumption and simplifies manufacturing compared to XLPE. These aspects make PP-based insulations very attractive as eco-friendly, sustainable alternatives for next-generation cables.

However, pure PP in its standard form has some performance limitations that impede its use as high-voltage insulation. PP is more crystalline and can be mechanically stiffer and more brittle than XLPE, especially at cold temperatures. High crystallinity can also lead to more pronounced electrical field inhomogeneities at spherulite boundaries, which was linked to lower breakdown strength in earlier tests of isotactic PP insulation. Furthermore, pure PP has a high melting point (~160 °C) but poor thermal flexibility. XLPE can withstand short-term emergency temperatures (~250 °C) without melting (due to its cross-linked structure). Thermoplastic

PP insulating materials must be carefully formulated to avoid deformation under extreme fault conditions. To overcome these challenges, recent research has focused on PP-based blends and nanocomposites [21, 22]. By blending PP with elastomers or adding nanofillers, engineers can toughen the polymer and reduce its crystallinity, thereby improving its mechanical ductility and electrical strength. For example, blending PP with a compatible thermoplastic elastomer has shown significant improvements. PP+SEBS (styrene-ethylene-butylene-styrene) blends exhibit reduced stiffness and brittleness and yield a finer, lower-crystallinity morphology that enhances dielectric breakdown performance. In one recent study, a PP–SEBS insulation with ~35% SEBS achieved an impressive AC breakdown strength of 66.4 kV/mm, comparable to or even exceeding the breakdown strength of traditional XLPE, while also demonstrating a long-predicted service life of ~42 years, as shown in Figure 3.2. The improved performance was attributed to the elastomer-induced reduction in crystalline size and fraction, which mitigated the initiation of electrical breakdown at crystal boundaries. Notably, beyond matching XLPE electrically, these PP-based blends retained the key sustainability advantage, full recyclability, and simpler processing without peroxides. The optimized PP/SEBS insulation outperformed conventional XLPE in both mechanical flexibility and dielectric

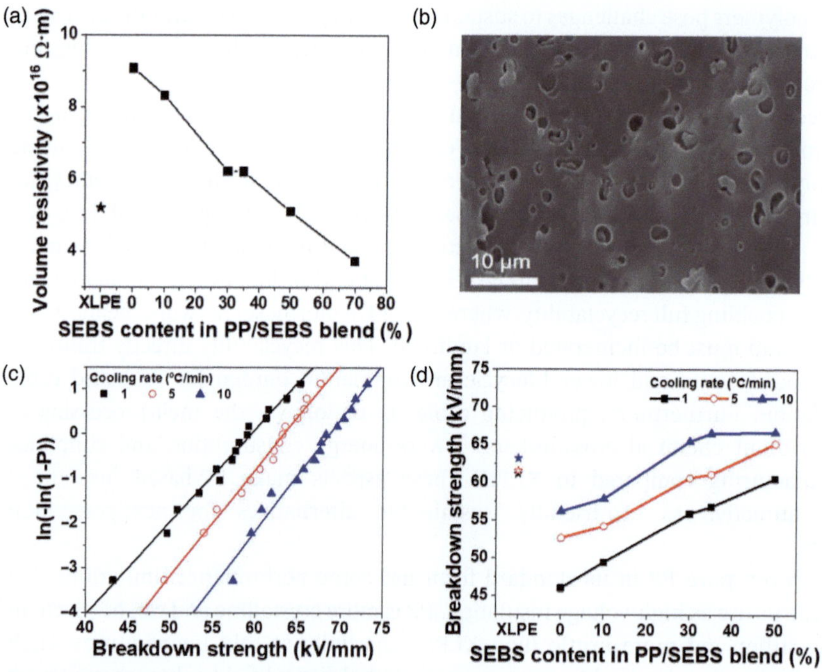

Figure 3.2 (a) Volume resistivity of PP/SEBS blends as a function of SEBS content. (b) SEM images of PP/SEBS blends with SEBS 35% content. (c) Weibull plots of AC breakdown strength of PP/SEBS blends as a function of cooling rate at 35% SEBS in blends and (d) breakdown strength of PP/SEBS blends with varying SEBS content. *Source:* [23] / MDPI / CC BY 4.0.

strength, highlighting its promise as a technologically and environmentally superior cable insulation [23].

Except for PP blends, other strategies for green cable insulation include nanocomposite dielectrics (e.g. PE or PP loaded with MgO or SiO_2 nanoparticles) to improve breakdown strength and suppress space charge accumulation [24, 25]. Nanocomposites can allow thinner insulation or higher voltage ratings with the same base polymer, potentially reducing material usage. Researchers have also explored copolymers and alternative thermoplastics, for instance, polyethylene copolymerized with rubbery units ethylene vinyl acetate (EVA) or polyolefin elastomers have been studied as drop-in replacements for XLPE [26, 27]. Some of these materials reached comparable dielectric properties, but issues like space charge in direct current (DC) fields and lower thermal limits are noted in earlier generations. The latest PP-based systems, by contrast, demonstrate very low dielectric loss and stable space charge behavior, making them strong candidates for high-voltage direct current (HVDC). The chemical simplicity (nonpolar backbone with no peroxide byproducts) of PP is inherently advantageous for HVDC insulation, as it reduces charge traps and electrical aging under DC stress. These improvements suggest that recyclable thermoplastic cable insulations can meet the rigorous reliability demands of underground transmission cables.

The thermal and mechanical performance of the new green cable polymers can be carefully engineered to match standards. A typical XLPE-insulated power cable is rated for continuous operation at 90 °C (with short-term emergency up to 105 °C) [28]. Thermoplastic PP insulation can also be rated 90 °C. In fact, certain PP grades or blends have higher softening points and can allow slightly higher operating temperatures before mechanical deformation. On the other hand, since PP melts at ~160 °C, short-circuit loads (which can raise conductor temperatures to 200–250 °C for a few seconds under fault conditions) are a design consideration. This can be compensated for by adjusting the cable structure or adding polymer additives that improve melt stability. For mechanical properties, unmodified PP is not as flexible as XLPE, but the introduction of elastomer modifiers (such as styrene ethylene butylene styrene or ethylene-octene copolymers) provides the necessary flexibility and impact resistance even at sub-zero temperatures. Recent formulations report elongation and tensile strength on par with XLPE, ensuring the cables can withstand installation bending and thermal expansion cycles [29]. Appropriately stable PP-based insulating materials exhibit excellent ageing properties and typically have a lower tendency to water treeing than XLPE, as the synthesis process for thermoplastics can be very pure (without cross-linking impurities). For example, lifetime modeling in one study showed the optimized PP/SEBS blend had an extrapolated life of more than 40 years at operating stress, comparable to XLPE service lifetime [30].

Green solid insulating materials for cables, particularly recyclable polyolefins (PP and advanced blends), have achieved levels of performance that are essentially equivalent to those of traditional XLPE in all key performance indicators. Another key performance characteristic of insulating materials is their insulating properties. Table 3.1 compares the performance of traditional XLPE cable

Table 3.1 Comparison of XLPE cable insulation and a next-gen recyclable PP blend insulation.

Property	XLPE (traditional)	Recyclable PP blend (green)
Dielectric strength (AC, 23 °C)	~30 kV/mm (thick sample); lab thin-film breakdown 60–70 kV/mm	~25–30 kV/mm; lab breakdown up to 65 + kV/mm (modification)
Dielectric constant (ε_r@50 Hz)	~2.3	~2.2
Dielectric loss (tanδ@90 °C)	~0.001	~0.001–0.002
Continuous operating temp.	90 °C (thermal class)	90 °C (similar rating; some blends up to 105 °C)
Water treeing resistance	High (with additives)	High (hydrophobic polymer)
Recyclability	Not recyclable	Fully recyclable

insulation with that of the new generation of recyclable PP insulation, highlighting the environmental benefits of the latter [23].

Another important aspect is fire safety and toxicity. In building wiring and transit cables, there is increasing demand for low-smoke and zero-halogen insulation to reduce hazardous fumes in the event of a fire. Traditional cable polymers like PVC emit hydrogen chloride gas and dense smoke when burning. Polyolefin-based insulations (PE, PP) are inherently halogen-free and can be compounded with mineral flame retardants (such as aluminum hydroxide or phosphorus/nitrogen-based flame retardants) to meet flame retardancy standards without halogens. These halogen-free flame-retardant cables dramatically reduce the emission of corrosive or toxic gases. Green cable insulations, being halogen-free by nature, readily fulfill these requirements. It is noted that some halogen-free formulations even offer better electrical and thermal stability than their halogenated counterparts. For example, replacing brominated flame retardants with phosphorus-based ones in PCB laminates has been shown to reduce the resin's polarity and increase its glass transition temperature, resulting in improved insulation resistance and thermal robustness [31]. Similarly, power cable compounds with non–halogen fire retardants can achieve high fire safety ratings without compromising dielectric performance. Thus, the move to halogen-free materials aligns with both environmental safety and electrical reliability.

It should be noted that cable systems also include accessories like joints and terminations, which use solid insulating materials (epoxies, silicone rubbers, etc.). Green innovations are appearing here as well. For instance, cold-shrink or elastomeric joint insulation made of silicone rubber prevents oil or resin from being poured on-site and is inherently recyclable or self-healable as polymer waste [32, 33]. Additionally, some suppliers are exploring bio-based or recycled materials for cable jackets and semi-conductive screens. Although these do not directly affect the dielectric function of the cable, they contribute to the overall sustainability of the cable system. The field of solid insulation for cables is undergoing a materials revolution. Recyclable thermoplastics, halogen-free formulations and nanocomposites enable cables to not only offer reliable electrical performance and a long service life, but also comply with the principles of the circular economy. The next sections will explore similar green material advancements in liquid and gaseous insulating media.

3.3 Green Liquid Insulating Materials for Transformers and Dielectric Fluids

High-voltage equipment such as power transformers, distribution transformers, high-voltage bushings, and capacitors relies on liquid dielectrics for insulation and cooling [34, 35]. The liquid insulation both provides electrical isolation between high voltage components and acts as a heat transfer medium to remove thermal losses. For over a century, the standard choice has been mineral oil (MO), a hydrocarbon oil refined from petroleum, owing to its good dielectric strength, low cost, and adequate cooling capability. MO has served extremely well in millions of transformers, but it has notable drawbacks from a sustainability and safety perspective. It is flammable (fire point ~160 °C), and can cause severe environmental contamination if leaked. MO is not readily biodegradable, so spills persist in soil/groundwater. Being petroleum derived, it contributes to fossil resource depletion. Furthermore, several catastrophic transformer fires in the past were fueled by MO, prompting stricter fire safety requirements for transformers located indoors or near sensitive installations. These concerns have intensified the search for green liquid insulations that are less hazardous and more eco-friendly, without sacrificing performance.

The leading green alternatives to mineral oil are ester-based insulating liquids, broadly classified into natural esters and synthetic esters. Natural esters (NE) are dielectric fluids derived from vegetable oils, common sources include soybean oil, rapeseed (canola) oil, sunflower seed oil, and others [36, 37]. Synthetic esters (SE) are engineered organic esters (often pentaerythritol tetra-fatty acid esters or similar) produced via chemical synthesis [38, 39]. They are not directly vegetable oil but can be made from either plant-derived or petrochemical feedstocks. Both types of esters are characterized by being biodegradable, fire-safe, and nontoxic, which gives them a much lower environmental risk profile than mineral oil. In fact, ester fluids have been classified as readily biodegradable and are often nontoxic to aquatic life. This means that natural esters leaked into the environment will decompose into harmless substances within a few weeks. Compared with MO, this greatly limits soil or water pollution. Many electric railways and wind farms have already adopted natural ester-insulated transformers to mitigate spill risks.

The electrical insulation performance of natural and synthetic esters is comparable to that of mineral oil, with some slight differences. In terms of dielectric strength, new ester fluids typically pass the same ASTM or IEC breakdown voltage tests required of mineral oil. For example, both natural and synthetic esters can achieve AC breakdown voltages of 50–60 kV (root mean square, RMS) in a standard 2.5 mm gap test (IEC 60156), often even higher if processed and dried properly. Table 3.2 provides a comparison of key properties for a representative mineral oil and ester fluids [40–42]. The breakdown voltage of esters is on par with or better than mineral oil when the fluids are dry and degassed. The relative permittivity (dielectric constant) of esters is slightly higher (~3.1–3.2 at room temperature) than mineral oil (~2.2). The dissipation factor (dielectric loss tangent) of new ester fluids is low, though generally a bit higher than ultra-refined mineral oil at ambient temperature. For instance, at 90 °C, a natural ester might have $\tan\delta$ on the order of 0.02–0.05, whereas a MO may be ~0.005–0.01. In practice, both fluid types easily meet insulation power-factor specifications (<0.5% at 20 °C, per IEC/ASTM limits).

Table 3.2 Properties of mineral oil and ester insulating liquids.

Property	Mineral oil	Natural ester	Synthetic ester
Dielectric strength (2.5 mm gap)	~45–60 kV	~50–60 kV	~50–65 kV
Relative permittivity (ε_r, 20 °C)	~2.2	~3.2	~3.2
Dielectric loss ($\tan\delta$@90 °C)	0.005–0.01	0.02–0.05	~0.005–0.01
Volume resistivity (Ω cm@20 °C)	~10^{12}–10^{13}	~10^{11}–10^{12}	~10^{12}
Flash point/fire point	~150 °C/~170 °C	~330 °C/~350 °C	~260 °C/~300 °C
Pour point	−40 °C to −60 °C	~ −20 °C to −30 °C	~ −50 °C
Water saturation @20 °C	~60 ppm	~1000 ppm	~250–300 ppm
Effect on paper moisture	Baseline	Strongly dries paper	Dries paper
Oxidation stability	Good	Moderate	Very good
Thermal conductivity (@20 °C)	~0.12 W/m·K	~0.15 W/m·K	~0.14 W/m·K
Fire safety class (IEC 61100)	O-class	K-class	K-class
Standards and guides	IEC 60296; IEEE C57.106	IEC 62770; IEEE C57.147	IEC 61099; IEEE C57.147

Thermal performance is another important performance indicator for esters. On the one hand, esters have higher viscosity than mineral oil (especially natural esters). At 20–25 °C, a natural ester can be approximately two to three times more viscous than mineral oil [43]. This means cooling flow is less efficient, and without design adjustments, an ester-filled transformer may run a few degrees hotter. Manufacturers counteract this by slightly overdesigning cooling radiators or allowing a bit higher temperature. On the other hand, esters' higher water tolerance results in drier paper insulation, which actually permits a higher safe hotspot temperature for the paper before aging accelerates. This can compensate for the higher fluid viscosity. Synthetic esters have viscosity closer to mineral oil and also pour points as low as −50 °C, making them suitable for cold climates where some natural esters would solidify without additives. Additive packages and certain formulated natural esters have improved low-temperature behavior, extending their usability in moderate cold environments. Nonetheless, for very cold regions, synthetic esters or special blends are often chosen due to their superior low-temperature fluidity.

One of the most lauded advantages of natural esters is their impact on insulation aging and moisture management. Transformers contain solid cellulose insulation (paper and pressboard) which degrades over time, especially under heat and in the presence of moisture and oxygen. Natural esters have a high saturation moisture content. They can dissolve significantly more water than mineral oil before reaching saturation. For example, at 20 °C, mineral oil might saturate at ~50 ppm moisture, whereas a natural ester can hold over 1000 ppm of water at the same temperature. This means that in an operating transformer, ester fluid actively pulls water out of the paper insulation, keeping the cellulose drier. It has been proven that transformers filled with natural esters experience slower paper aging and can safely operate at higher hotspot temperatures, effectively improving load capacity and extending the lifetime. Laboratory aging tests confirm that paper in natural ester can last as long or longer than in mineral oil at equivalent temperatures, thanks largely to the ester's water sequestration and the lower oxygen content in sealed designs [44]. The flip side is that natural esters themselves can oxidize more readily than mineral oil if exposed to air. Natural esters are triglycerides that will form acids and eventually polymerize in the presence of oxygen and moisture over time. Therefore, manufacturers often use sealed-tank designs or inert gas blankets for ester-filled transformers, or they add oxidation inhibitors. Synthetic esters generally have better oxidative stability than natural esters.

The use of natural ester dielectric fluid was pioneered in the late 1990s (notably, the commercial product Envirotemp FR3 was introduced around 1998). Since then, adoption has accelerated. Today, natural esters are widely used in distribution transformers (especially pole-top or pad-mounted transformers up to 36 kV). Many utilities have transitioned to buying new distribution transformers filled with ester fluid by default, for fire and environmental benefits. For power transformers (high voltage, large mega volt-ampere (MVA) units), ester use was initially slower due to conservative industry practices and the need for large-scale field validation. But in the past decade, numerous high-voltage ester-filled transformers have been successfully deployed. For example, a 420 kV–300 MVA autotransformer filled with natural ester was energized in 2018. Synthetic esters, while more expensive, have seen strong uptake in special applications such as offshore wind farm transformers and mobile transformers, where their combination of high fire safety and better low-temperature flow is desirable [45]. Both types of esters have also been used to retrofill existing mineral oil transformers, i.e. draining mineral oil and replacing it with ester to immediately improve fire safety and environmental profile. Retrofilling is an attractive life-extension and risk-reduction measure for older transformers, provided they are compatible.

Beyond esters, a few other green liquid insulation technologies deserve mention. Silicone oil (polydimethylsiloxane) has long been used in niche high-temperature transformers (it is stable up to 180 °C) and is less flammable than mineral oil [46]. Silicone fluid is inert and has relatively low environmental impact, but its very high cost and nonbiodegradability limit its widespread classification as "green." Still, silicone-filled transformers have been used in indoor and traction applications for safety. Another development is biodegradable hydrocarbon oils [47]. Recently, some manufacturers have introduced highly refined hydrocarbon fluids that are made to be readily biodegradable. These may be derived from gas-to-liquid processes or hydrotreated vegetable oils. They aim to combine the cooling performance of mineral oil with biodegradability. Research is also ongoing into nanofluids, insulating oils with nanoparticle additives (like magnetic nanoparticles and aluminum nitride), which can enhance thermal conductivity or breakdown strength [48]. Some studies have applied this to esters [49], finding that certain nanoparticles (e.g. TiO_2 or BN) can improve the dielectric strength of natural esters by capturing electrons or improving heat dispersion, as shown in Figure 3.3. While promising, nanofluids are not yet standard in industry practice due to concerns about long-term stability and compatibility.

The introduction of new insulating liquids has been accompanied by new standards. IEC and ASTM issued fluid specification standards. IEC 62770:2013 covers unused natural esters for transformers, and IEC 61099 (updated in 2010) covers synthetic organic esters. IEEE has a guide C57.147 (2018) for acceptance and maintenance of natural ester in transformers, and IEEE C57.154 covers the loading of high-temperature (ester-immersed) transformers. A recent IEC standard 62975:2021 provides guidelines for the maintenance and use of natural esters in equipment, reflecting accumulated field experience. Importantly, fire safety standards like IEC 61100 classify fluids by flammability (K-class fluids, must have fire point >300 °C).

Green liquid insulating materials, particularly natural and synthetic esters, have proven to be viable replacements for mineral oil in a wide range of transformer types. They meet industry dielectric performance requirements, and their use confers significant safety and environmental benefits: reduced fire risk, easier spill remediation, and use of renewable resources. These properties allow transformers to be sited in dense urban areas or sensitive sites with fewer concerns, and mitigate environmental damage from accidents. As grid operators move toward eco-friendly specifications, the role of ester liquids is poised to grow further. Green liquid dielectrics are now a mature technology, enabling the transformer sector to significantly lower its environmental footprint while maintaining reliability.

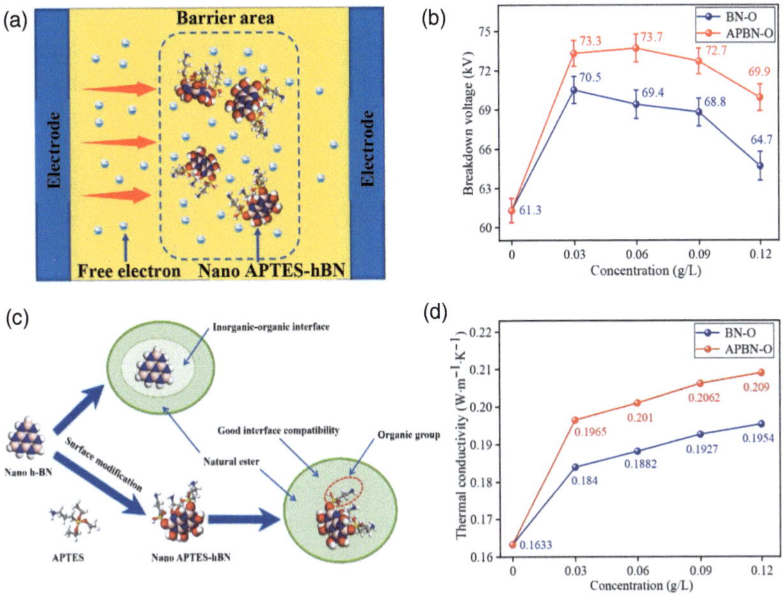

Figure 3.3 (a) Diagram of nanoparticles to improve the breakdown voltage of natural ester. (b) Breakdown voltage of nano-modified natural ester. (c) Diagram of nanoparticles to improve the thermal conductivity of natural ester. (d) Thermal conductivity of nano-modified natural ester. *Source:* [49] / with permission of Elsevier.

3.4 Green Gaseous Insulating Materials for Switchgear and Gas-Insulated Equipment

Gaseous dielectrics are used extensively in switchgear, circuit breakers, gas-insulated substations (GIS), and transmission lines to provide electrical insulation and arc-quenching [50–52]. For the last few decades, the virtually irreplaceable gas in high-voltage applications was sulfur hexafluoride (SF_6). SF_6 is an inorganic, colorless, inert gas with exceptional dielectric properties. At normal pressure, its dielectric strength is about two to three times that of air, and it can be compressed to achieve even higher insulation performance. It is also thermally stable and an excellent arc quencher, as it deionizes quickly after an arc by capturing electrons. These qualities made SF_6 the dominant insulation medium in medium-voltage (MV) and high-voltage (HV) switchgear since the 1960s, replacing bulkier alternatives like oil or air. However, SF_6 has a dark side. It is the most potent greenhouse gas known, with a 100-year GWP of approximately 23 500 (meaning 1 kg SF_6 has the warming effect of 23.5 tons of CO_2). Moreover, SF_6 has an extremely long atmospheric lifetime (~3200 years). Though nontoxic in its pure form and highly effective in service, any leakage or release of SF_6 contributes significantly to climate change. The electrical industry is estimated to be responsible for ~80% of global SF_6 emissions, primarily from equipment leaks and end-of-life handling. In recent years, this has led to an urgent push for SF_6-free insulating gas alternatives as part of global climate initiatives.

The quest for green insulating gases has centered on finding gases or gas mixtures that can approach the dielectric and interruption performance of SF_6 without its environmental impact. Several strategies have emerged. (1) Utilizing "natural" gases like dry air, nitrogen (N_2), and carbon dioxide (CO_2) as the insulating medium, often in combination with vacuum interrupters for breaking current. Dry air and N_2 have GWP = 0 and are nontoxic, making them very eco-friendly. (2) Developing new fluorinated molecules with much lower GWP than SF_6 but high dielectric strength. Leading candidates are the so-called fluoroketones and fluoronitriles developed in the 2010s. For interrupting current, vacuum interrupters (which extinguish arcs in a sealed vacuum bottle) are an established green alternative to SF_6 arc interruption in medium voltage. (3) Vacuum has no GWP and no byproducts. Modern designs are extending vacuum interrupters to higher voltages. However, vacuum alone is not an insulating medium for large GIS enclosures. A complete vacuum insulation of a large volume is impractical in HV equipment due to mechanical complexity and the risk of flashover if even minor partial vacuum or outgassing occurs. Thus, vacuum interruption is typically paired with a solid or gas insulation of the equipment.

Dry air or N_2 is inexpensive and harmless but has much lower dielectric strength, roughly only 30% of the insulating capability of SF_6 at 1 atm. To use air/N_2 in high voltage GIS, one must either accept a significantly larger equipment size or increase gas pressure. Some designs use pressurized dry air to reach adequate insulation levels. This increases the thickness and strength requirements of the gas enclosure but it is technically feasible. In medium-voltage switchgear, where equipment size is manageable, dry air insulation has been successfully implemented, often alongside vacuum breakers. These systems typically operate at a few bars of pressure and use epoxy or polymer insulators to shape the field. The benefit is zero greenhouse emissions and simple end-of-life handling. One example is Siemens' "Clean Air" technology, which uses filtered air at about 1.2–1.5 bar for insulation combined with vacuum interrupters. They have demonstrated this in GIS up to 145 kV in their blue portfolio [53]. Dry air-insulated switchgear exhibits slightly larger dimensions but provides reduced environmental impact, improved safety, and enhanced sustainability by eliminating SF_6.

The R152a (R152a is recognized as a chlorofluorocarbon) /CO_2 gas mixture represents a promising alternative to SF_6 for electrical insulation (as shown in Figure 3.4) [54]. This approach utilizes R152a, a gas with relatively high dielectric strength, blended with a buffer gas such as CO_2 to lower the mixture's boiling point and maintain its gaseous state under practical operating conditions. While pure R152a has a boiling point of −25 °C, which is higher than that of SF_6 (−64 °C) and limits its standalone use in colder climates, mixing it with CO_2 effectively reduces the liquefaction temperature. Experimental findings indicate that a mixture composed of 80% R152a and 20% CO_2 can achieve over 96% of the dielectric strength of SF_6. Since the environmentally impactful R152a constitutes only a portion of the blend, the resulting global warming potential (GWP) of the mixture is significantly lowered to approximately 117, representing a reduction of about 98% compared to pure SF_6. Consequently, the R152a/CO_2 mixture stands as a viable, cost-effective, and

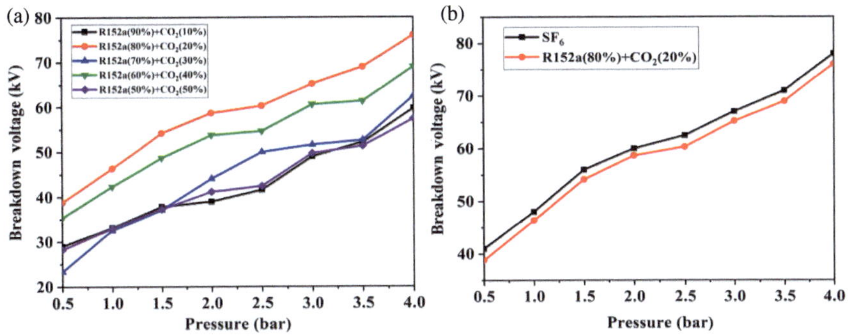

Figure 3.4 (a) AC power frequency breakdown voltage of R152a/CO_2 gas at varying mixture ratio and 8 mm electrode gap distance; (b) Breakdown characteristic comparison of R152a/CO_2 gas at 80%/20% mixture ratio and SF_6. *Source:* [54] / MDPI / CC BY 4.0.

environmentally friendly insulating medium suitable for applications in medium- and high-voltage equipment.

Unlike SF_6, substitute gases do not completely recombine after arc extinction, and the arc-extinguishing capacity of mixed gases is slightly lower. This is mitigated by design adaptations. For instance, slightly larger or differently shaped arcing contacts, and using a vacuum interrupter for current breaking whenever possible. Indeed, state-of-the-art HV GIS designs for 145 kV combine vacuum circuit breakers with the eco-gas for insulation. Another factor is temperature. As noted, these mixtures need higher fill pressure and careful control of temperature range. A high percentage of the fluorinated component yields better dielectric strength but raises the mixture's dew point. Manufacturers have to strike a balance between maximizing dielectric performance and ensuring the gas stays vaporized at the lowest ambient temperature expected. For very low-temperature climates ($< -30\,°C$), one might either heat the equipment or use a lower proportion of heavy gas. From a safety and environmental viewpoint, the new gases are designed to avoid toxic or long-lived residues. They contain no chlorine or bromine, so they won't produce ozone-depleting emissions. However, like SF_6, any arcing in these gases will produce some decomposition products (mainly HF, COF_2, and similar acidic gases from the breakdown of fluorocarbons). Therefore, handling procedures for alternative gases are similar to SF_6. To compare insulating gas options, Table 3.3 provides a qualitative overview of SF_6 and some prominent green alternatives [55–57].

Green gas insulating technology is at a transitional point where multiple major manufacturers have commercial offerings. For medium-voltage (up to 24 or 36 kV), SF_6-free switchgear is already widely available, using either vacuum + air (e.g. Schneider Electric's RM AirSeT uses air insulation, ABB's SafeRing Air uses air and vacuum, Siemens has 12–24 kV air-insulated panels) or using solid-encapsulated bus plus vacuum (Eaton's solid-insulated systems). At high voltage, pilot projects have proven the feasibility at 123, 145, and 170 kV levels. For example, the Koengen 145 kV GIS in Norway (2019) is the world's first SF_2-free

Table 3.3 Comparison of traditional SF_6 gas and green insulating gas alternatives.

Insulation medium	Dielectric strength (relative to SF_6 @1 bar)	Global warming potential	Key features and trade-offs
SF_6 (sulfur hexafluoride)	100% (baseline, excellent)	~23 500	Exceptional insulator and arc quencher; extremely high GWP, long atmospheric life
Dry air/N_2 (pressurized)	~30% at 1 atm (needs 3 × pressure)	0 (no greenhouse gas (GHG) impact)	Readily available, no global warming or toxicity issues; at HV, requires higher pressure (5–8 bar)
CO_2 (compressed)	~30–40% at 1 atm (slightly better than air)	~1 (baseline CO_2)	Nontoxic and cheap; poor dielectric
C_4 fluoronitrile + CO_2 mixture	~80–85% of SF_6 (at ~6–7 bar)	~300–500 (mixture)	GWP ~98% lower than SF_6 with comparable performance. Dielectric strength high but requires higher pressure
C_5 fluoroketone + Air mixture	~75–85% of SF_6 (at ~7 bar)	< 1 (mixture)	Ultra-low GWP (essentially zero) solution, good insulation performance at high pressure; Requires handling similar to SF_6 (not vented)
Vacuum + solid insulation	(Vacuum interrupter for arc; solid dielectric for insulation)	0	No gas used. Vacuum interrupters provide arc quenching; vacuum equipment requires hermetic sealing but has long-proven lifespan

GIS at that voltage, using clean air and vacuum breakers. Manufacturers like GE, Siemens Energy, and Hitachi Energy have all announced roadmaps to extend SF_6-free product ranges up to 245 and 420 kV by mid-2020s.

Green gaseous insulating materials are poised to replace SF_6 in the coming decade for most applications. The material characteristics that influence this shift are clear. GWP has become a crucial factor. This is a case where external environmental mandates have accelerated material innovation. The slight compromises (higher pressure, careful temperature management, use of vacuum interrupters) are generally considered worth the substantial environmental benefit. As the technology matures and economies of scale kick in, SF_6-free switchgear is expected to become the norm. Just as the PB oils were entirely phased out for toxicity reasons by the 1980s, we can anticipate SF_6 being largely phased out by the 2030s due to its climate impact, replaced by the greener gases and systems described above.

3.5 Green Insulating Materials in Electronics and Specialty Applications

In addition to the power transmission equipment sector, there is also a clear demand for sustainable insulation materials in the electronics and specialized electrical industries. Electronic insulation encompasses materials such as PCB laminates, wire enamels and coatings, potting and encapsulation compounds, as well as insulation within electronic modules. While the electrical stresses in electronics are much lower than in power systems, the sheer volume of electronic products and the mounting e-waste problem make the environmental profile of these insulating materials important. In recent years, significant efforts have focused on eliminating hazardous substances and improving recyclability in electronics insulation. Key themes include the move to halogen-free PCB materials, development of bio-based or recyclable substrate materials, and research into biodegradable electronics for specialized uses (like transient biomedical implants) [58–60].

Traditional PCB boards (flame retardant, FR-4) are made of fiberglass cloth impregnated with epoxy resin that contains brominated flame retardants tetrabromobisphenol-A (TBBPA). Brominated resins are highly effective in meeting flammability requirements by quenching combustion. However, if PCBs catch fire or are incinerated improperly during disposal, these brominated compounds release highly toxic fumes, including dioxins and furans. Although TBBPA is still allowed, there is a strong trend toward halogen-free PCB laminates in the industry. In halogen-free FR-4, brominated epoxy resins are replaced by phosphate-based or nitrogen-containing compounds. For example, as shown in Figure 3.5, a common

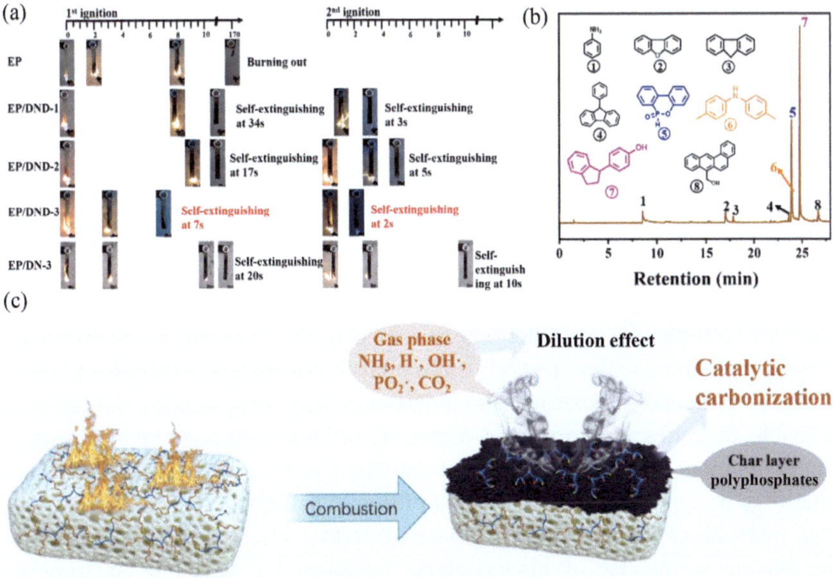

Figure 3.5 (a) Digital photos of EP and composites made of EP in UL-94 tests. (b) Pyrolysis gas chromatography/mass spectrometry (Py-GC/MS) of DND. (c) Possible flame-retardant action of DND in EP. *Source:* [61] / with permission of Elsevier.

approach is to incorporate a phosphorus-rich resin (DOPO-based nanophosphorus derivatives, DND), which chars into a protective layer when heated, or to use inorganic fillers like aluminum hydroxide that release water to cool the flame [61]. These halogen-free materials meet the required flame tests without emitting halogenated toxins. Halogen-free PCB substrates have been found to exhibit better electrical insulation properties and higher thermal stability in some cases. The reasoning is that phosphorus and nitrogen in the resin create less polar bonds than bromine, resulting in a resin system with lower dielectric constant and loss, as well as a higher glass transition temperature (T_g). Thus, halogen-free boards can have slightly improved high-frequency performance and better dimensional stability under heat. Overall, removing halogens from PCB insulation materials significantly reduces the generation of toxic smoke in fires and eases recycling, aligning with sustainability goals.

Traditional FR-4 boards are difficult to recycle. The glass fiber and thermoset epoxy are inseparably bonded, so recycling usually means shredding and incinerating to recover metals. To address this, researchers are exploring PCB substrates that are either biodegradable or more easily recyclable [62, 63]. One approach is the development of bio-based circuit boards using natural fibers and polymers. For instance, using polylactic acid (PLA) biopolymer as the matrix and cotton fiber cloth as reinforcement, mimicking the structure of FR-4 but with fully bio-sourced components. PLA is a compostable polyester derived from corn starch or sugar cane, and cotton is obviously biodegradable. The prototype boards showed promising mechanical and electrical performance for low-power circuitry, and could be thermoformed into shape (something rigid FR-4 cannot do). While such boards are not ready to replace high-density multilayer PCBs in advanced electronics; they hint at a future of electronics designed for environmental disassembly. After use, the board could theoretically biodegrade or be processed without harmful residues. Another is thermoplastic PCB substrates, such as those based on polycarbonate or poly(phenylene oxide) blends, which can be melted down and recycled [64]. Some commercial "high-frequency" laminates use poly tetra fluoro ethylene (PTFE) or liquid crystal polymer (LCP). These are not bio-based, but being thermoplastic, they could be recycled. We also see interest in paper-based electronics, where paper or cellulose films serve as the substrate and insulating layer. These are being explored for flexible and disposable electronics (radio-frequency identification, RFID tags, sensors) that could be easily recycled or composted.

Electronic components are often encased in potting compounds or conformal coatings for insulation and protection. Traditional potting materials include epoxies, urethanes, or silicones. Epoxies give a hard, permanent encapsulation but are not reclaimable. Silicones are more easily removed but have their own environmental cost in production. Efforts for greener encapsulation include using solvent-free and volatile organic compound (VOC)-free conformal coatings (to reduce emissions during application) and developing bio-based potting materials [65]. One interesting avenue is the use of biodegradable polymers for transient electronics.

For example, polyanhydride polymers have been studied as encapsulation layers that slowly hydrolyze over time, useful for devices intended to dissolve after functioning (like biomedical implants or environmental sensors) [66]. Although niche, these demonstrate that it's possible to insulate electronics with materials that will eventually disappear harmlessly. In consumer electronics, the emphasis is more on reworkability and recyclability. Companies are looking at adhesives that can be deboned with heat or specific chemicals, allowing components to be separated for recycling. There is also a push for thermally conductive but recyclable potting for power electronics. For instance, using thermoplastic polyurethane with ceramic fillers instead of thermoset epoxy for an inverter casing, so that at the end of life the potting can be melted off and the metal reclaimed [67].

Wires and cables in electronics also see materials evolution. PVC has been widely used for wire insulation in appliances and electronics, but as mentioned, PVC is being phased out in many contexts for health reasons. Alternatives include polyolefin-based insulations (PE, PP, or elastomer blends) that are halogen-free. These often require flame retardants. Non-halogen flame-retardant cables using magnesium hydroxide filler are now common, especially in IT and telecom equipment cabling. For high-temperature hookup wires, silicone rubber or cross-linked polyolefins are used, which avoid halogens. Some manufacturers have introduced bio-polymer jacketing (e.g. a TPE derived partly from castor oil) for low-voltage cables, boasting a reduced carbon footprint.

Electronic insulating materials are subject to various stress sources, primarily temperature cycling, humidity, and in some cases, UV exposure. Green materials must meet these reliability requirements. Halogen-free PCB resins typically have a higher T_g (approximately 150–170 °C, compared to 130–140 °C for standard FR-4), which effectively improves thermal reliability. Many halogen-free boards also have lower moisture absorption, which helps prevent moisture-related degradation. However, bio-based materials like PLA can have lower thermal limits (T_g ~60 °C) unless blended, so they are currently limited to low-temperature applications. Researchers are working to enhance biopolymers to improve their stability, for example, by adding cellulose nanofibers or blending them with more robust polymers.

For electronic insulation materials, greening means removing toxic additives (such as halogens and heavy metals), incorporating recyclable or renewable components, designing for longer service life (thereby extending equipment life and reducing waste), or using materials that can be recycled at the end of their service life. The progress is tangible, halogenated PCB materials are well on their way out, and some companies have even prototyped fully biodegradable electronics for specialized uses. Although the performance requirements of electronic products mean that not every bio-based experiment can be converted into large-scale production, continuous improvements in biopolymers and composite materials indicate that at least certain categories of electronic products can be made from bio-based insulating substrates in the future.

3.6 Conclusion and Outlook

This chapter investigates the prospects for next-generation solid, liquid, and gas green insulating materials, emphasizing how each type of material is being developed to meet the dual goals of high performance and sustainability. For every traditional insulating material that was once considered irreplaceable, whether it be XLPE in cables, mineral oil in transformers, or SF6 in switchgear, viable, eco-friendly alternatives have emerged through research and industry efforts. These green materials, including recyclable thermoplastics, natural ester fluids, fluorocarbon gas mixtures, and halogen-free laminates, have largely achieved technical performance equivalent to or even exceeding that of their predecessors. At the same time, they have greatly reduced environmental impact and improved safety.

A recurring theme is that material characteristics directly influence adoption in industrial practice. When a green material can demonstrate sufficient dielectric strength, heat resistance, and reliability, its recyclability or nonflammability becomes a powerful incentive for users to switch, sometimes even at a slightly higher initial cost. For example, the use of natural ester transformer oil is justified not only because of its biodegradability, but also because its high flash point and extended insulation life make any design adjustments or moderate cost increases reasonable. Similarly, PP cables are also being explored, as they simplify production (no cross-linking) and offer recycling options, which are in line with the circular economy goals of the power cable industry. In terms of switchgear, gases that replace SF_6 are now also being tested.

Green insulating materials are also gaining wider application due to their unique advantages. For example, ester fluids improve environmental safety and allow transformers to be designed for higher loads. Halogen-free PCBs eliminate toxins and provide better thermal performance for high-power electronic products. Therefore, even if a green insulating material requires moderate redesign or a premium, it is often advantageous to adopt it from a life cycle perspective and in terms of risk reduction.

From an industry perspective, there are still some challenges. One is the standardization of new materials and supply chain maturity. Green insulation materials must come from multiple suppliers with consistent quality. Technicians must receive training in their use. Currently, several oil companies are producing natural ester liquids. Polymer companies are developing recyclable compounds for cables. Gas suppliers are providing new mixed gases for switchgear. Another challenge is the retrofitting or upgrading of existing infrastructure. Some equipment can be retrofitted (e.g. oil can be retrofitted with esters), but other equipment may need to be replaced to accommodate different media.

Looking ahead, several interesting trends can be anticipated:

1) Insulation materials are green and recyclable, meaning they can be recycled or made from recycled materials. If high-voltage cables can be recycled, it will create a huge polymer waste cycle. Future PCB materials may be designed to be

soluble or separable for material recycling. Addressing the issue of electronic waste at the material level is important, but it is essential to ensure that encapsulation or adhesives do not hinder recycling.

2) Nanocomposites can achieve the same functionality using less insulating material. Using less material is inherently more sustainable. If nanofilled polymers allow for thinner insulation in motors or cables, then less polymer is needed overall. Combining this with green components (such as bio-based nanofillers or polymers) can result in the next generation of ultra-efficient, eco-designed insulation systems.
3) With the application of new materials, the industry will closely monitor their long-term behavior. Intelligent monitoring has been used to construct operational data, such as dissolved gas analysis of esters or density monitoring of new gases. Over time, familiarity with the operation of green materials will increase, reducing any perceived risks between new and old materials.
4) More and more utilities and manufacturers will specify environmental standards in their procurement. This will strengthen the business case for green materials and effectively incorporate sustainability into management policies.

Insulation materials have traditionally played a crucial role in the field of electrical engineering and are now clearly at the forefront of sustainable innovation. The materials discussed in this chapter can significantly reduce carbon dioxide emissions. They also enhance human safety (by reducing fire and toxicity risks) and minimize ecological harm. Importantly, this transformation has occurred without compromising technical performance, thanks to the collective efforts of the global research community and industry stakeholders. As technology continues to advance, we can anticipate even more creative solutions. State-of-the-art green insulating materials have provided the electrical industry with a pathway to become more low-carbon, environmentally friendly, and sustainable, truly aligning with the needs of next-generation power systems and electronic devices.

Acknowledgments

This work was financially supported by the China National Postdoctoral Program for Innovative Talents (No. BX20240194), China Postdoctoral Science Foundation (No. 2025M770495).

References

1 Sun, P., Li, C., Sima, W. et al. (2021). Super electrical insulating materials based on honeycomb-inspired nanostructure: high electrical strength and low permittivity and dielectric loss. *Advanced Electronic Materials* 8 (4): 2100979.

2 Liu, J. and Nicolosi, V. (2024). Electrically insulating electromagnetic interference shielding materials: a perspective. *Advanced Functional Materials* 35 (18): 2407439.

3 Li, S., Yu, S., and Feng, Y. (2016). Progress in and prospects for electrical insulating materials. *High Voltage* 1 (3): 122–129.

4 Zhan, Y., Chen, G., Hao, M. et al. (2020). Space charge measurement and modelling in cross-linked polyethylene. *Energies* 13 (8): 1906.

5 Yoon, H., Chen, G., and Hosier, I.L. (2023). Influence of thermal aging on dielectric properties of natural ester oil-impregnated paper. *IEEE Transactions on Dielectrics and Electrical Insulation* 30 (3): 1337–1344.

6 Wang, G., Kim, W.-H., Kil, G.-S. et al. (2019). Green gas for a grid as an eco-friendly alternative insulation gas to SF_6: from the perspective of partial discharge under AC. *Applied Sciences* 9 (4): 651.

7 Yang, Y., Gao, K., Ding, L. et al. (2021). Review of the decomposition characteristics of eco-friendly insulation gas. *High Voltage* 6 (5): 733–749.

8 Yao, Y., Yang, Z., Zhang, B. et al. (2024). Discovery of a novel binary azeotrope with positive synergistic insulation strength as eco-friendly SF_6-alternative. *Journal of Physics D: Applied Physics* 58 (3): 035503.

9 Wang, S., Dong, L., Peng, R., and Tang, H. (2025). Green synthesis of trifluoromethanesulfonyl fluoride as an eco-friendly alternative to SF_6 gas insulation and analysis of its acute inhalation toxicity. *Molecules* 30 (10): 2241.

10 Hetimy, S., Megahed, N., Eleinen, O.A., and Elgheznawy, D. (2024). Exploring the potential of sheep wool as an eco-friendly insulation material: a comprehensive review and analytical ranking. *Sustainable Materials and Technologies* 39: e00812.

11 Pourrahimi, A.M., Mauri, M., D, Auria, S. et al. (2024). Alternative concepts for extruded power cable insulation: from thermosets to thermoplastics. *Advanced Materials* 36 (52): 2313508.

12 Dai, X., Hao, J., Montanari, G.C. et al. (2022). Influence of thermal ageing on high-field polarisation characteristics and conductivity behaviour of submarine polymeric cables insulation. *High Voltage* 8 (1): 138–148.

13 Zhang, X., Hu, S., Huang, S. et al. (2024). Structure-performance relationship of polypropylene/elastomer/carbon black composites as high voltage cable shielding layer. *Composites Part A: Applied Science and Manufacturing* 185: 108334.

14 Hou, L., Liu, Y., Yang, L. et al. (2024). Concentration measurement of antioxidant 1010 in cross-linked polyethylene by terahertz time-domain spectroscopy. *IEEE Transactions on Plasma Science* 52 (10): 5024–5030.

15 Chen, X., Hong, Z., Dai, C. et al. (2022). Voltage stabilizer and nanofiller in enhancement of insulation properties of cross-linked polyethylene. *Journal of Applied Polymer Science* 139 (23): e52314.

16 Csanyi, G.M., Bal, S., and Tamus, Z.A. (2020). Dielectric measurement based deducted quantities to track repetitive, short-term thermal aging of polyvinyl chloride (PVC) cable insulation. *Polymers* 12 (12): 2809.

17 Tamus, Z.Á. (2024). Determination of trap distribution based on voltage response measurement for condition monitoring of nuclear power plant I&C cables. *IEEE Transactions on Instrumentation and Measurement* 73: 1–11.

18 Sadroddini, M., Razzaghi-Kashani, M., Miranzadeh, M., and Kassaee, M.Z. (2016). Controlling dielectric permittivity and dielectric loss by starch-coated

silver nanoparticles in ethylene–propylene rubber. *Polymer Composites* 39 (4): 1303–1310.

19 Liu, F., Shen, C., You, F. et al. (2023). Enhancing the sound and thermal insulation properties of polypropylene foam by preparing high melt strength polypropylene. *Macromolecular Rapid Communications* 44 (20): e2300344.

20 Li, Z., Zhong, Z., and Du, B. (2019). Dielectric relaxation and trap-modulated dc breakdown of polypropylene blend insulation. *Polymer* 185: 121935.

21 Hu, S., Wang, W., Dang, B. et al. (2020). Thermal properties and space charge behavior of thermally aged polypropylene/elastomer blends nanocomposite. *IEEE Transactions on Dielectrics and Electrical Insulation* 27 (2): 521–527.

22 Li, J., Gao, Y., Liu, B. et al. (2024). Electrical conduction properties of PP-based nanocomposite with high thermal conductivity for HVDC cable insulation. *Journal of Physics D: Applied Physics* 57 (16): 165301.

23 Nam, C.Y., Lee, J.H., Kim, M.A., and Yoon, H.G. (2025). High performance and recyclable polypropylene/styrene-ethylene-butylene-styrene blends for next generation cable insulation with enhanced breakdown strength through controlling crystallinity. *Polymers* 17 (10): 1361.

24 Gao, Y., Huang, X., Min, D. et al. (2018). Recyclable dielectric polymer nanocomposites with voltage stabilizer interface: toward new generation of high voltage direct current cable insulation. *ACS Sustainable Chemistry & Engineering* 7 (1): 513–525.

25 Plesa, I., Notingher, P.V., Stancu, C. et al. (2018). Polyethylene nanocomposites for power cable insulations. *Polymers* 11 (1): 24.

26 Luo, Z., Ye, H., Hu, J. et al. (2020). Synthesis of a pyrene-functionalized hyperbranched polyethylene ternary copolymer for efficient graphite exfoliation in chloroform and formation of ethylene-vinyl acetate/graphene nanocomposites. *Journal of Applied Polymer Science* 137 (43): e49320.

27 Baffie, F., Boyron, O., Lansalot, M. et al. (2023). Polyethylene and poly(ethylene-co-vinyl acetate) star polymers by iodine transfer polymerization. *Polymer Chemistry* 14 (38): 4419–4428.

28 Benato, R., Balanuye, İ., Köksal, F. et al. (2018). Installation of XLPE-insulated 400 KV submarine AC power cables under the dardanelles strait: a 4 GW Turkish grid reinforcement. *Energies* 11 (1): 164.

29 Ouyang, Y., Pourrahimi, A.M., Lund, A. et al. (2021). High-temperature creep resistant ternary blends based on polyethylene and polypropylene for thermoplastic power cable insulation. *Journal of Polymer Science* 59 (11): 1084–1094.

30 Yuan, C., Jiang, B., Zhu, Y. et al. (2024). Enhanced electrical insulation properties of 2-isopropenyl-2-oxazoline grafted polypropylene for high voltage direct current power cables. *Journal of Physics D: Applied Physics* 57 (41): 415503.

31 Singh, I. and Sivaramakrishna, A. (2024). Phosphorus-based polymeric flame retardants–recent advances and perspectives. *ChemistrySelect* 9 (26): e202401485.

32 Xia, R., Ouyang, B., Wang, Y. et al. (2023). Effect of temperature and thermal ageing of cable silicone rubber accessories on interface pressure of cable joints. *Applied Sciences* 13 (18): 10406.

33 Su, D. (2024). Silicone rubber thermal aging performance for cables and accessories. *Journal of Materials Science: Materials in Electronics* 35 (5): 328.

34 Rogkas, N., Manios, A., Pelekis, M. et al. (2024). Coupled EMAG-CFD-thermal analysis of a novel SST system utilizing forced-liquid cooling with biodegradable dielectric fluid. *Applied Thermal Engineering* 257: 124293.

35 Khan, S.A., Tariq, M., Khan, A.A. et al. (2022). Influence of area and volume effect on dielectric behaviour of the mineral oil-based nanofluids. *Energies* 15 (9): 3354.

36 Jaroszewski, M., Beroual, A., and Dziedziul, M. (2018). Electro-convection in natural ester based transformer fluid. *IEEE Transactions on Dielectrics and Electrical Insulation* 25 (6): 2398–2404.

37 Raymon, A. (2024). Transesterification approaches to natural esters for transformer insulating fluids: a review. *IEEE Transactions on Dielectrics and Electrical Insulation* 31 (2): 607–614.

38 Mahidhar, G.D.P., Sarathi, R., Taylor, N., and Edin, H. (2020). Dielectric properties of silica based synthetic ester nanofluid. *IEEE Transactions on Dielectrics and Electrical Insulation* 27 (5): 1508–1515.

39 Stuchala, F., Rozga, P., Malaczek, M., and Staniewski, J. (2024). Comparative analysis of selected insulating liquids including bio-based hydrocarbon and GTL in terms of breakdown and acceleration voltage at negative lightning impulse. *IEEE Transactions on Dielectrics and Electrical Insulation* 31 (5): 2443–2450.

40 García, B., Ortiz, A., Renedo, C. et al. (2021). Use performance and management of biodegradable fluids as transformer insulation. *Energies* 14 (19): 6357.

41 Mehta, D.M., Kundu, P., Chowdhury, A. et al. (2016). A review on critical evaluation of natural ester vis-a-vis mineral oil insulating liquid for use in transformers: Part 1. *IEEE Transactions on Dielectrics and Electrical Insulation* 23 (2): 873–880.

42 Frimpong, G., Hajek, J., and Kulasek, K. (2019). Eco-friendly ester liquid transformer designs. *Transformers Magazine* 4 (6): 33–36.

43 Lyutikova, M., Korobeynikov, S., and Konovalov, A. (2021). Evaluation of the properties of mixtures of aromatic mineral oil and synthetic ester for high-voltage equipment. *IEEE Transactions on Dielectrics and Electrical Insulation* 28 (4): 1282–1290.

44 Przybylek, P. (2023). Determination of mineral oil concentration in the mixture with synthetic ester using near-infrared spectroscopy. *Energies* 16 (17): 6381.

45 Liu, J., Chen, H., Yang, M. et al. (2024). Comparative study of natural ester oil and mineral oil on the applicability of the immersion cooling for a battery module. *Renewable Energy* 224: 120187.

46 Wen, X., Yuan, X., Lan, L. et al. (2017). Effect of transformer oil on room temperature vulcanized silicone rubber. *IEEE Transactions on Dielectrics and Electrical Insulation* 24 (4): 2337–2343.

47 Liu, F., Tu, T., Li, S. et al. (2019). Relationship between plankton-based beta-carotene and biodegradable adaptablity to petroleum-derived hydrocarbon. *Chemosphere* 237: 124430.

48 Adnan, M.M., Tveten, E.G., Glaum, J. et al. (2018). Epoxy-based nanocomposites for high-voltage insulation: a review. *Advanced Electronic Materials* 5 (2): 1800505.

49 Chen, R., Qiu, Q., Peng, X., and Tang, C. (2023). Surface modified h-BN towards enhanced electrical properties and thermal conductivity of natural ester insulating oil. *Renewable Energy* 204: 185–196.

50 Jonsson, E., Runde, M., Dominguez, G. et al. (2013). Comparative study of arc-quenching capabilities of different ablation materials. *IEEE Transactions on Power Delivery* 28 (4): 2065–2070.

51 Wang, J. and Bi, J. (2020). Protective effect study of solid-phase arc-quenching device on superimposed lightning strike. *Electric Power Systems Research* 182: 106195.

52 Wang, W., Gao, K., Yan, X. et al. (2025). Numerical study of the arc-quenching performance of eco-friendly $C_4F_7N/CO_2/O_2$ gas mixtures. *Journal of Physics D: Applied Physics* 58 (18): 185207.

53 Song, L., Zhou, J., Wang, C. et al. (2022). Airborne pathogenic microorganisms and air cleaning technology development: a review. *Journal of Hazardous Materials* 424 (Pt B): 127429.

54 Kharal, H., Kamran, M., Ullah, R., Saleem, M., Alvi, M. (2019). Environment-Friendly and Efficient Gaseous Insulator as a Potential Alternative to SF_6. *Processes* 7 (10): 740.

55 Pan, B., Wang, G., Shi, H. et al. (2020). Green gas for grid as an eco-friendly alternative insulation gas to SF_6: a review. *Applied Sciences* 10 (7): 2526.

56 Owens, J., Xiao, A., Bonk, J. et al. (2021). Recent development of two alternative gases to SF_6 for high voltage electrical power applications. *Energies* 14 (16): 5051.

57 Deng, J., Peng, M., Gao, Z. et al. (2020). Synthesis and dielectric properties of the eco-friendly insulating gas thiazyl trifluoride. *RSC Advances* 10 (5): 2740–2746.

58 Zhan, M. and Wool, R.P. (2013). Design and evaluation of bio-based composites for printed circuit board application. *Composites Part A: Applied Science and Manufacturing* 47: 22–30.

59 Lili, M., Sood, B., and Pecht, M. (2011). Effect of moisture on thermal properties of halogen-free and halogenated printed-circuit-board laminates. *IEEE Transactions on Device and Materials Reliability* 11 (1): 66–75.

60 Jeong, S., Kim, D.-H., Song, J. et al. (2019). Smartwatch strap wireless power transfer system with flexible PCB coil and shielding material. *IEEE Transactions on Industrial Electronics* 66 (5): 4054–4064.

61 Zheng, S., Shi, S., Zeng, K. et al. (2025). A compound containing phosphorus-rich arcuate structure towards improving fire safety and toughness, and maintaining transparency of epoxy resins. *Composites Communications* 56: 102386.

62 Liu, Y., Ahmad, M., Venditti, R.A. et al. (2024). Sustainable soft electronics combining recyclable metal nanowire circuits and biodegradable gel film substrates. *Advanced Electronic Materials* 10 (6): 2300792.

63 Zhu, Y., Wang, Z., Chen, Z. et al. (2024). Highly stretchable, biodegradable, and recyclable green electronic substrates. *Small* 20 (3): e2305181.

64 Li, X., Zhou, D., Yao, X. et al. (2024). Modified poly(phenylene oxide)/(CASR)TiO_3 composites with ultrahigh permittivity and low loss for microwave substrates. *Journal of Alloys and Compounds* 970: 172533.

65 Zhou, X., Fu, F., Shen, M. et al. (2023). Research progress of bio-based polybenzoxazine materials. *Advanced Sustainable Systems* 8 (2): 2300372.

66 Wu, X., Babi, M., Moran-Mirabal, J., and Pelton, R.H. (2024). Grafting polyanhydride polymers to cellulose nanofibers. *Cellulose* 31 (10): 6193–6210.

67 Chen, T., Xie, Y., Wang, Z. et al. (2021). Recent advances of flexible strain sensors based on conductive fillers and thermoplastic polyurethane matrixes. *ACS Applied Polymer Materials* 3 (11): 5317–5338.

4

Milestone of Insulating Plastics

Wenye Zhang[1], Hongzhe Zhang[1], and Jun-Wei Zha[1,2]

[1] *University of Science and Technology Beijing, School of Chemistry and Biological Engineering, No. 30 Xueyuan Road, Haidian District, Beijing 100083, P. R. China*
[2] *North China Electric Power University, State Key Laboratory of Alternate Electrical Power System with Renewable Energy Sources, School of Electrical and Electronic Engineering, No. 2 Beinong Road, Huilongguan, Changping District, Beijing 102206, P. R. China*

4.1 Introduction

The use of polymeric materials for electrical insulating has a rich history spanning over a century. Early electrical systems relied on natural insulators like gutta-percha and treated rubbers, but the advent of synthetic plastics in the 20th century revolutionized the field. Milestone developments in polymer science enabled the replacement of traditional paper, oil, and rubber insulation with durable, lighter, and more efficient insulating plastics [1–3]. Over time, a series of insulating plastics rose to prominence, each addressing the limitations of its predecessors. For example, polyethylene (PE) emerged in the mid-1900s as a superior insulator with high dielectric strength and low dielectric loss. Around the same period, polypropylene (PP) also gained attention for its low dielectric loss and higher thermal endurance, although its brittleness at low temperatures initially limited its widespread adoption [4, 5]. Meanwhile, polyvinyl chloride (PVC) offered flexibility and flame resistance for wiring applications [6]. By the 1960s, chemically cross-linked polyethylene (XLPE) became the material of choice for high-voltage cables, marrying the electrical excellence of PE with enhanced thermal stability [7, 8].

These classic materials set benchmarks in performance and reliability. However, new demands, ranging from higher operating temperatures and voltages to sustainability and smart functionality, have driven innovation beyond the classics. Researchers have also pioneered dynamic polymer networks, e.g., covalent adaptable networks (CANs) or vitrimers, that behave like traditional thermosets in operation yet are reprocessable or recyclable, addressing end-of-life challenges of cross-linked insulators [9, 10]. Furthermore, "smart" insulating materials capable of self-healing electrical damage or providing self-diagnostic feedback, such as a color change upon degradation, have recently been reported, signaling a new era of intelligent insulation systems [11–15].

Next-Generation Electrical Insulating Materials: Low-Carbon, Eco-Friendly, and Sustainable Solutions, First Edition. Edited by Jun-Wei Zha.

To begin with, the development, key properties, and historical usage of classic insulating plastics such as PVC, PE, PP, and XLPE in electrical systems will be discussed in detail. Following this, attention will shift to advanced solutions, including the emergence of dynamic/recyclable polymer networks, and smart polyolefin-based insulators with self-healing or sensing capabilities [13, 16–18]. Throughout these sections, significant breakthroughs in formulation, such as copolymerization, cross-linking techniques, and nano-additives, as well as in processing methods like extrusion and curing, will be highlighted, along with enhancements in electrical and mechanical performance. Furthermore, considerations of sustainability, e.g., recyclability, upcycling, and carbon footprint, will be integrated into the discussion, aligning with the global drive for eco-friendly materials.

In conclusion, by examining both the historical evolution and the cutting-edge innovations, including recent findings from 2020 to 2025, this chapter offers a comprehensive understanding of how insulating plastics have reached their current stage. From the widespread adoption of PE and PVC in the mid-20th century, through the refinement of XLPE and the introduction of PP-based thermoplastics, to the forefront development of recyclable and smart insulators, the trajectory of insulating plastics reflects a continuous pursuit of enhanced performance, reliability, and sustainability. The insights provided not only recount past milestones but also pave the way for future advancements in next-generation electrical insulating materials.

4.2 Development and Evolution of Traditional Insulating Plastics

Electrical insulation technology has been dominated by a few classic polymer materials that achieved widespread use due to their excellent dielectric properties and practical processability [19]. This section examines four cornerstone insulating plastics, PVC, PE, PP, and XLPE, outlining their development history, properties, and the key milestones that established their roles in the electrical industry (Table 4.1).

4.2.1 Polyethylene

Polyethylene was one of the first synthetic polymers to make a major impact on electrical insulating materials. Discovered in the 1930s, PE is a semicrystalline thermoplastic polymer with excellent insulating characteristics. Low-density polyethylene (LDPE) in particular, with its highly branched structure, offers a very low dielectric constant (~2.3) and low dielectric loss, making it an ideal insulator for high-frequency and high-voltage applications [20, 21]. A milestone achievement was the use of PE in radar coaxial cables during World War II, where its low loss tangent significantly improved signal transmission compared to earlier rubber insulations. By the 1950s, PE had been adopted in power cables, with

Table 4.1 Comparison of traditional insulating plastics (properties and usage).

Material	Dielectric constant @1 MHz	Dielectric strength (kV/mm)	Thermal limit (°C)	Recyclability	Typical applications
PVC	~3.2	10–20	70–90	Limited (thermoplastic, but Cl content)	Low-voltage wire insulation, cable jackets, appliance cords (good flame retardance);
PE (low-density polyethylene, LDPE)	~2.3	~75	~80	Yes (thermoplastic)	Coaxial cables, medium-voltage cables (with XLPE), dielectric films (low-loss applications);
XLPE	~2.3–2.4	~50	90	No (thermoset)	Medium-voltage and high-voltage power cables (extruded insulation), joints, and terminations;
PP	~2.2	~35	105–110	Yes (thermoplastic)	Capacitor films (BOPP), high-voltage cables (thermoplastic insulation), appliance, and automotive wires (heat-resistant, halogen-free).

LDPE-insulated cables proving superior to paper-oil cables in ease of installation and moisture resistance.

The dielectric strength of pure PE is remarkably high; laboratory values for LDPE can exceed 70–100 kV/mm for thin samples, owing to its chemically inert, nonpolar structure and the absence of significant dipoles or fillers that could initiate breakdown. This high breakdown field, combined with volume resistivity on the order of 10^{16} Ω/cm, established PE as a benchmark insulator in the mid-20th century. However, a drawback of PE, especially LDPE, is its relatively low melting point (~110 °C) and softening temperature. As power equipment and cables operate at elevated temperatures, often reaching up to 90 °C under normal service conditions, pure PE insulation risks deformation or melting under load. This limitation motivated one of the most important innovations in insulating plastics, the cross-linking of PE to create XLPE.

Another challenge with early PE was its mechanical softness and tendency to environmental stress cracking under long-term stress or exposure to certain chemicals. Over the years, formulations were improved, for instance, high-density polyethylene (HDPE), which has a more linear structure and higher crystallinity, offers greater mechanical strength and a higher melting point (~130 °C) at the cost of slightly higher dielectric constant [22–24]. HDPE is found to be used in specialty insulators and also as a component in blends for cable insulation [25, 26].

By the 1960s, antioxidant additives were introduced into PE insulation formulations as a milestone in improving longevity; these additives trap free radicals formed during electrical aging or thermal exposure, thereby preventing dielectric degradation over decades of service.

PE exhibited inherent advantages, especially in low-loss dielectric films. While biaxially oriented polypropylene (BOPP) eventually surpassed PE in capacitor films due to the superior temperature resistance of PP, PE film remained essential in certain high-frequency cable dielectrics and as an inner layer in coaxial cables [27, 28]. The concept of enhancing the insulation property of PE with structural improvements continued to evolve, leading to the development of XLPE and various copolymers. In summary, polyethylene laid the foundation for modern polymeric insulators. Its introduction marked a turning point, demonstrating that plastics could meet and even exceed the electrical performance of traditional insulators. The need to overcome the thermal limitation of PE directly led to further advancements, including the development of XLPE. Today, PE remains a core material in cable insulation, often in its cross-linked form, and in many composite insulating structures, owing to its excellent dielectric properties and chemical simplicity.

4.2.2 Polypropylene

PP is a polyolefin similar to PE, but it exhibits higher crystallinity and a higher melting point. Isotactic PP, introduced in the late 1950s, crystallizes into a strong and relatively rigid material, with a melting point of approximately 160–170 °C. Initially, the potential of PP as an insulator was recognized due to its very low dielectric loss, comparable to PE, and its slightly lower dielectric constant (~2.2). In fact, one of the landmark applications of PP in electrical engineering is in metallized film capacitors, where BOPP films serve as the dielectric material [29, 30]. From the 1960s and increasingly through the 1970s and 1980s, BOPP film capacitors became ubiquitous in alternating current (AC) motor run capacitors, power factor correction, and direct current (DC) link capacitors in power electronics [28, 31]. The key milestone was that PP film offered extremely high breakdown strength, with several hundred kV/mm in ultra-thin films, and a low dielectric loss, enabling the production of compact capacitors. Meanwhile, these capacitors can self-heal after dielectric breakdown in a small area by locally vaporizing the metal film [32–34].

Although PP was successful in capacitors, its use as bulk cable insulation presented greater challenges. Since the glass transition temperature (T_g) is around 0 °C, pure isotactic PP is too rigid and brittle at low temperatures, meaning that in cold climates, a PP-insulated cable could crack when flexed. Additionally, the crystallinity of PP can lead to the formation of space charge under DC high-voltage, which may induce dielectric stress. Throughout the 1970s to the 1990s, XLPE remained the dominant high-voltage insulator. However, research on PP did not cease. A major driver for revisiting PP was its high thermal endurance. PP can maintain its form up to ~105 °C continuously or even higher for short periods, significantly above the 90 °C limit for XLPE cables [35, 36]. By the

2010s, the demand for fully recyclable thermoplastic cable insulation systems led to renewed development of PP, culminating in the introduction of commercial high-voltage PP-based cable insulation systems. For example, the first thermoplastic high-voltage cable, often referred to as P-Laser by Prysmian, became available, using a PP-based blend that included a small percentage of carefully formulated additives and a dielectric fluid to fill microvoids. This was a milestone in that it demonstrated a viable alternative to XLPE, which could be melted and recycled at the end of life.

Two primary strategies have been employed to make PP suitable for cable insulation: copolymerization or blending with elastomers and nucleation or crystal control. The first strategy yields materials often referred to as thermoplastic polyolefins (TPOs) [37–39]. By introducing a rubbery phase, such as ethylene-propylene rubber (EPR), into PP or using random PP copolymers with a small percentage of ethylene, the resulting material gains toughness and flexibility. For instance, adding ~2% ethylene comonomer reduces the stiffness of PP and allows considerable elongation without fracture, while still preserving a high melting point around 150 °C. This means that a copolymer PP insulation can handle thermal stress while being less brittle, which is a crucial characteristic for HV cables that may experience mechanical strain. The second strategy, nucleation and additives, tackles another issue related to the electrical conductivity of PP under high fields and elevated temperatures. Isotactic PP has exceptionally low DC conductivity (~10^{-15} S/m at 70 °C), but further reduction in conductivity and suppression of space charge can be achieved with additives. Studies have shown that adding β-nucleating agents to modify crystalline morphology or certain nanoparticles, e.g., metal oxides, can further reduce the electrical conductivity of PP.

By the early 2020s, these innovations bore fruit: PP-based thermoplastic insulating materials demonstrated dielectric performance comparable to, or in some cases exceeding, that of XLPE. In particular, several PP copolymers and blends exhibited superior performance at elevated temperatures, where XLPE tends to show a steeper increase in DC conductivity (Figure 4.1). Moreover, PP generates no byproducts or gases during processing, and end-of-life materials can be remelted and reused directly, offering clear advantages in terms of recyclability and circular economy potential.

In summary, the milestone achievements of PP in insulating plastics are twofold: the widespread use of PP films in capacitors and the emerging success of PP-based thermoplastic power cable insulation. Decades of research have addressed the initial shortcomings of PP through copolymerization and additives, resulting in materials that combine the high thermal stability, a melting temperature (T_m) up to ~170 °C, with adequate flexibility and excellent electrical properties. PP is poised to play a larger role in high-voltage insulation as the industry seeks sustainable, reprocessable alternatives to thermosets. Its journey exemplifies how the inherent strengths of the materials, such as low dielectric loss, high T_m, can be leveraged, and its weaknesses, such as brittleness and space charge, can be mitigated through innovative polymer chemistry and formulation.

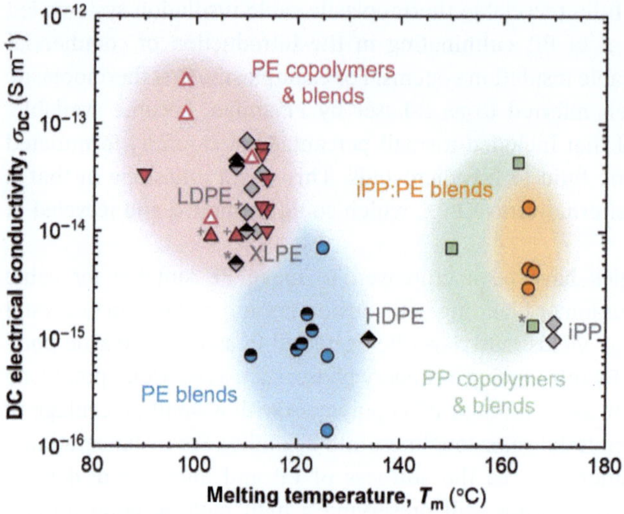

Figure 4.1 Values of DC electrical conductivity of various polyolefin insulating materials at 70 °C. *Source:* [40] / John Wiley & Sons / CC BY 4.0.

4.2.3 Polyvinyl Chloride

PVC is another historically important insulating plastic, first synthesized in the early 20th century and commercialized by the 1930s. The significance of PVC lies in its versatility and flame retardance. Unlike polyolefins, PVC contains chlorine atoms in its polymer structure, which confer inherent flame-retardant behavior, an essential property for insulation in building wires and appliance cords where fire safety is critical [41]. By the 1950s, flexible PVC, achieved by adding plasticizers, had become a standard insulation and jacketing material for low-voltage electrical wires, coaxial cables, and harnesses. A milestone was the replacement of rubber insulation in household wiring with PVC in many countries due to its superior durability and self-extinguishing characteristics.

In terms of dielectric properties, PVC has a higher dielectric constant (~3.0–4.0) and higher dielectric losses than nonpolar polymers like PE. Its dielectric strength is generally in the range of 10–20 kV/mm, which is adequate for low and moderate voltages but lower than that of PE. Moreover, the electrical resistivity of PVC is slightly lower than that of polyolefins due to the polar chlorine, but still high enough for most insulation needs. For these reasons, PVC was traditionally not used as a high-voltage insulation for power cables, instead, it found its application in 600 V class insulation, appliance wiring, and telecommunications cables. A classic application is PVC-insulated copper building wire, which can operate up to about 70 °C; heat-stabilized formulations with special additives can reach 90 °C ratings for thermoplastic high heat-resistant nylon-coated (THHN) wires, representing a milestone in extending the thermal performance [6, 42].

However, PVC insulation has faced challenges and changes over time. One issue is that PVC releases hydrochloric acid (HCl) gas when it thermally degrades or burns, which can corrode metals and harm electronics. This has led to the development of zero-halogen cable materials for sensitive environments, which are often polyolefin-based compounds that avoid chlorine entirely [43]. Another issue is plasticizer migration; the additives that make PVC flexible can slowly evaporate or leach out, especially under heat, causing the insulation to become brittle with age. To address this, modern PVC cable compounds use more stable plasticizers and polymer blending. Milestones were also achieved in the stabilization of PVC against heat and UV. Early PVC used heavy metal stabilizers like lead compounds, which have since been replaced in many regions by calcium-zinc or organic stabilizers for health and environmental reasons [44].

Recycling PVC from cables has historically been difficult due to contamination and the evolution of HCl gas when remelting PVC. Recent research efforts are noteworthy in this context. For example, research by Guo et al. demonstrated room-temperature co-upcycling of PVC and PP into useful hydrocarbons by using special ionic liquids to remove chlorine [43, 45]. As shown in Figure 4.2, this innovative chemical recycling approach can convert mixed PVC/PP waste into valuable products under mild conditions, highlighting the push toward sustainable end-of-life solutions for PVC-based insulators. While such processes are not yet industry standards, they represent important steps toward resolving the legacy environmental issues of PVC.

In summary, the milestone contribution of PVC to insulating plastics is its widespread use in flame-resistant, flexible insulation for low-voltage applications. It proved that plastics could be made inherently flame-retardant and mechanically robust enough for everyday electrical use. Though gradually being supplemented or replaced by halogen-free alternatives in some applications, PVC remains in extensive use globally. Its development spurred advances in formulating polymer compounds for fire safety and longevity, and modern sustainability initiatives continue to find ways to mitigate the environmental impact of PVC while leveraging its advantageous properties.

4.2.4 Cross-Linked Polyethylene

XLPE represents one of the most significant milestones in the evolution of insulating plastics. Developed in the 1960s to overcome the thermal limitations of PE, XLPE is produced by chemically bonding linear PE chains at various points along their backbone. This cross-linking is typically achieved using peroxide additives, such as dicumyl peroxide (DCP), which decompose at elevated temperatures to generate radicals that induce covalent bonding between chains [46]. As a result, the material does not exhibit a distinct melting point. Instead, it possesses a broader softening range and maintains mechanical integrity up to and beyond 120 °C. This thermal resilience enabled power cables insulated with XLPE to operate safely at conductor temperatures of 90 °C, with short-term emergency ratings above 100 °C. These properties marked a substantial improvement over the 70–80 °C limit typical of thermoplastic PE cables. By around 2010, these improvements culminated in XLPE high-voltage direct current (HVDC) cables rated up to 320 kV and eventually 525 kV, which was

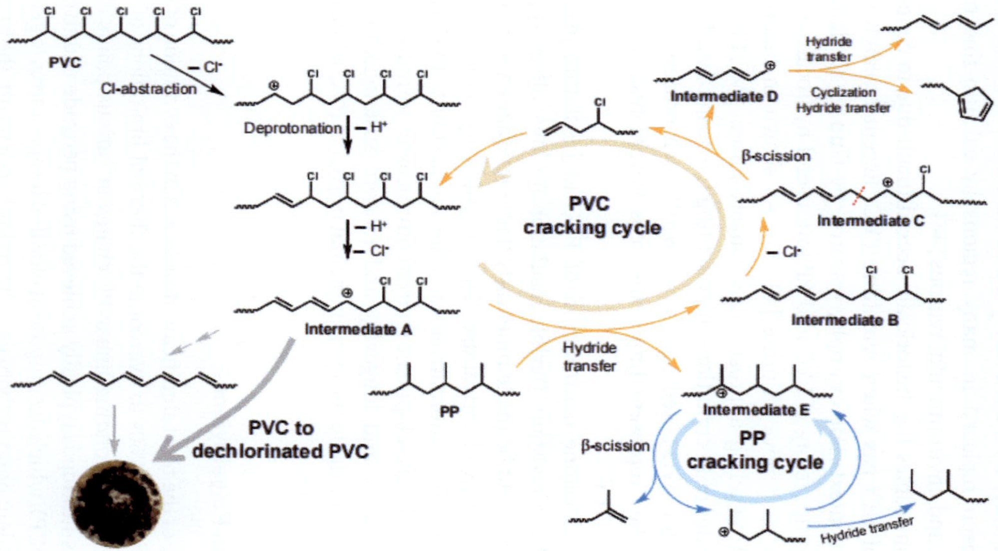

Figure 4.2 Proposed mechanism for the PP-PVC co-conversion. *Source:* [45] / Springer Nature.

a major industry milestone, as extruded polymeric cables competed with traditional lapped paper-oil cables in HVDC power transmission.

A historical challenge in XLPE production involved the presence of cross-linking byproducts. Peroxide curing generates small volatile molecules, including methane, acetophenone, and cumyl alcohol, which can become trapped within the polymer matrix (Figure 4.3) [40]. If not adequately removed, these byproducts may create voids or increase dielectric loss. A major technological advancement in the 1990s addressed this issue through vacuum degassing, wherein cables are held in heated chambers postcuring to diffuse out residual volatiles. This significantly improved reliability, particularly in HVDC applications, where even trace impurities can elevate conduction. Despite these improvements, XLPE faced further limitations in HVDC systems due to the risk of space charge accumulation under DC stress, which can distort the electric field distribution. To mitigate this, innovations such as

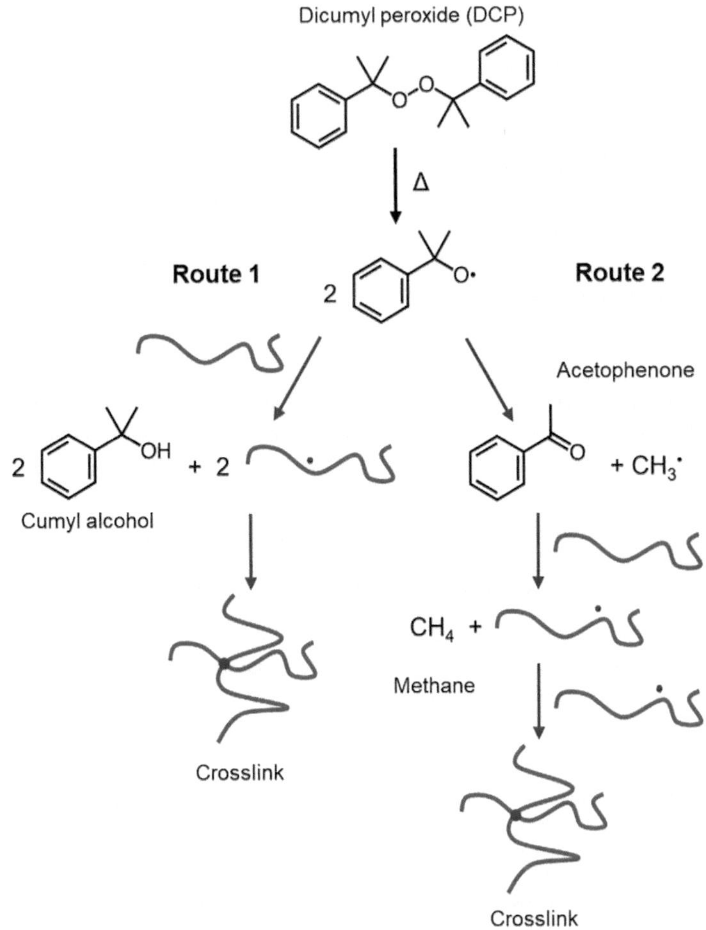

Figure 4.3 Above 160 °C, DCP decomposes into radicals that cross-link polyethylene via hydrogen abstraction. This produces either cumyl alcohol (Route 1) or acetophenone and methane (Route 2) through distinct degradation pathways. *Source:* Adapted from [40].

super-clean insulation with parts-per-billion-level purity and the incorporation of specific additives were introduced. For example, the inclusion of nanoparticles like magnesium oxide (MgO) or aluminum oxide (Al_2O_3), as well as organic molecules, has been shown to suppress space charge and reduce DC conductivity, tailoring XLPE for HVDC performance [5, 47–49].

From a manufacturing perspective, XLPE established the standard cable production model combining extrusion with in-situ curing. Modern cable fabrication typically involves triple-layer co-extrusion, including the conductor shield, insulation, and insulation shield, followed by curing in a high-pressure nitrogen tube. This continuous process, refined over several decades, reflects the industrial maturity of XLPE technology. Nevertheless, the thermoset nature of XLPE presents a major sustainability challenge. Since the material cannot be remelted, recycling is difficult [50, 51]. Waste XLPE, whether from production or decommissioned cables, is often landfilled, used as inert filler, or pyrolyzed for energy recovery. The inability to reprocess XLPE into new insulating products has driven research into thermoplastic alternatives, such as PP-based systems discussed earlier.

It is important to note that not all cross-linking methods for PE rely on peroxides. An alternative technique introduced in the 1970s is silane cross-linking (the Sioplas process), in which PE is grafted with silane groups and subsequently cured by moisture. This water-induced condensation reaction forms -Si-O-Si- cross-links and allows curing at lower temperatures without generating peroxide byproducts. Although this method has seen extensive use in low-voltage applications and plumbing, it has not replaced peroxide-cured XLPE in high-voltage environments. More recently, researchers have explored click-chemistry-based cross-linking methods that produce no volatile byproducts.

In conclusion, XLPE remains a foundational insulating material, enabling the widespread adoption of solid dielectric cables across voltage classes. Its introduction resolved the thermal limitations of PE and set the benchmark for high-performance insulation. However, the trade-off between its excellent electrical properties and limited recyclability continues to shape the search for next-generation insulating materials. Many emerging developments, particularly those involving recyclable thermoplastics or novel dynamic networks, are directly motivated by the desire to retain the advantages of XLPE while overcoming its sustainability limitations. With a proven record across countless kilometers of installed infrastructure, XLPE continues to define the standard for reliable high-voltage insulation.

In summary, classic insulating plastics such as PE, PVC, PP, and XLPE (Figure 4.4) have each made unique contributions to the development of electrical insulation. PE ushered in the era of low-loss, high-strength insulating polymers, while PVC provided flame-retardant, flexible wiring solutions. PP, though initially limited by its brittleness, became crucial in the capacitor field and has recently resurfaced as a recyclable cable insulating material. XLPE addressed critical thermal issues, enabling polymer insulation to be used at high voltages. Despite recycling challenges, its superior overall performance has made it a primary material for power cables. The development of these materials has laid a solid foundation for the next

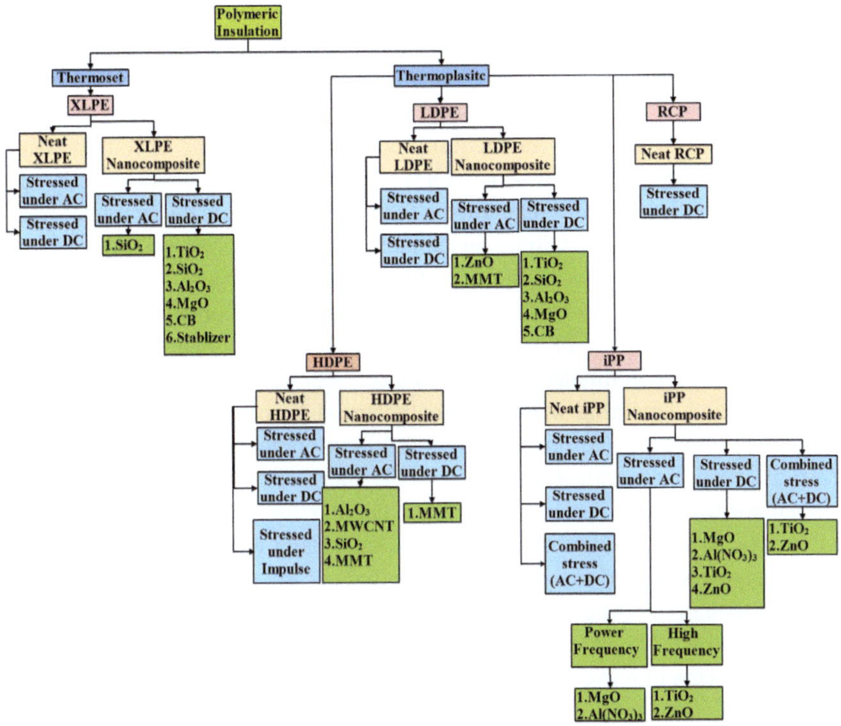

Figure 4.4 Typical classification of plastic insulation. *Source:* [2] / John Wiley & Sons.

generation of novel insulating materials, maintaining foundational properties while incorporating sustainable and smart characteristics, to be discussed in the following sections.

4.3 Development of Next-Generation Insulating Plastics

Electrical insulating plastics are essential in power cables, transformers, and equipment, but traditional options like XLPE, PVC, and other petrochemical-derived polymers present significant environmental challenges [52]. In recent years, growing environmental obligations, resource scarcity, and waste management concerns have driven researchers and industry stakeholders to seek more sustainable, low-carbon, and eco-friendly insulating materials. Traditional cable insulations, like XLPE used in high-voltage cables and PVC used in wiring, are nonbiodegradable, often difficult to recycle, and can persist in landfills for decades or longer. The manufacturing and end-of-life incineration of these plastics also contribute to carbon emissions and pollution, for instance, burning halogenated PVC can release toxic dioxins and corrosive acids. As the world transitions toward a more sustainable and circular economy, insulating plastics must be re-engineered to reduce their life-cycle carbon footprint and overall environmental impact [53]. This involves developing recyclable or biodegradable materials, or materials derived from renewable resources, that are free of harmful additives, while maintaining or even enhancing the electrical properties required for safe and reliable operation. The shift toward eco-friendly insulating plastics is being accelerated by multiple driving forces: increasingly stringent regulations (e.g., halogen- and heavy metal-free requirements for cables), corporate sustainability commitments, and a growing recognition that sustainability and high performance must go hand in hand.

In this section, we explore the sustainable development of mainstream insulating plastics, particularly XLPE, PP, and PVC, under the growing demands of decarbonization and environmental responsibility. Key technologies, functional mechanisms, and emerging materials are reviewed to provide an overview of this evolving landscape. Current research and development in this field focuses on several converging strategies: (1) recyclable thermoplastic insulation materials to replace thermoset designs, enabling a circular life cycle; (2) halogen-free and low-toxicity formulations to improve fire safety and end-of-life disposal; (3) biobased polymers and green synthesis routes that cut reliance on fossil fuels and even sequester carbon; (4) enhanced durability and self-healing features to extend service life; and (5) life-cycle assessment (LCA) to quantitatively guide material choices and process improvements.

4.3.1 Low-Carbon, Eco-Friendly, and Sustainable Advances in Insulating Plastics

4.3.1.1 Challenges with Conventional Insulating Plastics

XLPE, a semicrystalline polymer obtained from PE, features compact chain segments with crystalline and amorphous regions formed by ordered and disordered molecular structures, respectively. Owing to its excellent dielectric strength, low dielectric loss, and favorable thermal and mechanical properties, XLPE has long served as a primary insulating material in high-voltage power cables. Compared with unmodified PE, XLPE exhibits significantly improved mechanical strength and thermal stability, raising its long-term operating temperature from 70 to 90 °C. As shown in Figure 4.5, XLPE is typically synthesized through radiation cross-linking, silane cross-linking, and peroxide cross-linking [40, 46]. Figure 4.5a illustrates the radiation cross-linking mechanism, where irradiation generates active hydrogen atoms that attack PE chains, creating free radicals. These radicals react with radical sites on PE chains, resulting in cross-linking. In silane cross-linking (Figure 4.5b), peroxides first decompose under heat to generate radicals that extract hydrogen from PE, forming long-chain radicals. These then react with silane double bonds to produce grafted PE. In moist environments, the silane groups hydrolyze and undergo condensation to form stable siloxane bonds. As shown in Figure 4.5c, peroxide cross-linking, the most common method, relies on peroxide decomposition to generate radicals that abstract hydrogen from PE, transferring radical sites to the chains and promoting cross-linking into XLPE. However, this

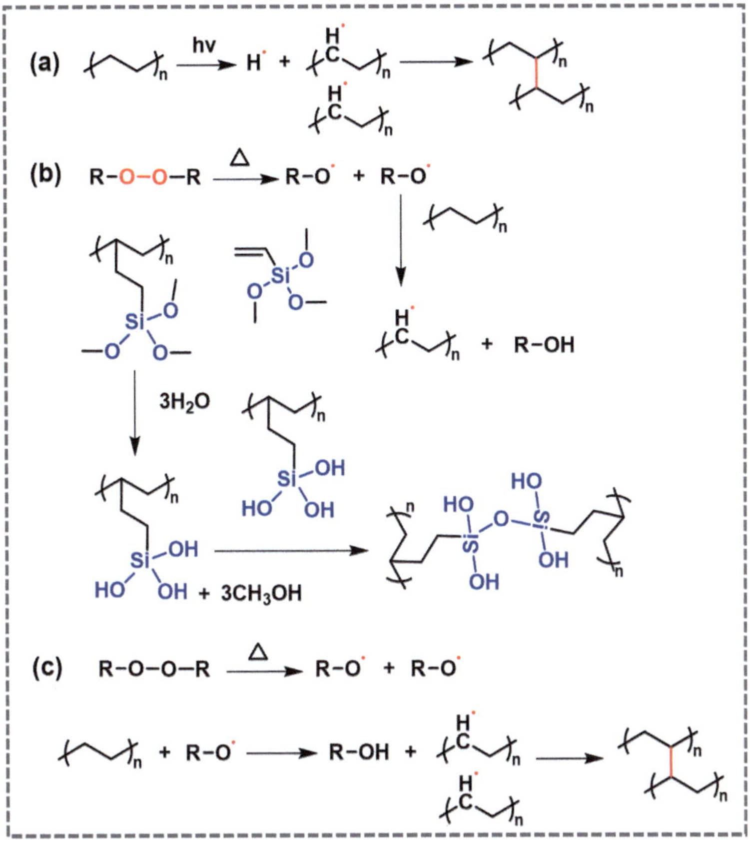

Figure 4.5 Mechanisms of XLPE synthesis. (a) Radiation cross-linking; (b) silane cross-linking; (c) peroxide cross-linking. *Source:* [46] / Acta Physica Sinica.

permanently cross-linked molecular structure renders XLPE nonrecyclable after extrusion and curing. It cannot be remelted or easily reshaped, making its recycling extremely difficult. End-of-life XLPE cable insulation and manufacturing scrap are traditionally disposed of through landfilling or incineration. Due to its environmental persistence and extremely slow degradation rate, landfilled XLPE contributes to long-term plastic pollution. Although incineration can recover some energy, it leads to carbon dioxide emissions and potentially hazardous combustion byproducts. Thus, despite its outstanding performance, XLPE faces serious sustainability challenges at the end of its service life, hindering its future development under environmentally conscious frameworks.

Similarly, PVC, which is widely used as insulation and sheathing for low-voltage cables, also raises environmental and health concerns [54]. PVC contains approximately 30% chlorine by weight. When burned, either in accidental fires or waste incineration, hydrogen chloride gas is released, and persistent organic pollutants such as dioxins are produced [55]. During fire events, PVC cables emit corrosive gases and dense smoke, posing severe risks to human safety and equipment.

While PVC is technically recyclable through melt processing, practical recycling is often economically and technically challenging due to the presence of additives such as plasticizers and stabilizers, including lead compounds, as well as issues with material heterogeneity and separation [56]. Consequently, PVC cable insulation tends to have a low recycling rate in many regions, with large quantities ending up as landfill waste. Moreover, both XLPE and PVC are derived from petrochemical processes, specifically vinyl and chlorine, respectively, which result in significant greenhouse gas emissions. These environmental drawbacks have motivated researchers to explore cleaner, more sustainable alternatives for electrical insulation materials.

4.3.1.2 Toward Sustainable Plastic Insulating Materials

One major advancement toward sustainability in electrical insulating materials is the replacement of nonrecyclable thermosets, such as XLPE, with recyclable thermoplastic polymers that can match or even exceed the performance of their thermoset counterparts. PP has emerged as a leading candidate for next-generation high-voltage cable insulation. Unlike XLPE, PP is a linear thermoplastic material that softens and melts when heated, which means that end-of-life PP insulation can be reprocessed and potentially recycled into new products. Moreover, PP exhibits superior thermal stability compared to XLPE, with an isotactic PP melting point of approximately 160 °C, compared to around 110 °C for XLPE. This advantage potentially enables cable operation at higher temperatures or under greater current loads without premature aging [47]. PP also offers excellent dielectric properties, with a low relative permittivity (~2.2–2.3, comparable to PE) and extremely low dielectric loss. In addition, PP does not produce byproducts, such as methane, during manufacturing, unlike XLPE, which is often cross-linked using peroxides that release volatile compounds that require control. PP is also halogen-free and does not release toxic halogenated smoke during combustion, making it significantly safer than PVC or other chlorine-containing materials in fire scenarios. Because of these advantages, PP is regarded as an eco-friendly insulating material for both AC and DC cable applications. In fact, several cable manufacturers have begun developing high-voltage thermoplastic PP-based insulation systems. For example, the EU-funded HVDC amp project, along with companies such as Prysmian and Borealis, has been working on PP-based high-voltage DC cable systems that eliminate the need for cross-linking. As of 2021, at least one 525 kV HVDC cable insulated with thermoplastic PP has successfully passed type testing, demonstrating the feasibility of PP in demanding high-voltage environments.

However, PP insulation materials also face technical challenges, particularly the tendency to accumulate space charge under DC stress, which may lead to electric field distortion in HVDC cables. Extensive research is addressing this issue through material modifications, such as incorporating nanoparticle fillers or chemically grafting polar groups to inhibit charge transport [57, 58]. For instance, grafting 4-methylstyrene onto PP has been shown to significantly enhance DC breakdown strength and suppress space charge accumulation, bringing the performance of modified PP close to that of XLPE. In parallel, studies have demonstrated the feasibility of mechanical recycling for hybrid systems combining XLPE with thermoplastics. Pioneering work by NKT (a major cable producer based in Denmark, Nordiske Kabel Traadfabriker) and its Nordic partners showed that production waste or post-consumer XLPE from cables can be cryogenically ground and blended with virgin polyolefins to produce new materials. For example, composites containing 25–50% recycled XLPE mixed with PP were found to retain good mechanical properties and impact strength, making them suitable for injection-molded products. Aging tests further indicated that long-term stability could be achieved by adding stabilizers such as antioxidants and metal deactivators. This circular approach uses waste XLPE as a filler in new polymer products, which not only reduces waste generation but also extracts value from materials once considered nonrecyclable. In 2020, NKT announced the industrialization of an XLPE mechanical recycling process at its Stenlille facility in Denmark, converting cable insulation waste into plastic pellets and molding them into new products such as cable drums. The environmental benefit is significant: recycling 1 ton of XLPE into new plastic products reduces carbon dioxide emissions by approximately 2 tons compared to using virgin materials. This clearly demonstrates how innovations in material design, such as the blending of thermoplastics and thermosetting plastics, and recycling technology can dramatically lower the carbon footprint of insulating plastics.

Another important aspect of sustainable insulation design is the elimination of halogens and other hazardous substances. PVC, owing to its excellent mechanical properties, electrical insulation capabilities, processability, and low cost, has long served as a key material for insulation and sheathing in small-diameter cables. However, PVC is difficult to degrade after its service life, and its recycling is often accompanied by the release of harmful additives such as plasticizers and lead-based stabilizers, posing environmental and health risks [54, 56]. Additionally, disposal by incineration or landfilling not only contributes to greenhouse gas emissions but may also lead to the release of persistent organic pollutants like dioxins. These challenges have driven the development of low smoke zero halogen (LSZH) cables, which use alternative polymers and flame retardants to meet fire safety standards without chlorine or bromine. Halogen-free insulation materials include PE, ethylene-vinyl acetate (EVA) with mineral fillers, and specialty thermoplastics like polyolefin elastomers and modified polyphenylene ether (mPPE). These materials do not emit corrosive gases or dense smoke upon combustion. For instance, Alpha Wire's EcoWire uses Noryl® mPPE instead of PVC. It is free from halogens, phthalates, and heavy metals, compliant with the Restriction of Hazardous Substances (RoHS) and Waste Electrical and Electronic Equipment (WEEE) directives, and offers high dielectric strength, enabling thinner insulation. Compared to PVC, it has 25% lower density and is thermoplastically recyclable. While material costs are higher, growing regulatory pressure and environmental awareness are increasing its appeal.

PVC recycling has improved in Europe through initiatives like VinylPlus, with rising mechanical recycling rates. However, challenges remain: many processes lead

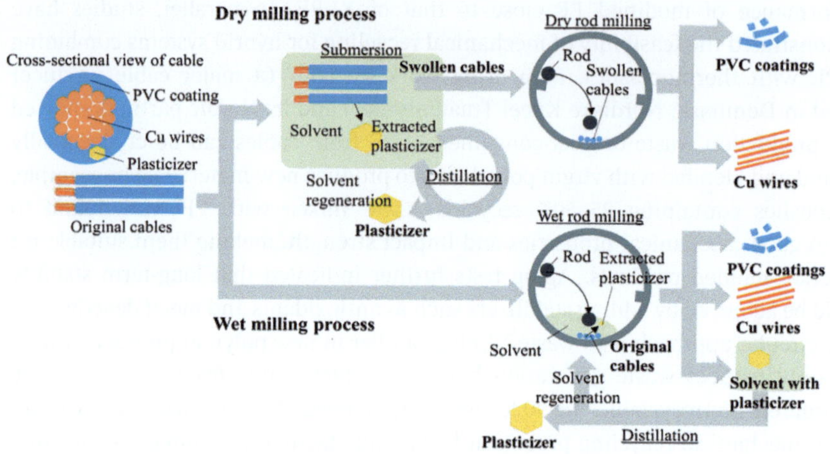

Figure 4.6 Schematic illustrations of the dry and wet rod milling processes. *Source:* [54] / Elsevier / CC BY 4.0.

to downcycled materials, and the presence of legacy additives complicates reuse. As shown in Figure 4.6, typical PVC recycling from end-of-life cables involves: (1) mechanical pretreatment including shredding and separation; (2) purification using methods such as electrostatic or solvent separation; and (3) advanced recovery includes techniques like pyrolysis and chemical dechlorination to enhance quality and reduce toxic residues [54].

In response to the above issues, researchers have focused on technical upgrades for PVC in the context of sustainable development, concentrating on areas such as thermal degradation behavior analysis, functional enhancement composites, service life monitoring, and green recycling, and have made significant progress. Aerts et al. [59] used a Hansen solubility parameter-based solvent system (γ-valerolactone and water) to separate PVC from fillers like TiO_2 and $CaCO_3$. Applied to commercial waste, this method achieved >95% PVC recovery and >99% filler removal without significant degradation, offering a green, bio-based solution. Muniz et al. [60] proposed using infrared thermography to monitor the surface emissivity of PVC cables for nondestructive aging prediction. This method shows good correlation with conventional metrics like elongation-at-break and thermogravimetric analysis, offering potential for in-situ lifespan monitoring. In addition, studies show that specific microbes (*Pseudomonas*, *Bacillus*) can enzymatically degrade phthalate plasticizers like Di-2-ethylhexyl phthalate into nontoxic small molecules through staged hydrolysis and ring cleavage, supporting eco-friendly pretreatment during recycling [43]. Although this does not degrade the PVC backbone, it helps reduce environmental toxicity from additives.

In summary, phasing out PVC and halogenated plastics from insulation materials and developing cleaner, more efficient recycling technologies will help strengthen fire safety, reduce pollution, and improve waste management, which are crucial to sustainable development.

4.3.1.3 Recyclable Insulating Plastics Enabled by Dynamic Covalent Bonding

To address the recycling challenges posed by the irreversible cross-linked structures of traditional thermoset insulating materials, as well as the resulting resource waste and environmental burden, researchers have proposed the introduction of dynamic covalent bonds (DCBs) to construct reversible cross-linked networks. This approach imparts materials with recyclability, self-healing, and other advanced functionalities. As illustrated in Figure 4.7, several types of dynamic covalent reactions have been commonly employed, including transesterification, transcarbonation, acetal and imine exchange, vinyl-urea, Diels-Alder reactions, and disulfide exchange [9, 10, 61–63]. These reactions can undergo reversible bond exchange under thermal, photochemical, chemical, or catalytic stimuli, enabling insulating plastics to be reprocessed by heat or reconfigured in solution.

Lee et al. [64] proposed a nitrene C—H insertion strategy using perfluorophenyl azide (PFPA) groups to in situ introduce DCBs into thermoplastics such as LDPE and PP, thereby forming CANs that are recyclable thermosets (Figure 4.8a,b). This method applies to low-functionality polyolefin systems, where nitrenes generated from the thermal decomposition of PFPA undergo insertion into polymer C—H bonds, linking cross-linkers bearing dynamic bonds, such as disulfide, acetal, or imine bonds, to construct reversible topological network structures, as shown in Figure 4.8c. Among them, the LDPE cross-linked with disulfide (LDPE-SS) vitrimers system, utilizing disulfide exchange, exhibits good self-healing ability, thermal/mechanical recyclability, and tunable stress relaxation behavior.

Notably, the CANs exhibit excellent reprocessability and chemical recyclability. Through mechanical hot pressing, the material can be repeatedly reused, as shown in Figure 4.9a. Although repeated processing slightly reduces the cross-linking density and mechanical properties, LDPE-SS still retains high mechanical performance after five cycles (tensile strength: 9.9 ± 0.3 MPa; elongation at break: $101\% \pm 28\%$, as shown in Figure 4.9b). More importantly, by taking advantage of the reversibility of the dynamic bonds, the polymer network can be de-cross-linked in the presence of a catalyst, such as trioctylphosphine via solvent-mediated processes using dimethyl

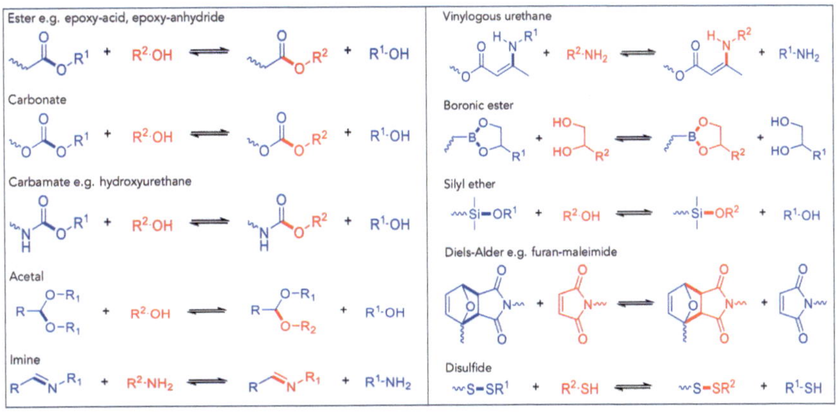

Figure 4.7 Types of dynamic covalent bond. *Source:* Adapted from [9, 10, 62, 63].

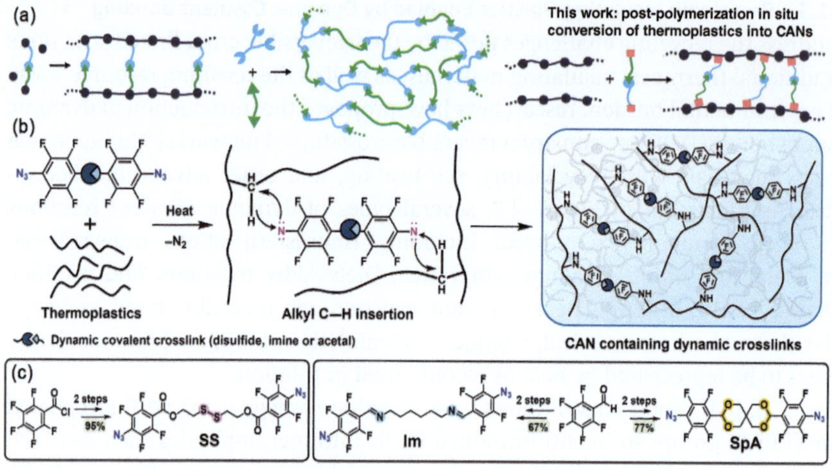

Figure 4.8 Previous work and this approach for cross-linking thermoplastics to form CANs using PFPA-based dynamic covalent cross-linkers. (a) Strategies to synthesize CANs. (b) Thermal activation of PFPA groups forms reactive nitrenes that undergo alkyl C–H insertion with thermoplastics to form permanent amine bonds; during this process, DCBs are incorporated to form a CAN. (c) Synthesis of SS, Im, and SpA cross-linkers in two steps each. *Source:* [64] / with permission of Elsevier.

disulfide and toluene, and then re-cross-linked under vacuum thermal treatment. The reconstructed material not only maintains its mechanical performance but also exhibits a more homogeneous network structure. Furthermore, the authors demonstrated the applicability of this strategy to postconsumer LDPE/PP mixed plastic waste (Figure 4.9c), thereby expanding the potential for high-value utilization of waste polyolefins.

In addition, Neidhart et al. [65] achieved the conversion of polyolefins, particularly PE, from thermoplastic to thermosetting materials through a C—H bond functionalization strategy, ultimately realizing their reprocessability and recyclability, and providing an innovative molecular design pathway for the closed-loop use of XLPE materials. Conventional polyolefin chains lack reactive functional groups, making them difficult to cross-link or de-cross-link in a controlled manner. In this work, transition metal catalysis was used to activate inert C—H bonds, allowing the subsequent introduction of functional groups capable of forming DCBs, such as alkenyl or azide moieties. This approach enables the precise construction of cross-linking sites without altering the polymer backbone, thereby laying the foundation for the formation of reversible networks. As shown in Figure 4.10a, the study employs a three-step strategy of C—H bond activation–grafting–cross-linking. First, reactive bifunctional small molecules are grafted onto the PE main chain, followed by the introduction of reversible covalent cross-linking structures such as imine bonds, ester exchange, and disulfide bonds to construct dynamic covalent networks (DCPNs). Unlike traditional irreversible DCP cross-linking, this method preserves the mechanical strength of the network while allowing for thermal or chemical de-cross-linking. This enables controlled degradation and multiple rounds

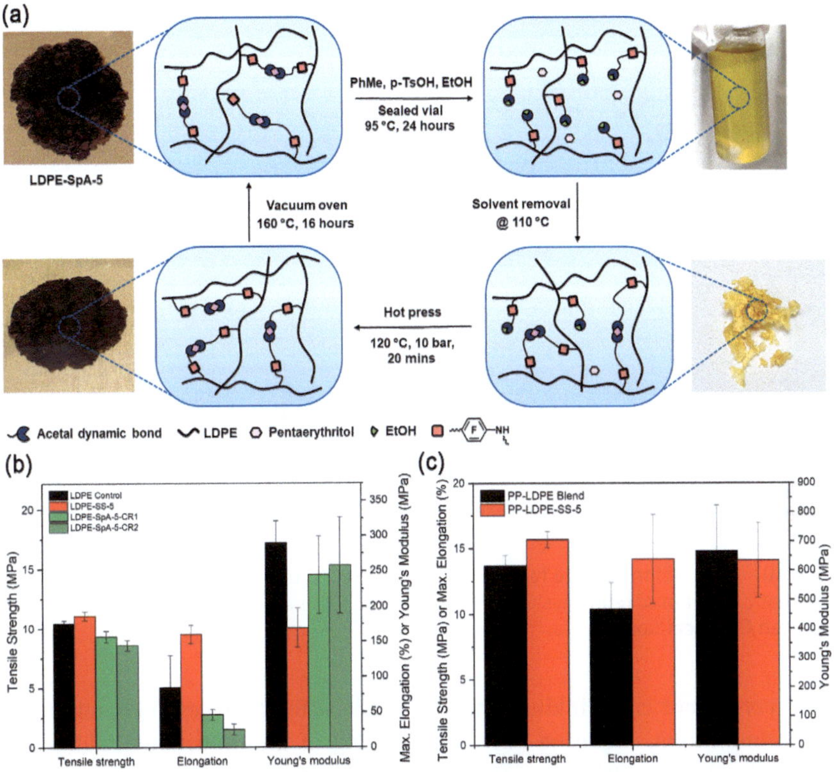

Figure 4.9 Recycling process of LDPE-SpA-5 and cross-linking process of PP-LDPE-SS-5 and their mechanical properties. (a) Illustration and photographs of the chemical recycling process of LDPE-SpA-5. (b) Comparison of tensile strength, elongation, and Young's modulus of LDPE control, LDPE-SS-5, LDPE-SpA-5-CR1, and LDPE-SpA-5-CR2. (c) Comparison of tensile strength, elongation, and Young's modulus of PP-LDPE blend and PP-LDPE-SS-5. *Source:* [64] / with permission of Elsevier.

of reprocessing (see Figure 4.10b). Through precise molecular design and dynamic cross-linking chemistry, the authors successfully created polyolefin materials that exhibit both thermoset-level performance and thermoplastic-like processability. This dual-functionality platform provides theoretical and technical support for the development of high-performance, recyclable cable insulation materials.

4.3.1.4 Recyclable Insulating Plastics Enabled by Dynamic Noncovalent Bonding

Another effective strategy for greening insulating plastics is the construction of supramolecular networks based on reversible noncovalent interactions, such as hydrogen bonding or ionic associations. For instance, molecules capable of forming hydrogen-bonding networks can be incorporated into PE or PP [66]. These reversible interactions act as dynamic cross-links that enhance mechanical performance under service conditions while dissociating upon heating to restore processability. Such systems have been explored for cable insulation applications

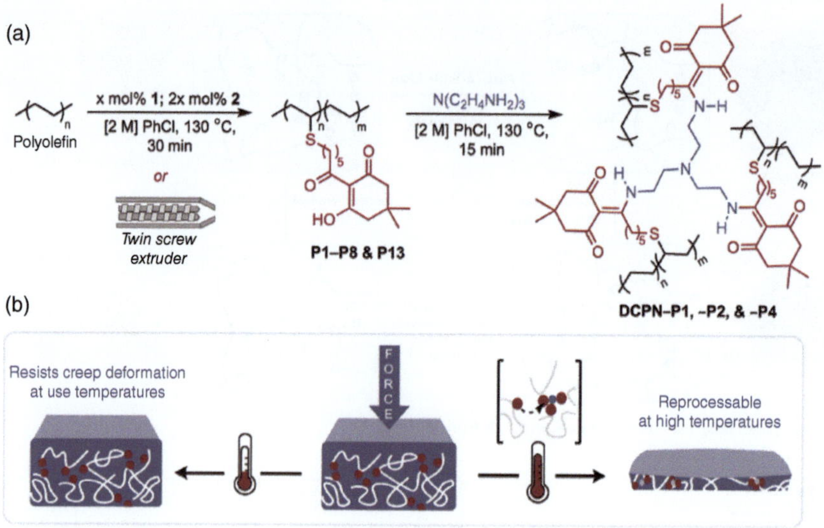

Figure 4.10 (a) Reaction scheme for C—H functionalization to provide DCPNs. (b) Schematic diagram of reprocessability characteristics. *Source:* [65] / with permission of American Chemical Society.

to avoid permanent cross-linking. They are intrinsically reprocessable and often easier to implement, requiring only the blending of multifunctional additives. However, the thermal stability of these networks is generally lower than that of covalent bonds. Therefore, careful molecular design, such as employing multiple hydrogen bonds per cross-link, is necessary to ensure adequate performance at operating temperatures.

In the development of recyclable XLPE materials, dynamic noncovalent bonds, particularly hydrogen bonds, have emerged as promising alternatives to permanent covalent cross-links due to their reversibility, catalyst-free property, and strong thermal responsiveness. Unlike conventional XLPE, hydrogen-bonded networks can undergo dissociation and reformation under thermal stimuli, enabling thermal reconfiguration and reprocessing. For instance, Mao et al. [18], constructed a multiple hydrogen-bonding network composed of amide-triazole-carboxylic acid motifs by reacting 3-amino-1,2,4-triazole (ATA) with maleic anhydride grafted polyethylene (PE-g-MAH), achieving efficient dynamic cross-linking under metal-free and solvent-free conditions, as shown in Figure 4.11a. The resulting material exhibited stable melt rheology between 170 and 190 °C and allowed multiple rounds of mechanical recycling via simple hot-pressing. After two recycling cycles, its mechanical properties remained nearly unchanged, and both stress relaxation behavior and differential scanning calorimetry results confirmed the reversibility of the cross-linked network. These findings demonstrate that PE-based physical cross-linked networks formed by multiple hydrogen bonds can provide mechanical integrity and creep resistance, while also enabling recyclability through network reconstruction.

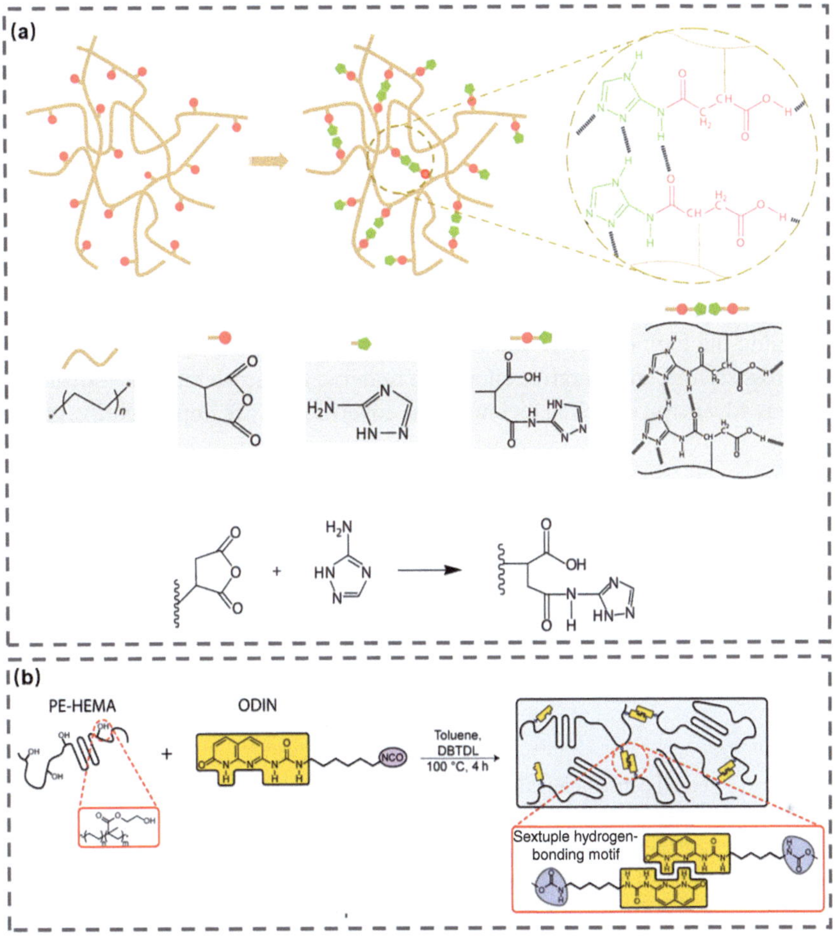

Figure 4.11 (a) Schematic illustration of formation of H-bonds cross-linking via amide triazole ring-carboxylic acid units in PE. *Source:* [18] / Springer Nature. (b) Synthesis of hydrogen bonding functionalized poly(ethylene-co-(2-hydroxyethyl methacrylate)) (PE-HEMA); cartoon depicts semicrystalline nature of resulting polymer cross-linked via multiple hydrogen bonding moieties. *Source:* [67] / with permission of American Chemical Society.

In comparison, highly ordered hydrogen bonding arrays, such as the hexameric hydrogen bond array (6HBA) reported by Tellers et al. [67], utilized TAZ-PZ structural units to create noncovalent cross-linking sites with strong synergistic effects in PE (Figure 4.11b). This system significantly enhanced the storage modulus, which increased to $\sim 1.2 \times 10^5$ Pa, stress relaxation efficiency, and mechanical response rate. Although the recyclability and thermal reprocessability were not directly evaluated, the observed fast dynamic response indicates structural potential for constructing high-performance, responsive insulation networks.

Hence, dynamic noncovalent (hydrogen-bonded) XLPE systems offer a balanced combination of mechanical properties and thermal processability, providing a viable route toward closed-loop recycling of thermoset insulating materials. In cable and electrical equipment applications, these materials, characterized by excellent electrical insulation, deformation responsiveness, and reprocessability, hold promise as sustainable and low-carbon alternatives to conventional nonrecyclable XLPE.

In summary, the field of insulating materials is undergoing a transformation aligned with global sustainability goals. Table 4.2 summarizes and compares the advantages and disadvantages of traditional and new sustainable insulating plastics, clearly showing the future direction of their development. Next-generation insulating plastics will be characterized by circularity, including recyclability or biodegradability, low-carbon manufacturing, and safe end-of-life behavior, all while maintaining the high electrical and mechanical performance required during service. Across sectors, collaborative efforts are underway to accelerate this transition: industry is advancing recycling technologies, governments are implementing green

Table 4.2 Overview of conventional vs. sustainable insulating plastic approaches, showing the shift toward recyclable, halogen-free, and low-carbon solutions.

Aspect	Conventional	Sustainable
Base polymer	XLPE: not recyclable PVC: emits toxins	PP (thermoplastic) and dynamic cross-linked polyethylene: recyclable Halogen-free polyolefins (PE, EVA, mPPE): cleaner combustion
Feedstock source	Fossil petrochemicals (100% oil/gas)	Bio-based plastic from renewable crops (captures CO_2)
Fire safety	PVC releases HCl smoke and dioxins; XLPE is halogen-free but produces smoke when burning	Low-smoke zero-halogen cables (polyolefin-based): minimal toxic emissions
End-of-life disposal	Landfill (persists for decades) Incineration (energy recovery with emissions)	Mechanical recycling into new products (e.g. XLPE → plastic parts) Chemical recycling (pyrolysis to fuels) under development Molecular structure design enables recycling and reuse
Life-cycle GHG emissions	High (fossil origin and waste incineration CO_2)	Reduced: recycling saves ~2 tons CO_2 per ton XLPE
Functional additives	Lead-based stabilizers (old PVC): toxic Peroxide cross-linker (XLPE): energy-intensive cure	Nontoxic stabilizers (Ca/Zn for PVC) Nitrogen or mineral flame retardants (no bromine) Reversible cross-link bonds for reprocessing (research stage)
Durability and reliability	Susceptible to aging (thermal oxidation, electrical trees) requiring frequent maintenance/replacement	Enhanced with nanoparticles and self-healing polymers to extend service life Condition monitoring to optimize lifetime use

material policies, and researchers are developing high-performance, sustainable insulating polymers. Together, these initiatives are driving the continued progress of low-carbon, eco-friendly, and sustainable insulating materials.

4.3.2 Smart Insulators: Self-Healing and Self-Adaptive Plastics

4.3.2.1 Extrinsic Approaches to Self-Healing

During long-term service of polyolefin-based insulating materials, electrical breakdown and mechanical damage are key factors limiting their reliability. To extend material lifespan and ensure system safety, developing self-healing PE insulation materials has become an important research focus in recent years. According to different healing mechanisms, existing strategies can be classified into extrinsic and intrinsic approaches. This section focuses on the application of extrinsic self-healing strategies in insulating plastics, highlighting their significance in promoting low-carbon, eco-friendly, and sustainable development of insulating materials.

PE, a typical nonpolar thermoplastic polymer, plays a critical role in cable insulation, packaging, and high-voltage devices. However, due to the lack of reversible cross-linking sites or polar-responsive units in its molecular structure, PE exhibits limited self-healing capability when subjected to micro-damages such as electrical tree discharge or mechanical cracking. To enhance its service reliability, extrinsic self-healing systems have been explored as early strategies to compensate for electrical degradation and prolong material lifespan. Among them, microcapsule-based self-healing has been one of the earliest and most effective approaches validated in PE insulation [68]. This strategy embeds microcapsules containing healing monomers and catalysts into the polymer matrix. Upon damage, the capsules rupture and release their contents, initiating localized polymerization to restore structural integrity. This method offers advantages such as simple structural design and low energy requirements for thermal activation.

As illustrated in Figure 4.12, researchers incorporated microcapsules with a urea-formaldehyde shell encapsulating dicyclopentadiene (DCPD) into LDPE, along with dispersed Grubbs' second-generation catalyst. Upon crack formation in LDPE, the capsules rupture and release DCPD, which undergoes ring-opening metathesis polymerization (ROMP) under thermal activation to form highly cross-linked poly(dicyclopentadiene) (PDCPD), thereby filling the crack and restoring the damaged structure (see Figure 4.13a for the mechanism). Figure 4.13b presents optical micrographs of scratched pure PE and PE/microcapsule composites before and after heating at 60 °C for 30 min, showing negligible change in scratch width for neat PE but a pronounced narrowing and partial closure for the composite due to microcapsule-triggered self-healing [70]. These images provide direct evidence of the heat-triggered responsiveness and repair effectiveness of the microcapsule-based system in addressing mechanical damage in PE.

In addition, Zhang et al. [71] systematically investigated the electrical performance recovery of LDPE insulation materials under mechanical damage, with particular focus on the influence of thermal activation conditions on healing efficiency. Similar to previous studies, this work employed urea-formaldehyde resin as

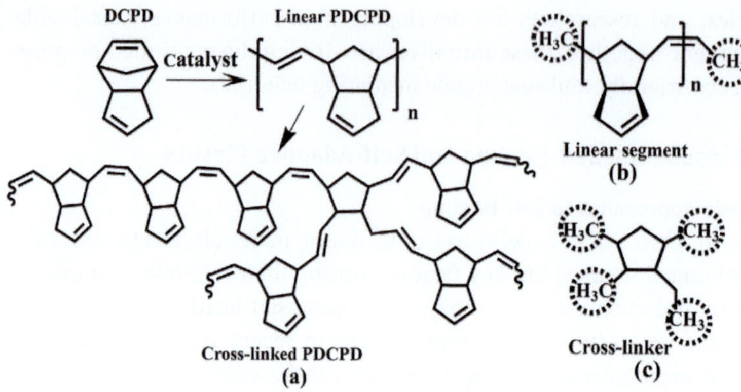

Figure 4.12 (a) Schematic diagram of the ring-opening polymerization mechanism of DCPD. (b) The linear segment and (c) cross-linker to construct a simulated PDCPD network. The circles represent carbon atoms that form cross-link bonds during the simulated polymerization process. *Source:* [69] / with permission of Elsevier / CC BY-NC-ND 4.0.

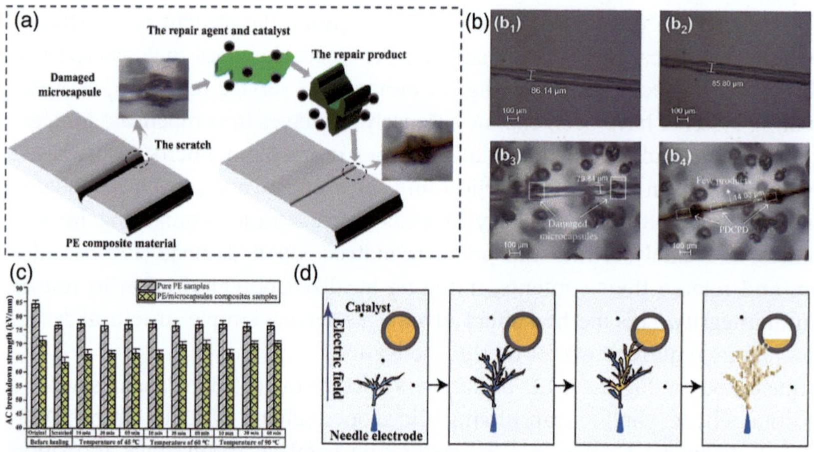

Figure 4.13 The self-healing mechanism and performance of LDPE/microcapsule composite insulation. (a) Schematic diagram of self-healing behavior of PE/microcapsules composite material. (b) Morphology of the scratch. (b_1) Unheated pure LDPE, (b_2) heated pure LDPE, (b_3) unheated composite, and (b_4) heated composite. (c) AC breakdown strength of polyethylene samples before and after scratch and repair. *Source:* [70] / Springer Nature. (d) The electrical tree self-healing process of composite material. *Source:* [69] / with permission of Elsevier / CC BY-NC-ND 4.0.

the shell material and DCPD as the healing agent, with Grubbs' second-generation catalyst dispersed in the matrix. As shown in Figure 4.13c, before heating, the initial breakdown strengths of pure PE and PE/microcapsule composites were approximately 85.5 and 71.5 kV/mm, respectively. The slightly lower strength of the composite is attributed to the presence of microcapsules disrupting the matrix homogeneity and interfacial structure. After introducing mechanical notches, the breakdown strengths decreased to 75.2 kV/mm for pure PE and 66.3 kV/mm for composite, indicating that the electric field concentration at microcracks is the primary cause of dielectric failure. At this stage, no healing occurred, as the microcapsules remained intact. Upon thermal treatment at 45, 60, and 90 °C for varying durations (10, 30, and 60 min), DCPD was gradually released and polymerized into cross-linked PDCPD under the action of the dispersed Grubbs catalyst, effectively filling the damaged pathways. After heating at 90 °C for 60 min, the breakdown strength of the composite recovered to ~78.4 kV/mm, approaching its original value and confirming the effectiveness and completeness of the healing process. In contrast, pure PE showed only slight thermal relaxation effects under the same conditions, with no substantial recovery of breakdown strength. The potential of this extrinsic self-healing system to mitigate electrical tree-induced micro-damage was further validated. Figure 4.13d illustrates the coupling between electrical tree propagation and microcapsule-triggered healing [69]. Starting from a needle electrode, electrical tree channels expanded outward under a high electric field. As the growing tree reached the dispersed microcapsules, localized stress ruptured the capsules, releasing DCPD. In the presence of Grubbs catalyst, rapid ROMP occurred, forming dense PDCPD networks that filled and coated the tree paths, thereby restoring insulation and preventing further breakdown.

These findings confirm that microcapsule-based extrinsic healing strategies in PE insulation have been thoroughly explored, from macro-scale performance recovery to micro-scale healing mechanisms. This approach offers a promising path for enhancing the reliability and lifetime of polyolefin insulation systems. However, several limitations must be acknowledged. First, microcapsule healing is inherently single-use; once the capsules rupture and release their contents, they cannot respond to subsequent damage. Second, uneven dispersion or aggregation of microcapsules may introduce local defects, undermining dielectric uniformity and long-term performance. Moreover, high filler content can compromise the mechanical and dielectric properties of the composite, leading to reduced breakdown strength or increased dielectric loss. Finally, the healing efficiency is constrained by the proximity of microcapsules to damage sites and by reaction conditions such as temperature or electric field intensity. Incomplete triggering or insufficient filling may occur. Therefore, although microcapsule-based extrinsic systems provide an effective initial solution for damage management in PE insulation, further optimization is necessary to ensure long-term reliability and repeated healing capability in practical applications.

Additionally, PP, a typical TPO insulating material, is widely used in HVDC cables, insulation layers, and capacitor packaging due to its excellent electrical insulation, thermal resistance, and low cost. In addition, its good thermal processability and recyclability make PP a promising candidate for sustainable electrical insulation applications. However, the lack of intrinsic dynamic functional groups in its molecular structure hinders spontaneous self-healing following multiscale electrical damage, such as electrical treeing, space charge accumulation, and localized breakdown, which often leads to rapid degradation or even failure of insulation performance, limiting the long-term reliability of PP-based systems.

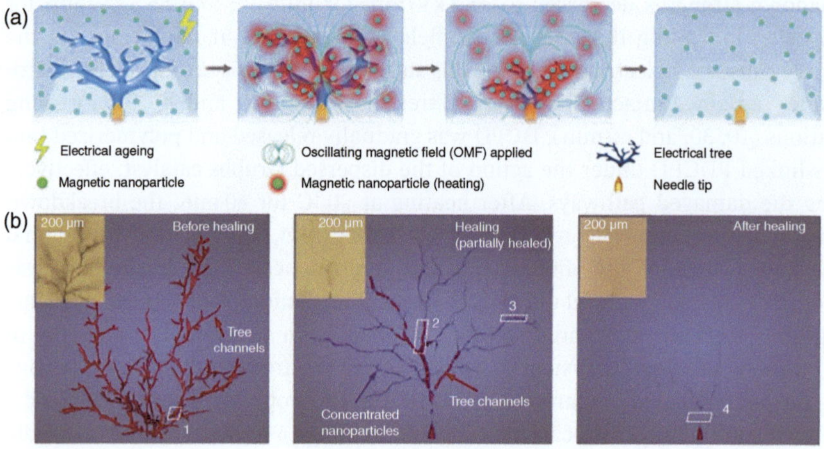

Figure 4.14 Self-healing of electrical tree damage in thermoplastics via defect-targeted superparamagnetic heating. (a) Schematics of the magnetic healing of electric trees, starting from tree formation, application of an OMF, and migration of particles to produce healing. (b) 3D micro-X-ray tomography of electrical trees in PP with 0.09 vol% SPNPs before and after healing. *Source:* [17] / Springer Nature.

To address this challenge, extrinsic self-healing strategies have been explored, aiming to achieve localized repair of damaged regions via functional additives, external energy input, or interfacial reconstruction. In this context, a defect-oriented superparamagnetic heating-assisted self-healing strategy has been proposed [17]. Surface-functionalized superparamagnetic nanoparticles (SPNPs) are uniformly dispersed into the PP matrix. Upon damage, the application of an oscillating magnetic field (OMF) induces nanoscale local heating by activating SPNPs near the damage site, as shown in Figure 4.14a. This targeted thermal effect softens or even melts the polymer locally, promoting chain mobility and structural reconstruction that effectively seals the damaged channels and restores electrical insulation. Unlike conventional bulk thermal healing approaches, this method leverages the preferential migration and enrichment of SPNPs along discharge-induced defect paths, allowing energy to be precisely focused at the damaged zones (Figure 4.14b). As a result, global energy consumption and undesired thermal effects are minimized. Experiments showed that SPNP-doped PP samples exhibited clear structural recovery and restored dielectric strength after simulated electrical tree discharge. The breakdown voltage increased from 14 to 17.5 kV post-repair, and infrared imaging combined with density distribution analysis confirmed effective local healing. Moreover, the process demonstrated excellent repeatability, maintaining stable breakdown strength and insulation performance across multiple damage-repair cycles, indicating robust damage memory and network stability. Dielectric spectroscopy further revealed that with SPNP concentrations between 0.05 and 0.18 vol.%, the dielectric constant of PP slightly increased to approximately 2.3–2.6, while the dielectric loss remained below 10^{-3}, well within the safety threshold for insulating applications. These results highlight the feasibility of externally activated, particle-assisted localized heating as an effective pathway for the precise healing and performance recovery of polyolefin-based insulation materials, offering practical prospects for use in HVDC cables, capacitor packaging, and other high-voltage systems.

In summary, extrinsic self-healing strategies represent one of the earliest and most effective approaches applied to plastic insulation materials, particularly in thermoplastic dielectric systems such as PE. These strategies achieve damage-responsive healing primarily by introducing functional healing units, e.g. microcapsules, fibers, or liquid-filled cavities, into the polymer matrix. A typical system involves the combination of urea-formaldehyde microcapsules encapsulating monomers such as DCPD and dispersed Grubbs' catalyst. When mechanical or electrical damage occurs, such as cracking or electrical treeing, the microcapsules rupture under stress, releasing the healing agent. In the presence of the catalyst, in situ polymerization takes place, forming a cross-linked polymer network that fills cracks and blocks discharge pathways, thereby restoring insulating performance.

4.3.2.2 Intrinsic Approaches to Self-Healing

In the preceding section, extrinsic self-healing approaches in polyolefin insulating materials were discussed, emphasizing the incorporation of external healing agents encapsulated within microcapsules or vascular networks. While extrinsic strategies effectively repair macroscopic damage, they typically offer limited repeatability and depend heavily on external reservoirs and precise damage-triggered mechanisms, presenting challenges in maintaining long-term healing functionality.

In contrast, intrinsic self-healing has emerged as a highly promising alternative approach, enabling polymeric insulating materials to autonomously repair internal damage through dynamic chemical bonds or reversible interactions embedded within their molecular structures. Intrinsic self-healing relies on reversible covalent bonds, supramolecular interactions, hydrogen bonding, or dynamic coordination bonds, allowing the material itself to repeatedly and autonomously restore its structural integrity and dielectric performance without the need for external healing agents [10, 72–74]. Recent studies have demonstrated that intrinsically self-healing polymers can significantly enhance the durability and reliability of insulating plastics under mechanical stress, thermal cycling, and electric field fluctuations. This self-contained healing mechanism offers unique advantages such as repeatable healing, reduced system complexity, and the potential for seamless integration into existing manufacturing processes, making it particularly suitable for advanced electrical insulation applications. This section discusses the key intrinsic healing mechanisms, highlights representative polymer systems, and explores their practical significance in achieving long-term stable and sustainable insulation performance.

Extrinsic strategies rely on encapsulated microcapsules, liquid healing agents, or thermally triggered flowable phases that are released upon damage to fill cracks. Although these systems offer high healing efficiency, they suffer from drawbacks such as depletion of healing agents and compromised dielectric homogeneity. In contrast, intrinsic self-healing strategies introduce reversible dynamic bonds,

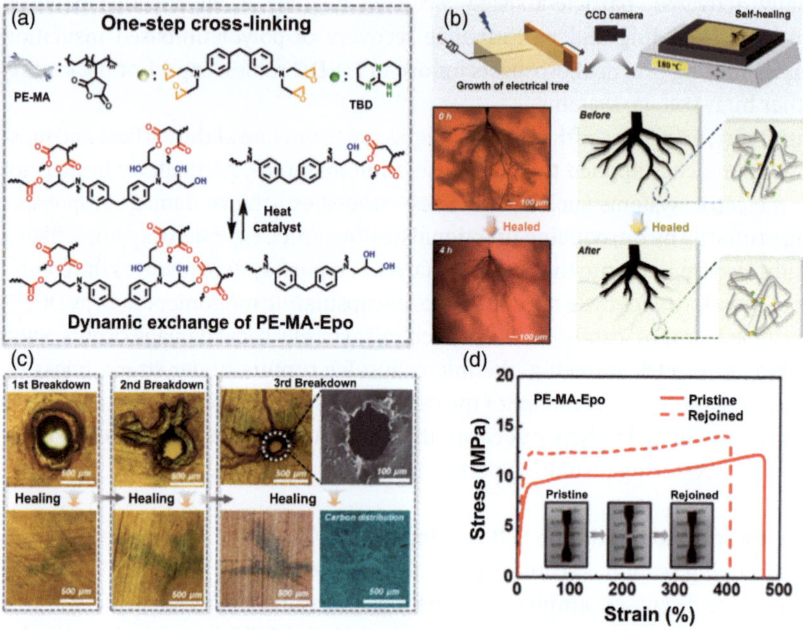

Figure 4.15 (a) Dynamic exchange mechanism between ester bonds and free hydroxyl groups. (b) Dynamic exchange mechanism between ester bonds and free hydroxyl groups. (c) Cyclic electrical breakdown/self-healing properties of PE-MAEpo. (d) The mechanical self-healing properties of PE-MAEpo. *Source:* [75] / with permission of American Chemical Society.

either dynamic covalent or noncovalent, into the polymer network, enabling reversible bond dissociation and reformation. These materials offer continuous and repeatable healing, thermal reprocessability, and improved performance under complex service conditions, making them more suitable for high-performance insulation applications.

As shown in Figure 4.15a, Li et al. [75] constructed an XLPE network with dynamic ester bonds through molecular design. In this approach, maleic anhydride (MA) was grafted onto a polyethylene backbone to introduce reactive ester groups. These groups then reacted with polar epoxy monomers via ring-opening to form side chains containing ester linkages. The resulting material, PE-MA-epo, retained the excellent crystallinity and mechanical strength of PE while significantly enhancing the dielectric constant ($\varepsilon_r > 7$) due to the introduction of polar groups. More importantly, the dynamic nature of ester bonds endowed the material with intrinsic self-healing functionality. During electrical damage recovery, exchange reactions of ester bonds enabled rapid chain mobility and network reconfiguration at 180 °C, efficiently repairing surface scratches, interfacial fractures, and electrical breakdown sites (Figure 4.15b–d). After two electrical breakdown-repair cycles, the breakdown strength retained 80% of its initial value, and breakdown holes were restored to near-original integrity (Figure 4.15c), indicating excellent durability. Additionally, the tensile strength of the healed material reached 90% of the initial value (Figure 4.15d), and the dielectric properties remained largely unchanged. This intrinsic self-healing mechanism, enabled by dynamic ester bonds, enhances thermal stability and dielectric performance while improving operational reliability and lifespan in high-electric-stress environments such as high-voltage cables and capacitors, highlighting its great potential for next-generation polyolefin-based insulating materials.

4.3.2.3 Self-Adaptive Insulating Plastics with Enhanced Performance

With the rapid advancement of electrical and electronic systems toward higher voltage, power density, and complexity, traditional insulating plastics often struggle to reliably meet the demands posed by fluctuating operating conditions, mechanical stresses, thermal cycling, and electrical aging. Conventional polymeric insulating materials, such as PE, PP, and epoxy resins, typically lack adaptability to dynamically changing environmental conditions, potentially leading to premature insulation failure, material degradation, and compromised device reliability.

To address these challenges, recent research has focused on the development of adaptive insulating plastics. These materials can autonomously regulate their properties in response to environmental stimuli or internal stresses [16, 76, 77]. These innovative materials integrate features such as dynamic chemical bonds, self-healing mechanisms, stimulus-responsive fillers, or smart polymer architectures, enabling functionalities including enhanced dielectric performance, improved thermal conductivity, and increased mechanical toughness. By automatically responding to changes in temperature, mechanical stress, electric fields, or chemical environments, adaptive insulating plastics offer a promising solution for achieving higher reliability, longer service life, and greater sustainability. This section reviews recent advances in the development of adaptive insulating plastics, elucidating their adaptive mechanisms, key material innovations, and practical applications in modern electrical systems.

Li et al. [16] proposed a novel organic semiconductor-modified insulating material – anthraquinone/cross-linked polyethylene (ET/XLPE) – based on the "Adaptive Charge Capture (ACC)" mechanism, aiming to significantly enhance cable insulation performance, as shown in Figure 4.16. Unlike conventional static doping methods, the polar organic small molecule anthraquinone (ET) undergoes electrophoretic migration under an electric field, enabling active enrichment in regions with the strongest electric field and weakest insulation, thereby achieving precise reinforcement at local defect sites. This "adaptive localization–charge regulation" strategy effectively establishes a reverse built-in electric field, which suppresses charge injection and accumulation, significantly reduces the internal electric field strength of the insulating material with a maximum reduction of 18%, and increases the electrical tree inception voltage by 123%, demonstrating superior performance among organic semiconductor-modified dielectrics. Furthermore, ET shows a stronger ability to form deep traps compared to conventional anthracene (EN), combined with its higher electron affinity and narrower bandgap, enabling enhanced capture and dissipation of high-energy electrons to inhibit electrical tree development at an early stage of insulation damage. This work not only

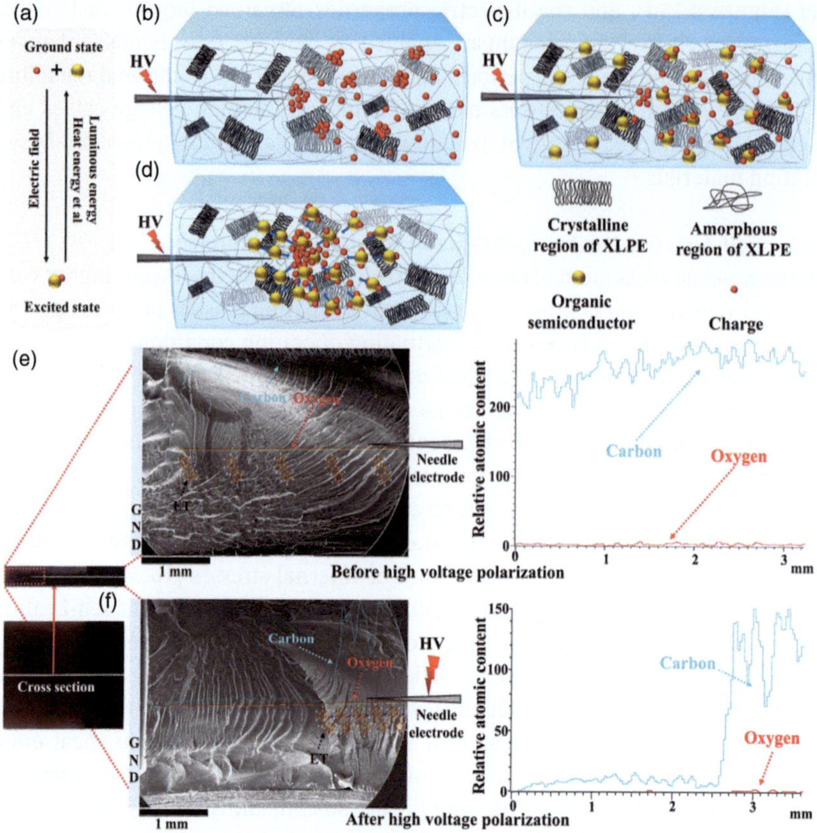

Figure 4.16 Mechanism of electric tree suppression in organic semiconductor/XLPE systems. (a) Schematic of electron energy dissipation. Charge transport behavior under high voltage in XLPE (b), EN/XLPE (c), and ET/XLPE (d). In XLPE, rapid charge injection causes damage. EN captures charges and slows transport. ET accumulates near the needle tip, forming a reverse field that suppresses local electric field intensity. SEM-EDS analysis before (e) and after (f) polarization. ET shows uniform distribution initially, but accumulates near the needle electrode after polarization, as indicated by increased carbon and oxygen signals. *Source:* [16] / Springer Nature / CC BY-NC-ND 4.0.

elucidates the microscopic mechanism of adaptive functionalization of polar organic semiconductors in dielectrics but also demonstrates their great potential in constructing intelligent insulation systems, providing a novel approach for the design of ultra-high-voltage cable materials.

4.4 Conclusion and Outlook

This chapter reviews the development history of electrical insulating plastics, highlighting a series of notable milestones, each of which addressed the critical challenges of its time and paved the way for subsequent innovations. From the emergence of PE and PVC in the mid-20th century, which established polymers as reliable insulating materials, to the development of XLPE, enabling polymer insulation in high-voltage environments, and on to the current emergence of sustainable and smart materials, this process has never ceased. Key performance metrics, including breakdown strength, thermal stability, and dielectric loss, have been enhanced or optimized for specific needs, such as the high breakdown strength of PE and PP, and the flame-retardant properties of PVC. Meanwhile, concerns about recyclability and environmental impact in modern society are being addressed through novel material chemistry, such as dynamic network structures, and rethinking traditional concepts.

A clear trend is that interdisciplinary innovation is driving the advancement of insulating materials. Chemistry contributes new polymers and cross-linking chemistries, such as CANs and vitrimers; materials science introduces nano-additives and processing techniques, such as nano-dielectrics and advanced extrusion techniques; and electrical engineering sets performance targets, such as higher voltage tolerance and smart grid compatibility. The intersection of these fields is where future milestones may emerge. For instance, improving the sustainability of polyolefin-based insulation will increasingly depend on advances in catalysis and polymer molecular design, combined with materials engineering approaches to preserve high performance. This suggests that future insulating materials may be designed through "molecular engineering" to incorporate self-healing, recyclable, adaptive, or self-monitoring properties without compromising their core insulating functionality.

References

1 Bjurström, A., Edin, H., Hillborg, H. et al. (2024). A review of polyolefin-insulation materials in high voltage transmission; from electronic structures to final products. *Advanced Materials* 36 (52): 2401464.

2 Angalane, S.K. and Kasinathan, E. (2022). A review on polymeric insulation for high-voltage application under various stress conditions. *Polymer Composites* 43 (8): 4803–4834.

3 Pleşa, I., Noţingher, P., Schlögl, S. et al. (2016). Properties of polymer composites used in high-voltage applications. *Polymers* 8 (5): 173.

4 Meng, P., Zhou, Y., Yuan, C. et al. (2019). Comparisons of different polypropylene copolymers as potential recyclable HVDC cable insulation materials. *IEEE Transactions on Dielectrics and Electrical Insulation* 26 (3): 674–680.

5 Yu, S., Lee, S.H., Han, J.A. et al. (2020). Insulative ethylene-propylene copolymer-nanostructured polypropylene for high-voltage cable insulation applications. *Polymer* 202: 122674.

6 Badi, N.S.A., Makmud, M.Z.H., Jamain, Z. et al. (2024). Enhancement of structural and dielectric properties of PVC/CNC nanocomposites as electrical insulation materials. *Polymers and Polymer Composites* 32: 1–12.

7 Gao, Y., Li, J., Li, Z., and Du, B. (2023). Polymer composites with high thermal conductivity for HVDC cable insulation: a review. *IEEE Transactions on Dielectrics and Electrical Insulation* 30 (6): 2444–2459.

8 Li, S., Yu, S., and Feng, Y. (2016). Progress in and prospects for electrical insulating materials. *High Voltage* 1 (3): 122–129.

9 Kamarulzaman, S., Png, Z.M., Lim, E.Q. et al. (2023). Covalent adaptable networks from renewable resources: crosslinked polymers for a sustainable future. *Chem* 9 (10): 2771–2816.

10 Zheng, N., Xu, Y., Zhao, Q., and Xie, T. (2021). Dynamic covalent polymer networks: a molecular platform for designing functions beyond chemical recycling and self-healing. *Chemical Reviews* 121 (3): 1716–1745.

11 Yang, X., Huang, W., Dong, H., and Zha, J.W. (2025). Smart polydimethylsiloxane materials: versatility for electrical and electronic devices applications. *Advanced Materials* 37 (17): e2500472.

12 Lugger, S.J.D., Houben, S.J.A., Foelen, Y. et al. (2021). Hydrogen-bonded supramolecular liquid crystal polymers: smart materials with stimuli-responsive, self-healing, and recyclable properties. *Chemical Reviews* 122 (5): 4946–4975.

13 Habib, S., Qureshi, A., Shakoor, R.A. et al. (2022). Corrosion inhibition performance of polyolefin smart self-healing composite coatings modified with Zno@B-cyclodextrin hybrid particles. *Journal of Materials Research and Technology* 21: 3371–3385.

14 Wan, B., Pu, T., Xing, Y., and Zha, J.W. (2023). Triple-chain segments regulation enable shape memory and photo deformation of smart polyimide. *Journal of Polymer Science* 62 (7): 1370–1377.

15 Zhang, Q., Qu, D.-H., Feringa, B.L., and Tian, H. (2022). Disulfide-mediated reversible polymerization toward intrinsically dynamic smart materials. *Journal of the American Chemical Society* 144 (5): 2022–2033.

16 Li, K., Zhou, L.-M., Zhu, X.-Q. et al. (2025). Anthrone/XLPE: an adaptive charge capture intelligent insulation material for advanced electric power transmission. *Communications Engineering* 4 (1): 38.

17 Yang, Y., He, J., Li, Q. et al. (2018). Self-healing of electrical damage in polymers using superparamagnetic nanoparticles. *Nature Nanotechnology* 14 (2): 151–155.

18 Mao, H.-D., Zhang, T.-T., Guo, Z.-Y. et al. (2022). A cross-linked polyethylene with recyclability and mechanical robustness enabled by establishment of multiple hydrogen bonds network via reactive melt blending. *Chinese Journal of Polymer Science* 41 (7): 1104–1114.

19 Wang, T.-Y., Mao, J., Zhang, B. et al. (2024). Polymeric insulating materials characteristics for high-voltage applications. *Nature Reviews Electrical Engineering* 1 (8): 516–528.

20 Song, S., Zhao, H., Yao, Z. et al. (2020). Enhanced electrical properties of polyethylene-graft-polystyrene/LDPE composites. *Polymers* 12 (1): 124.

21 Ouyang, Y., Mauri, M., Pourrahimi, A.M. et al. (2020). Recyclable polyethylene insulation via reactive compounding with a maleic anhydride-grafted polypropylene. *ACS Applied Polymer Materials* 2 (6): 2389–2396.

22 Nilsson, F., Karlsson, M., Gedde, U.W. et al. (2021). Nanocomposites and polyethylene blends: two potentially synergistic strategies for HVDC insulation materials with ultra-low electrical conductivity. *Composites Part B: Engineering* 204: 108498.

23 Zhang, K., Zhong, L., Gao, J. et al. (2019). Temperature dependence of crystalline structure and DC performance in LLDPE/HDPE blending material. *IEEE Transactions on Dielectrics and Electrical Insulation* 26 (3): 754–759.

24 Vidakis, N., Petousis, M., Kalderis, D. et al. (2024). Reinforced HDPE with optimized biochar content for material extrusion additive manufacturing: morphological, rheological, electrical, and thermomechanical insights. *Biochar* 6 (2): 37.

25 Chen, X., Jiang, C., Hou, Y. et al. (2019). Polyethylene blends with/without graphene for potential recyclable HVDC cable insulation. *IEEE Transactions on Dielectrics and Electrical Insulation* 26 (3): 851–858.

26 Andersson, M.G., Hynynen, J., Andersson, M.R. et al. (2017). Highly insulating polyethylene blends for high-voltage direct-current power cables. *ACS Macro Letters* 6 (2): 78–82.

27 Shi, M., Li, X., Jiang, Y. et al. (2023). 2D nanosheet spray coating for scalable processing of high-energy-density dielectric polymer films. *Advanced Electronic Materials* 9 (7): 2300187.

28 Zhang, T., Yu, H., Jung, Y.H. et al. (2022). Significantly improved high-temperature energy storage performance of BOPP films by coating nanoscale inorganic layer. *Energy & Environmental Materials* 7 (2): e12549.

29 Qiao, Y., Yin, X., Zhu, T. et al. (2018). Dielectric polymers with novel chemistry, compositions and architectures. *Progress in Polymer Science* 80: 153–162.

30 Huang, B., Yu, J., Dong, J. et al. (2024). Improving charge storage of biaxially-oriented polypropylene under extreme electric fields by excimer UV irradiation. *Advanced Materials* 36 (52): 2311713.

31 Chen, K., Du, B.X., Liu, H., and Xiao, M. (2025). Enhanced high-temperatures energy storage performance of BOPP film via surface grafting modification. *Chemical Engineering Journal* 505: 159724.

32 Ren, W., Yang, M., Guo, M. et al. (2024). Metallized stacked polymer film capacitors for high-temperature capacitive energy storage. *Energy Storage Materials* 65: 103095.

33 Zhang, Y., Khanbareh, H., Roscow, J. et al. (2020). Self-healing of materials under high electrical stress. *Matter* 3 (4): 989–1008.

34 Mai, W., Yu, Q., Han, C. et al. (2020). Self-healing materials for energy-storage devices. *Advanced Functional Materials* 30 (24): 1909912.

35 Yang, X., Zhao, H., Li, C. et al. (2022). Improved wide-temperature-range insulation properties of block polypropylene by UV-irradiated cografting of maleic anhydride and 4-tert-butylstyrene. *ACS Applied Polymer Materials* 5 (1): 247–258.

36 Li, X., Liu, T., Chen, X. et al. (2024). Electrical properties of MAH-G-PP modified PP/SEBS matrix semi-conductive composite materials. *Composites Science and Technology* 253: 110657.

37 Sun, M., Xiao, Y., Liu, K. et al. (2023). Synthesis and characterization of polyolefin thermoplastic elastomers: a review. *The Canadian Journal of Chemical Engineering* 101 (9): 4886–4906.

38 Zhou, Y., Yuan, C., Li, C. et al. (2019). Temperature dependent electrical properties of thermoplastic polypropylene nanocomposites for HVDC cable insulation. *IEEE Transactions on Dielectrics and Electrical Insulation* 26 (5): 1596–1604.

39 Kar, G.P., Saed, M.O., and Terentjev, E.M. (2020). Scalable upcycling of thermoplastic polyolefins into vitrimers through transesterification. *Journal of Materials Chemistry A* 8 (45): 24137–24147.

40 Pourrahimi, A.M., Mauri, M., D'Auria, S. et al. (2024). Alternative concepts for extruded power cable insulation: from thermosets to thermoplastics. *Advanced Materials* 36 (52): 2313508.

41 Iwashita, T., Keller, M.R., Hagimoto, Y., and Sugawa, O. (2016). Leakage currents precede short circuits in PVC-insulated cable when exposed to external radiant heat. *Fire and Materials* 41 (4): 339–348.

42 Lewandowski, K. and Skórczewska, K. (2022). A brief review of poly(vinyl chloride) (PVC) recycling. *Polymers* 14 (15): 3035.

43 Ait-Touchente, Z., Khellaf, M., Raffin, G. et al. (2023). Recent advances in polyvinyl chloride (PVC) recycling. *Polymers for Advanced Technologies* 35 (1): e6228.

44 Tsunekawa, M., Ito, M., Yuta, S. et al. (2011). Removal of lead compounds from polyvinylchloride in electric wires and cables using cation-exchange resin. *Journal of Hazardous Materials* 191 (1–3): 388–392.

45 Gao, Z., Wang, Y., Yuan, L. et al. (2024). Room-temperature co-upcycling of polyvinyl chloride and polypropylene. *Nature Sustainability* 7 (12): 1691–1698.

46 Wang, J.-Q., Li, W.-K., Zhang, W.-Y. et al. (2024). Aging and life control of cross-linked polyethylene as cable insulation material. *Acta Physica Sinica* 73 (7): 078801.

47 Gao, Y., Huang, X., Min, D. et al. (2018). Recyclable dielectric polymer nanocomposites with voltage stabilizer interface: toward new generation of high voltage direct current cable insulation. *ACS Sustainable Chemistry & Engineering* 7 (1): 513–525.

48 Yang, J., Li, R., Hu, L., and Wang, W. (2024). Influence of thermal aging on space charge characteristics and electrical conduction behavior of cross-linked polyethylene cable insulation. *Polymers* 16 (11): 1600.

49 Hu, S., Zhou, Y., Yuan, C. et al. (2020). Surface-modification effect of MgO nanoparticles on the electrical properties of polypropylene nanocomposite. *High Voltage* 5 (3): 249–255.

50 Miyake, G., Liu, K., Zhao, Y. et al. (2025). Repurposing post-consumer polyethylene to access cross-linked polyethylene with reprocessability, recyclability, and tunable properties. *Angewandte Chemie International Edition* 64 (19): e202502641.

51 D'Auria, S., Pourrahimi, A.M., Favero, A. et al. (2023). Polyethylene based ionomers as high voltage insulation materials. *Advanced Functional Materials* 33 (36): 2301878.

52 Hirschler, M.M. (2000). *Electrical Insulating Materials: International Issues.* ASTM International.

53 Bryce, T. and Nicholls, B.P.F. (2024). Closing the loop on thermoset plastic recycling. *Science* 384 (6692): 156–157.

54 Kumar, H., Kumagai, S., Saito, Y., and Yoshioka, T. (2024). Latest trends and challenges in PVC and copper recovery technologies for end-of-life thin cables. *Waste Management* 174: 400–410.

55 Tian, Y., Han, M., Gu, D. et al. (2024). PVC dechlorination for facilitating plastic chemical recycling: a systematic literature review of technical advances, modeling and assessment. *Sustainability* 16 (19): 8331.

56 Lu, L., Li, W., Cheng, Y., and Liu, M. (2023). Chemical recycling technologies for PVC waste and PVC-containing plastic waste: a review. *Waste Management* 166: 245–258.

57 Yuan, C., Zhou, Y., Zhu, Y. et al. (2022). Improved high-temperature electrical properties of polymeric material by grafting modification. *ACS Sustainable Chemistry & Engineering* 10 (27): 8685–8693.

58 Kwon, T., Lee, S.H., Kim, J.H. et al. (2023). Polypropylene nanocomposites doped with carbon nanohorns for high-voltage power cable insulation applications. *Advanced Composites and Hybrid Materials* 6 (5): 167.

59 Aerts, A., Gándara, S.C., Rubens, M. et al. (2025). Sustainable physical recycling of rigid PVC waste: a biobased solvent approach for the recovery of PVC, TiO_2, and $CaCO_3$. *Waste Management* 202: 114798.

60 Muniz, P.R., Teixeira, J.L., Santos, N.Q. et al. (2017). Prospects of life estimation of low voltage electrical cables insulated by PVC by emissivity measurement. *IEEE Transactions on Dielectrics and Electrical Insulation* 24 (6): 3951–3958.

61 Lei, Z., Chen, H., Huang, S. et al. (2024). New advances in covalent network polymers via dynamic covalent chemistry. *Chemical Reviews* 124 (12): 7829–7906.

62 Van Lijsebetten, F., Holloway, J.O., Winne, J.M., and Du Prez, F.E. (2020). Internal catalysis for dynamic covalent chemistry applications and polymer science. *Chemical Society Reviews* 49 (23): 8425–8438.

63 Chakma, P. and Konkolewicz, D. (2019). Dynamic covalent bonds in polymeric materials. *Angewandte Chemie International Edition* 58 (29): 9682–9695.

64 Lee, Z.Y., Kamarulzaman, S., Rasyiddin, R. et al. (2025). Dynamic crosslinking of thermoplastics via perfluorophenyl nitrene C—H insertion to form recyclable thermosets. *Chem* 102479.

65 Neidhart, E.K., Hua, M., Peng, Z. et al. (2023). C—H functionalization of polyolefins to access reprocessable polyolefin thermosets. *Journal of the American Chemical Society* 145 (50): 27450–27458.

66 Leone, G., Palucci, B., Zanchin, G. et al. (2022). Dynamically cross-linked polyolefins via hydrogen bonds: tough yet soft thermoplastic elastomers with high elastic recovery. *ACS Applied Polymer Materials* 4 (5): 3770–3778.

67 Tellers, J., Canossa, S., Pinalli, R. et al. (2018). Dynamic cross-linking of polyethylene via sextuple hydrogen bonding array. *Macromolecules* 51 (19): 7680–7691.

68 Yang, Y., Dang, Z.M., Li, Q., and He, J. (2020). Self-healing of electrical damage in polymers. *Advanced Science* 7 (21): 2002131.

69 Zhang, Y., Wang, Y., Li, Y., and Zheng, R. (2022). Self-healing of electrical tree damage of polyethylene/microcapsules insulation composite material. *Journal of Materials Research and Technology* 19: 1614–1626.

70 Wang, Y., Li, Y., and Zhang, Y. (2019). Effect of doping microcapsules on typical electrical performances of self-healing polyethylene insulating composite. Applied Sciences 9 (15): 3039.

71 Zhang, Y., Wang, Y., Li, Y. et al. (2021). Self-healing of mechanical damage of polyethylene/microcapsules electrical insulation composite material. *Journal of Materials Science: Materials in Electronics* 32 (22): 26329–26340.

72 Wang, S. and Urban, M.W. (2020). Self-healing polymers. *Nature Reviews Materials* 5 (8): 562–583.

73 Ratwani, C.R., Kamali, A.R., and Abdelkader, A.M. (2023). Self-healing by Diels-Alder cycloaddition in advanced functional polymers: a review. *Progress in Materials Science* 131: 101001.

74 Ren, Y. and Dong, X. (2024). Dynamic polymeric materials via hydrogen-bond cross-linking: effect of multiple network topologies. *Progress in Polymer Science* 158: 101890.

75 Li, Z., Wen, Y., Song, Z. et al. (2023). Dynamic cross-linked polyethylene networks with high energy storage and electrical damage self-healability. *ACS Macro Letters* 12 (10): 1409–1415.

76 Yang, X., Hu, J., Wang, S. et al. (2022). A dielectric polymer/metal oxide nanowire composite for self-adaptive charge release. *Nano Letters* 22 (13): 5167–5174.

77 Shen, Z., Jia, Z., Hao, Y. et al. (2022). A fabrication method for adaptive dielectric gradient insulating components. *High Voltage* 8 (1): 59–69.

5

Milestone of Insulating Resin

Yang Zhang[1] *and Jun-Wei Zha*[1,2]

[1] *University of Science and Technology Beijing, School of Chemistry and Biological Engineering, No. 30 Xueyuan Road, Haidian District, Beijing 100083, P. R. China*
[2] *North China Electric Power University, State Key Laboratory of Alternate Electrical Power System with Renewable Energy Sources, School of Electrical and Electronic Engineering, No. 2 Beinong Road, Huilongguan, Changping District, Beijing 102206, P. R. China*

5.1 Introduction

Under the profound transformation of the global energy landscape and the accelerated implementation of the "dual carbon" strategy, the power equipment manufacturing sector faces unprecedented challenges in both environmental protection and technological upgrading. According to international energy forecasts, global electricity demand will increase by 60% by 2040, with expanding markets for renewable energy generation equipment, ultra-high voltage transmission systems, and intelligent power electronic devices [1, 2]. As a core material in power equipment, insulating resins critically influence energy conversion efficiency, operational lifespan, and the carbon footprint across the entire industry chain. Widely used epoxy resin-based insulating materials emit large amounts of volatile organic compounds such as bisphenol A (BPA) during production [3]. Their curing process requires sustained heating at 150–180 °C for 4–6 hours, resulting in high energy consumption per unit [4]. More critically, traditional thermosetting resins are difficult to degrade after equipment decommissioning. Approximately 1.2 million tons of insulating resins from global electrical waste are disposed of via landfill or incineration annually, causing secondary pollution from persistent organic pollutants [5]. This "high pollution, high energy consumption, and poor recyclability" severely hinders the power industry's green transition.

Against this backdrop, the development of next-generation electrical insulating resins has become a global research focus. These include bio-based resins with low-carbon feedstocks, as well as low-carbon, self-healing, and recyclable insulating resins [6]. Bio-based insulating resins replace petroleum-derived raw materials with biomass sources such as lignin and cellulose derivatives, effectively reducing resource consumption and carbon emissions associated with traditional petrochemical production [7]. Additionally, the naturally occurring polar groups in

Next-Generation Electrical Insulating Materials: Low-Carbon, Eco-Friendly, and Sustainable Solutions, First Edition. Edited by Jun-Wei Zha.

biomass molecular structures facilitate optimized dielectric properties. Self-healing and recyclable resins employ innovative designs based on dynamic covalent bonds and supramolecular interactions, endowing materials with damage repair capabilities during service and closed-loop recyclability after decommissioning, fundamentally addressing the solid waste pollution challenge of thermosetting resins [8]. The synergistic development of low-carbon feedstock and end-of-life solutions marks a new stage in electrical insulating resins characterized by both superior performance and environmental sustainability. Therefore, this chapter focuses on the current research challenges of next-generation electrical insulating resins: achieving green, low-carbon, and sustainable insulation materials.

5.2 Bio-Based Insulating Resins

In the context of low-carbon transformation of novel power systems, conventional petroleum-based insulating resins face dual challenges of resource unsustainability and ecological pollution [9]. The synthesis process of fossil-derived conventional insulating materials (e.g. epoxy resins and silicone rubber) involves high energy consumption with substantial carbon emissions and releases volatile toxic byproducts such as benzene derivatives [10]. Furthermore, the three-dimensional network structures formed during cross-linking curing render these materials recalcitrant to degradation and recycling after their service life expires, thereby exacerbating secondary environmental issues like microplastic pollution [11]. Concurrently, the interfacial polarization losses and electrical treeing degradation behavior of insulating media under high-temperature and high-humidity conditions highlight the performance limitations of traditional materials regarding corona resistance and hygrothermal stability. Addressing these environmental constraints, bio-based insulating resins demonstrate ecological advantages through their renewable nature and cleaner manufacturing processes [12]. Derived from forestry/agricultural byproducts, these bio-based feedstocks are converted into insulating resin monomers via catalytic synthesis and molecular reconstruction technologies, achieving significant carbon emission reductions compared with petrochemical routes. Moreover, rational molecular topology design enables bio-based insulating resins to exhibit ultralow intrinsic dielectric loss characteristics, effectively enhancing their electrical insulation performance [13].

Bio-based compounds have recently attracted widespread attention and are extensively used in the preparation of biodegradable thermosets. Common bio-based feedstocks include cellulose, vegetable oils, cashew phenols, proteins, vanillin, furfural, and others. Importantly, the bio-based compounds employed for resin synthesis are not limited to their native structures but also encompass derivatives obtained through chemical transformation [14]. Here, we present various bio-based monomers such as alcohols (phenols), acids, and other derivatives, and summarize their synthetic routes for constructing bio-based monomers or preparing bio-based curing agents.

5.2.1 Alcohols and Phenols

Alcohols and phenols are key components of resins and can be converted into epoxides via simple etherification reactions or polymerized stepwise to form polyesters or polyurethanes. Bio-based alcohols are typically derived from fructose, glucose, and sorbitol, while bio-based phenols are commonly sourced from vanillin, cashew phenol, eugenol, and tannic acid.

Vanillin is a representative bio-based phenol and aldehyde, primarily derived from lignin degradation products and vanilla bean extracts. Its molecular structure, featuring phenolic hydroxyl, methoxy, and aldehyde groups, offers multiple reactive sites, providing a unique platform for synthesizing novel bio-based insulating resins [15]. In epoxy resin synthesis, vanillin can cross-link with polyamine curing agents to form a three-dimensional network containing Schiff base structures. The phenolic hydroxyl groups of vanillin react with epichlorohydrin to yield epoxy monomers, resulting in resins with excellent mechanical properties and biodegradability [16]. In polyimide systems, vanillin serves as a renewable raw material; its phenolic hydroxyl groups efficiently construct hexasubstituted cyclotriphosphazene (HVP) monomers via nucleophilic substitution, while the aldehyde groups act as dynamic cross-linking sites, forming reversible imine bonds with diamines. This imparts thermal reprocessability and acid-degradable recyclability to the material, and the rigid aromatic structure significantly enhances the mechanical strength and thermal stability of the polyimide [17]. Additionally, leveraging the dynamic covalent nature of vanillin's aldehyde groups, polyurethane resin systems synthesized via Schiff base reactions with polyamines exhibit self-healing capabilities by activating reversible imine bond breakage and reformation under thermal stimulation at 80 °C [18]. For example, Zhang et al. [16] constructed the hyperbranched epoxy resin VanEHBP by proton transfer polymerization using bio-based vanillin as a raw material and utilized its abundant functional groups to form a dynamic imine and β-hydroxy ester bonded cross-linked network (as shown in Figure 5.1a). The hyperbranched topology not only provided energy dissipation cavities, but its high-density hydroxyl groups also catalyzed the ester exchange reaction, which reduced the activation energy of stress relaxation to 27.4 $kJ\cdot mol^{-1}$ (47.5 $kJ\cdot mol^{-1}$ for the control) and significantly enhanced the network rearrangement efficiency. It also exhibits excellent overall performance: tensile strength of 104.5 MPa, toughness of 3.58 $MJ\cdot m^{-3}$, and impact strength of 57.0 $kJ\cdot m^{-2}$. Zhang et al. [20] developed a strong, fully bio-based polyimide material (DACNF-PI-LCD) through dynamic imine bonds formed via a Schiff base reaction between dialdehyde cellulose nanofibrils (DACNF) derived from softwood pulp and plant oil-based aliphatic long-chain diamine (LCD) synthesized from castor oil derivatives. The design leveraged renewable resources and a catalyst-free, room-temperature process analogous to papermaking wet pressing. The resulting films exhibited exceptional mechanical properties, with a tensile strength of 83.74 MPa and Young's modulus of 3.90 GPa, significantly surpassing most conventional petroleum-based resins. This strength arose from synergistic hydrogen bonding and dynamic imine cross-linking, optimized by LCD's long carbon chain, enhancing structural density (0.89 g/cm^3). Critically, the materials demonstrated outstanding water barrier properties: high

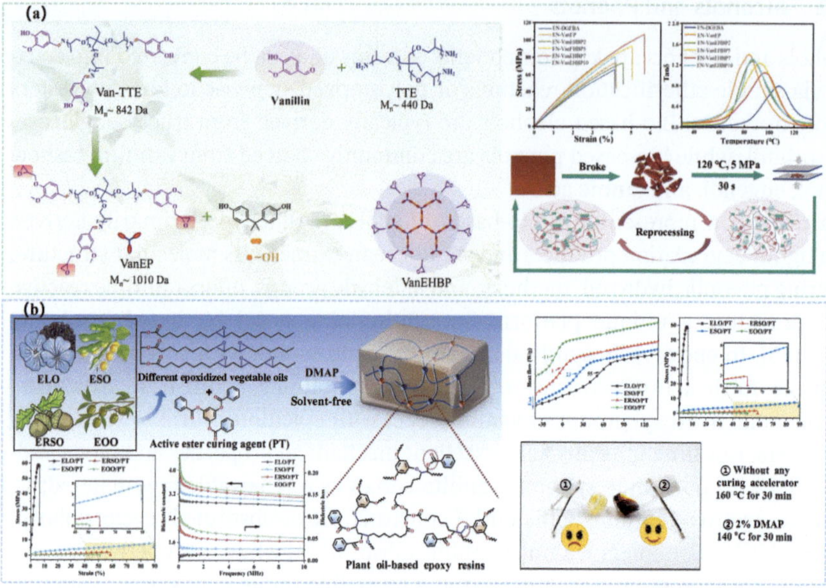

Figure 5.1 (a) Molecular design of hydrovanilloin-based resins and performance characteristics. *Source:* [16] / Springer Nature / CC BY 4.0. (b) Molecular design of plant oil-based resins and performance characteristics. *Source:* [20] / with permission of Elsevier.

hydrophobicity (increased contact angle), low water absorption (4.9%), and reduced water vapor transmission compared to pure cellulose films. They also displayed remarkable solvent resistance, maintaining integrity after 7 days in water, NaOH, and organic solvents. Furthermore, the dynamic imine bonds enabled thermally driven self-healing (72–87% property recovery) and closed-loop recyclability, while the fully bio-based composition ensured complete biodegradability within weeks in soil. These properties position DACNF-PI-LCD as a sustainable, high-performance alternative to nondegradable plastics, particularly in applications demanding mechanical robustness and environmental stability.

Plant oils are typical bio-based alcohols and represent major globally cultivated agricultural commodities, with an annual production exceeding 200 million tons. Palm oil and soybean oil alone account for over 50% of this output, ensuring a stable supply [21]. Their characteristic triglyceride structure imparts excellent lubricity, hydrophobicity, and film-forming ability. Through simple reactions such as transesterification and epoxidation, plant oils can be converted into high-performance resin monomers, plasticizers, and cross-linkers. In epoxy systems, the unsaturated double bonds in plant oils can be transformed into oxirane rings using various oxidants, including hydrogen peroxide, hypochlorous acid, and organic peracids [22]. For instance, Lyu et al. [19] demonstrated that the unsaturated bonds in triglycerides undergo electrophilic addition with peracids under a formic acid/H_2O_2 system to yield epoxidized plant oils with high epoxy values. Due to the unique low-polarity fatty acid chains of epoxidized plant oils, when cured synergistically with aromatic hardeners, the resulting resins exhibit ultra-low dielectric constants, high toughness (elongation at break of 87%), and excellent hydrophobicity. In polyesteramide (PEA) systems, the unsaturated fatty bonds in plant oils can be oxidized into dihydroxy monomers, whose hydrophobicity (water contact angle >100°) and low water absorption ensure stability in electronic packaging applications. High epoxy–value linseed oil-based resins achieve a modulus of 1.63 GPa and T_g of 65 °C through the formation of rigid networks. By minimizing hydroxyl content, this system significantly reduces polarizability, resulting in dielectric loss lower than conventional epoxy resins. It also overcomes mechanical performance limitations typical of bio-based materials, surpassing commercial resins and providing an environmentally friendly, customizable solution for high-frequency electronic device encapsulation (as shown in Figure 5.1b). Furthermore, Gao et al. [23] synthesized a C_{18}–OH dihydroxy monomer from *Sapium sebiferum* oil via catalytic oxidation, which was melt-polycondensed with hexamethylenediamine to construct a dual-dynamic cross-linked PEA. This material demonstrates triple advantages: Bio-based sustainability: entirely plant-derived monomers enable a green synthesis route replacing petrochemical plastics. Mechanical performance: synergistic dynamic hydrogen bonding (accounting for 79.6% of intermolecular interactions) and hydroxyl-ester covalent bonds, combined with flexible long alkyl chains, deliver breakthrough mechanical strength (58.0 ± 5.3 MPa) and toughness (130.1 ± 18.8 MJ/m^3), outperforming conventional polyamide engineering plastics. Dielectric properties: dense polar groups (hydroxyl, amide, ester) along molecular chains create strong dipole moments; their dynamic dissociation-recombination supports tunable dielectric response. The material features closed-loop recyclability with monomer recovery rates exceeding 70%, and maintains over 90% performance after five thermal compression cycles. Topological network activation energies (61.5–94.2 kJ/mol) confirm dynamic bond exchange mechanisms. As a high-performance adhesive, it exhibits shear strength up to 10.29 MPa and temperature stability within −20 to 80 °C (variation <15%), highlighting its innovative value in next-generation insulating resins. Plant oil-based insulating resins thus combine unique hydrogen-bond-rich dynamic networks and abundant flexible chain segments, delivering excellent mechanical strength and toughness. These properties enable effective application as encapsulants for transformer insulation and cable accessories, establishing plant oils as a prime source of high-quality bio-based resin monomers.

5.2.2 Acids

Bio-based acids represent another important class of structural units for constructing epoxy polymers (EP). They can be utilized to synthesize epoxides or epoxy oligomers [24].

Furfural and its derivatives are considered among the most attractive biomass platform chemicals. They are primarily sourced from agricultural byproducts such as corn cobs, cottonseed hulls, rice husks, and sugarcane bagasse. These raw materials can be upgraded into high-value-added chemicals including

2-methylfuran, tetrahydrofuran, furfuryl alcohol, 2,5-furandicarboxylic acid (FDCA), and 2-methyltetrahydrofuran and further transformed into functional materials such as synthetic resins [25]. From a molecular perspective, furfural is a multifunctional compound containing reactive groups including an aldehyde group, an oxirane (epoxy) ether ring, and conjugated olefinic bonds. Functionalization of furfural resins or the synthesis and subsequent polymerization of furfural-based monomers can both yield furfural-functional polymers. Due to the high reactivity of the aldehyde group, preparing functional monomers is relatively easy, making polymerization of these monomers a more straightforward route to furfural-based functional polymers. This simplicity in utilizing functional monomers for polymer synthesis is a distinctive characteristic in the development of furfural-based functional polymeric materials. Among furfural derivatives, FDCA is one of the twelve most representative bio-based chemicals identified by the U.S. Department of Energy and serves as a typical bio-based acid. FDCA can be converted into furandicarbonyl chloride via chlorosulfonyl chloride treatment under an inert atmosphere. Subsequently, it undergoes esterification with glycerol diglycidyl ether in dichloromethane (DCM) solvent, using triethylamine as an acid scavenger. After purification by silica gel column chromatography, the resulting diglycidyl furandicarboxylate (DGF) epoxy monomer is obtained. This monomer forms a three-dimensional cross-linked network through hydroxyl ring-opening curing, yielding a bio-based epoxy resin that exhibits both high thermal resistance ($T_g > 170\,°C$) and closed-loop recyclability [26]. Additionally, 2,5-furandimethanol, another furfural derivative, can directly react with epichlorohydrin to synthesize furan-based epoxy resins containing ether linkages [27]. Zhang et al. [28] developed a multifunctional castor-oil-based waterborne polyurethanes (WPUs) by incorporating dynamic disulfide bonds via controlled integration of dithiodiphenylamine (DTDA). The bio-based design utilized renewable castor oil polyols combined with DTDA's rigid aromatic structure and reversible S-S linkages. The resulting polymer networks achieved exceptional mechanical properties, with tensile strength tunable from 7.2 to 38.4 MPa and Young's modulus up to 628 MPa, significantly outperforming conventional plant-oil-based WPUs. These enhancements arose from synergistic effects: DTDA increased cross-linking density and hard-segment content, while the disulfide bonds enabled dynamic network reconfiguration. Crucially, the materials exhibited outstanding functional properties, including autonomous self-healing (100% mechanical recovery after 6 hours at 60 °C) and reprocessability (100% property retention after four recycling cycles at 140 °C). The broad T_g range (−10 to 83 °C) facilitated versatile shape-memory behavior, enabling dual-to-quadruple shape-memory effects with high fixity ($R_f \approx 100\%$) and recovery ($R_r > 92\%$). Although electrical insulation is not explicitly quantified, the material's robust dielectric characteristics are evidenced by its integration into flexible conductors; Ag nanowire (AgNW)–laminated films maintained stable conductivity even after 1500 bending cycles or repeated cutting/healing, confirming effective electrical insulation and matrix integrity. Surface properties included high pencil hardness (5–6H) and adhesion (5B grade), further supporting durability. This work demonstrates a sustainable strategy to reconcile high strength,

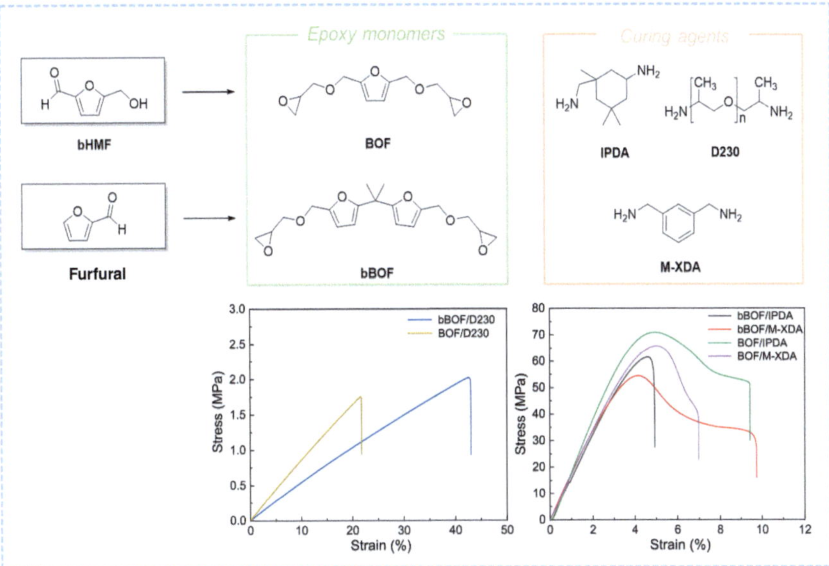

Figure 5.2 Synthesis pathways for 2-furoic acid bio-based resins as well as design of resin molecular structure and characterization of mechanical properties. *Source:* [27] / with permission of American Chemical Society.

reprocessability, self-healing, and multiresponsive functionality in bio-based WPUs, with applicability in smart coatings and flexible electronics.

As shown in Figure 5.2, Jiang et al. [27] reported the synthesis and performance evaluation of bis-furan epoxy resins (BOF and bBOF) derived from renewable resources, 5-hydroxymethylfurfural (HMF) and furfural derivatives. Through a mild and scalable synthetic route, the reduced product of HMF (bHMF) and the furfural derivative 5,5′-(propane-2,2-diyl)bis(furan-5,2-diyl)dimeth-anol (BFD) were reacted with epichlorohydrin to obtain high-purity epoxidized compounds. Using three types of amine curing agents, such as flexible aliphatic D230, cycloaliphatic isophorone diamine (IPDA), and aromatic *m*-xylylenediamine (M-XDA) epoxy/amine systems were constructed to systematically study the thermomechanical properties of the cured resins. The experimental results showed that resins cross-linked with rigid curing agents, such as IPDA or M-XDA, exhibited excellent mechanical properties. Specifically, the BOF/IPDA system achieved a tensile strength of 72.2 MPa, a storage modulus of 8124 MPa at 25 °C, and T_g of 80 °C, significantly outperforming conventional DGEBA-based resins. Thermogravimetric analysis indicated good thermal stability with a 5% weight loss temperature ($T_d = 5\%$) up to 300 °C. Furthermore, the resins demonstrated low water absorption (<2.6%) and good solvent resistance. This study confirmed that the rigid furan ring structure effectively enhances the comprehensive performance of epoxy resins, providing a new strategy for developing high-performance bio-based thermosetting materials.

Wu et al. [26] reported the design, preparation, and closed-loop recycling strategy of a fully bio-based epoxy-amine thermoset material (DGF/MBCA). This material is cross-linked via dynamic ester bonds formed between lignin-derived 4,4′-methylenebis(cyclohexylamine) (MBCA) and cellulose-based 2,5-furandicarboxylic acid dimethyl ester (DMFD) diglycidyl esters of 2,5-furandicarboxylic acid (DGF). It exhibits a glass transition temperature (T_g) of 170 °C, a maximum stress up to 100 MPa, and a storage modulus (E′) of 1.2 GPa at 25 °C, outperforming conventional BPA-based epoxy resins. By molecularly introducing dynamic ester bonds, the material undergoes catalyst-free ester exchange reactions in methanol at 70 °C, achieving complete depolymerization and recovering 90% of the DMFD monomer. Mechanistic studies indicate that the tertiary amine groups in the

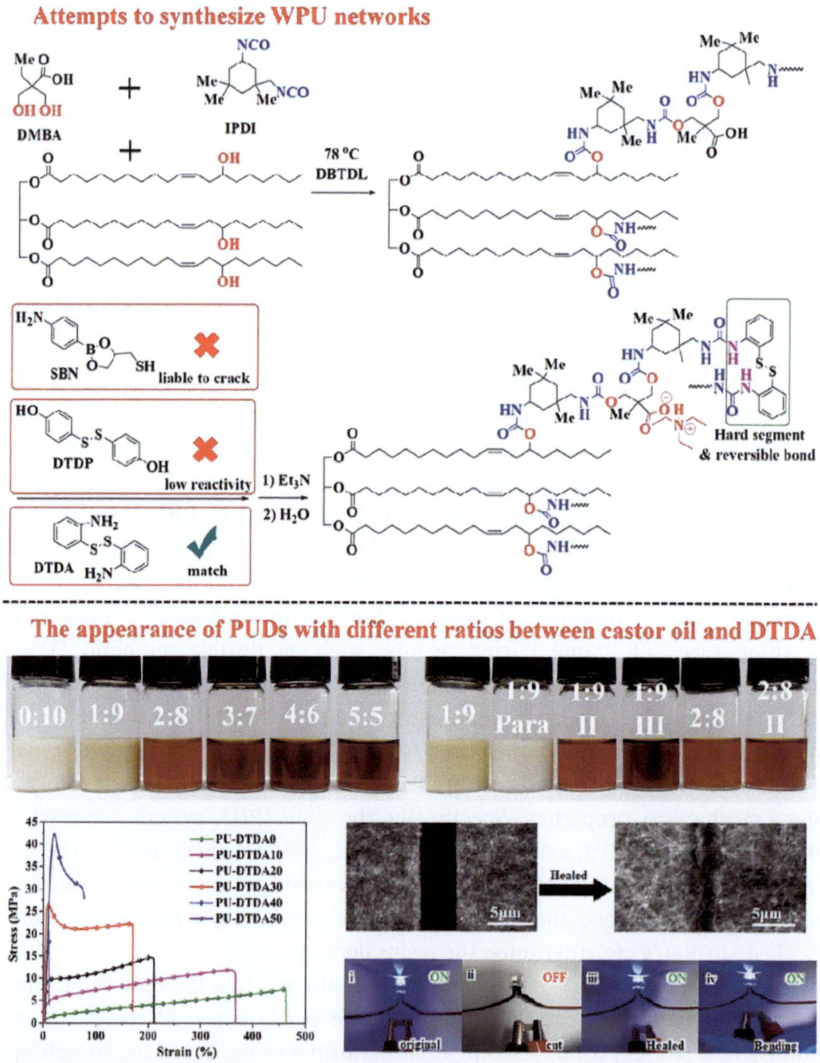

Figure 5.3 Synthesis of novel bio-based epoxy resins via castor-oil and their mechanical and methanol recovery properties. *Source:* [26] / The American Association for the Advancement of Science.

MBCA backbone form a hydrogen-bonding network with the furan ester structure, synergistically promoting nucleophilic attack by methanol on ester bonds. During recycling, the polyester side chains are cleaved via acetolysis, regenerating MBCA diamine and glycerol diglycidyl ether, thus forming a closed-loop cycle. The material was successfully applied in glass fiber (GF) and plant fiber reinforced composites. The GF composites allowed nondestructive recovery of fiber matrices (flexural modulus of 17.5 GPa) after methanol depolymerization (Figure 5.3). This work presents a new paradigm for developing high-performance, recyclable thermosets through the synergistic design of dynamic bonds and bio-based precursors.

5.3 Self-Healing Insulating Resins

In the construction of new power systems driven by the "dual carbon" strategy, electrical equipment is rapidly evolving toward higher efficiency, compactness, and extended service life, imposing stringent demands on the reliability and sustainability of insulation materials. Conventional insulating resins are prone to irreversible damage such as microcracks and electrical tracking under prolonged electromechanical stress, leading to increased failure risks, frequent downtime, and material replacement, which in turn cause resource wastage and environmental pollution. Self-healing insulating resins, featuring dynamic bond networks with intelligent responsiveness, can autonomously repair insulation defects, significantly extend material lifespan and reduce maintenance needs. This capability not only decreases resource consumption and waste generation over the entire lifecycle but also lowers the carbon footprint by prolonging equipment service cycles [29]. Compared to the passive failure of traditional materials, the active maintenance of self-healing resins positions them as a pivotal technology supporting the upgrade of new energy equipment and smart grids. They offer an innovative solution for building green, low-carbon energy infrastructure and serve as a critical technological foundation for sustainable development in the electrical field.

5.3.1 Classification of Self-Healing Insulating Resins

Self-healing resins are classified into intrinsic and extrinsic types based on their repair mechanisms. Intrinsic self-healing systems achieve damage repair through molecular-level dynamic bond network reconstruction, with the key advantage of requiring no external energy or agents during the healing process. These materials demonstrate exceptional repair efficiency in concealed components such as cable joint sealing layers and GIS insulators. Compared to extrinsic systems, their sustainable healing capability stems from the reversibility of dynamic bonds: the temperature-responsive Diels-Alder bonds accommodate the cyclic thermal expansion of cables during operation; redox-sensitive disulfide bonds enable spontaneous repair in chemically corrosive environments; and the ultra-fast reconfiguration of hydrogen bond networks (millisecond-scale response) effectively addresses high-frequency vibration scenarios [30].

Extrinsic self-healing systems rely on embedded healing agents within microcapsules or three-dimensional vascular networks. Although they offer short-term advantages with single-repair efficiencies of 85–95%, they suffer from inherent systemic drawbacks: microcapsule rupture creates localized weak zones that reduce insulation strength, causing an ~30% drop in breakdown field strength; the complex vascular structures induce dielectric anisotropy, leading to ±15% fluctuations in surface flashover voltage; photo- or electro-triggered resins require external energy supply devices, significantly increasing operational complexity. Experimental data show that after three repair cycles, extrinsic materials exhibit dielectric performance degradation of 40–60%, whereas intrinsic materials degrade only 8–12% over the same period [31]. This deterioration primarily results from the irreversible consumption of healing agents and phase separation phenomena, which under ultra-high voltage conditions may further cause interface electric field distortion and related secondary issues [32].

5.3.2 Dynamic Bond Types for Intrinsically Self-Healing Insulating Resins

In the advancement of modern insulation materials, molecular engineering of dynamic covalent bonds in intrinsic self-healing resins has progressively emerged as the central paradigm for achieving material self-repair functionality.

Diels-Alder (D-A) reaction, a cycloaddition between dienes and dienophiles, is well-known for its "click" characteristics, facilitating polymer design and synthesis. More importantly, the DA reaction is thermally reversible: upon heating to an appropriate temperature, the retro-D-A reaction occurs, and upon cooling, the D-A reaction reforms covalent D-A bonds. This crucial feature endows D-A cross-linked polymers with reprocessability and reparability. Heo and Sodano [33] developed a novel thermoresponsive self-healing polyurethanes (1 DATT and 1.5 DATT) leveraging thermally reversible D-A reactions between furan and maleimide moieties embedded in the polymer network. The dynamic D-A bonds enabled repeated healing cycles through retro-DA dissociation (≈130 °C) and subsequent D-A re-cross-linking (≈97 °C), as confirmed by variable-temperature nuclear magnetic resonance spectroscopy (NMR) and differential scanning calorimetry (DSC). Crucially, these materials integrate shape-memory functionality via a polyurethane backbone composed of triethanolamine (TEA) and hexamethylene diisocyanate (HDI), allowing autonomous crack closure without external pressure. Both polymers exhibited robust mechanical properties comparable to structural epoxies: tensile strengths of 76.8 MPa (1 DATT) and 87.3 MPa (1.5 DATT), Young's modulus ≈ 2.5 GPa, and fracture toughness values exceeding 1 $MPa{\cdot}m^{0.5}$. Healing efficiency, evaluated via compact tension tests, reached 84.1% for 1.5 DATT after the first cycle and retained 75.9% after the third cycle, outperforming existing DA-based systems. The shape-memory effect facilitated complete crack surface contact during healing, while the high thermal stability (decomposition >250 °C) ensured structural integrity during repeated thermal cycles. This synergistic design eliminates the need for external repair forces and provides a viable pathway for

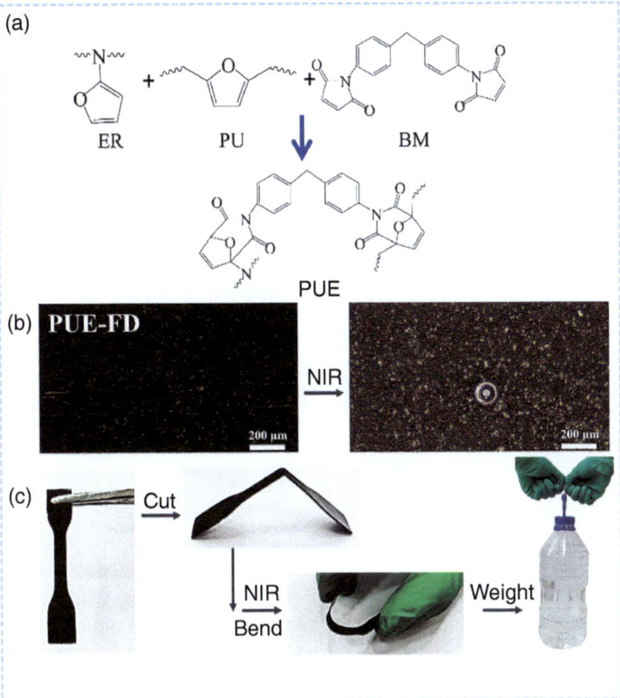

Figure 5.4 Synthesis pathways (a), scratch repair (b), and fracture repair characterization (c) of **D-A** dynamic resins. *Source:* [34] / with permission of Elsevier.

high-performance self-healing materials in structural applications. As shown in Figure 5.4, Tian et al. [34] incorporated dynamic D-A bonds into an epoxy-modified polyurea system (PUE-FD), imparting efficient photothermal self-healing capability to the coating. The thermal reversibility of DA bonds enabled crack repair within 60 seconds under near-infrared irradiation (0.5 W/cm^2), achieving a self-healing efficiency of 89.1%, which remained above 80% after three cycles. Concurrently, the multiple hydrogen bond network of functionalized reduced graphene oxide (FrGO) synergistically enhanced molecular chain rearrangement, significantly improving the coating's mechanical strength (tensile stress of 48.1 MPa) and substantially extending its service life through reversible healing.

Disulfide bond systems enable self-healing through redox reactions of sulfur bonds. Due to their relatively long bond length and low bond energy, disulfide bonds are more prone to cleavage compared to other bonds, allowing even room-temperature self-repair [35]. For instance, Zhang et al. [36] developed an intrinsic self-healing epoxy insulation material based on disulfide bonds. The incorporation of a small amount of disulfide bonds enhanced the tensile strength (up to 114.21 MPa) and elongation at break (20.39%) while optimizing dielectric properties by reducing dielectric constant and loss. The material demonstrated effective mechanical damage repair with an efficiency of 81.23%, alongside visible restoration of electrical treeing damage: disulfide bonds notably suppressed electrical tree

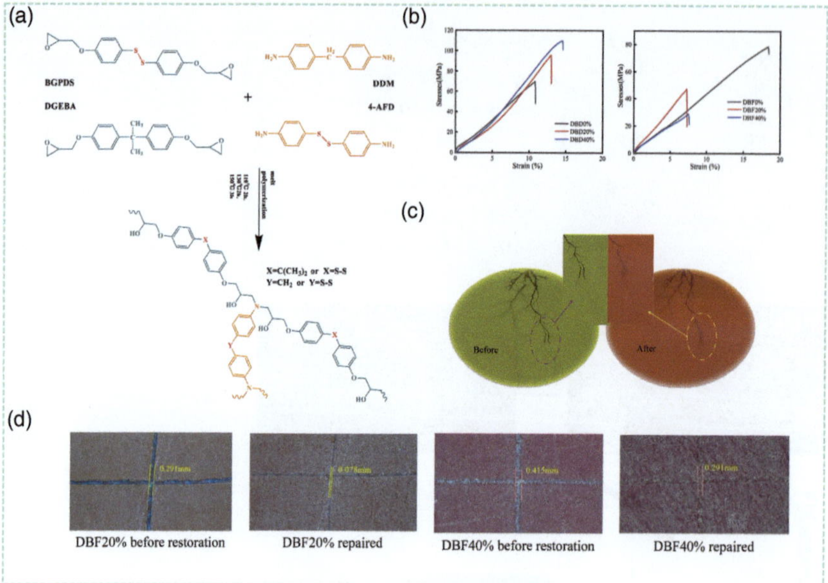

Figure 5.5 Structural design and performance characterization of **disulfide** dynamic resins. *Source:* [36] / with permission of Elsevier. (a) molecular structure design; (b) mechanical property characterization; (c) electrical dendrite repair characterization; (d) scratch repair characterization.

branching (reducing dendritic structures) and limited damage extent. After thermal treatment at 120 °C, carbonization traces of electrical trees diminished and partial discharge intensity decreased, confirming the reparability of electrical damage (Figure 5.5).

Zhu et al. [29] engineered an ultra-tough, room-temperature self-healing epoxy/MXene composite through hierarchical integration of multidynamic bonds and a biomimetic inverse-artificial nacre structure. Inspired by mussel nacre and byssus, the design incorporated: (1) flexible side-chain quadruple hydrogen bonds (H-bonds), (2) dynamic disulfide bonds, and (3) interfacial H-bonds via UPy-modified MXene nanosheets. This multidynamic bond system enabled exceptional mechanical performance, achieving a record-high toughness of 210.75 MJ/m^{-3} (exceeding spider silk by 30%), tensile strength of 34.50 MPa, and elongation at break of 864.72%. The aligned MXene nanosheets within the epoxy matrix formed a gas-impermeable barrier, reducing oxygen transmission rate (OTR) to 21.7 $cm^3/m^{-2}\cdot d^{-1}$ and water vapor transmission rate (WVTR) to 554.2 $g/m^{-2}\cdot d^{-1}$ orders of magnitude lower than unmodified epoxy. Critically, the dynamic bonds facilitated rapid autonomous repair: 90.4% recovery of tensile strength within 2 hours at 25 °C, with efficient healing maintained at 0 °C (81.5%) and underwater due to hydrophobic protection of bond mobility. The composite retained 90.2% of its mechanical strength after five damage-healing cycles, demonstrating robust recyclability. Additionally, the biomimetic network enhanced interfacial adhesion and enabled stable underwater flexible sensing. This work resolves the traditional conflict between strength, toughness, and self-healing in thermosets, positioning the material for demanding applications in marine corrosion protection and flexible electronics.

5.3.2.1 Schiff Base Bond

The incorporation of dynamic **Schiff base bond** offers an innovative solution for acid–base environmental adaptability. This system leverages the reversible hydrolysis of imine (C=N) bonds, enabling spontaneous self-healing in acidic conditions and effectively restoring insulation resin materials. Liu et al. [37] developed a multifunctional imine epoxy polymer (EPCN) by constructing dynamic imine bonds via a one-pot synthesis that combines inexpensive industrial precursors, terephthalaldehyde and BPA diglycidyl ether cured with D230. The imine exchange reaction facilitated efficient self-healing within 8 minutes at 150 °C, completely eliminating cracks and achieving a welding efficiency of 67%. The material also supports multiple reprocessing cycles via hot pressing while maintaining robust mechanical properties. Additionally, it exhibits thermally adaptive shape memory behavior and excellent ultraviolet shielding.

Zhou et al. [38] successfully synthesized a novel imine epoxy polymer (BDSER) through the reaction of a disulfide-containing difunctional epoxy monomer with 4,4′-diaminodiphenyl disulfide (4-AFD). The high density of dynamic disulfide bonds within the cross-linked network enabled exceptional bond exchange characteristics, evidenced by an ultra-short stress relaxation time of 9 seconds at 200 °C without catalysts and a low activation energy (38 kJ/mol), facilitating rapid network reconfiguration. Mechanically, BDSER exhibited robust performance comparable to conventional epoxies, with a high T_g (133 °C), a storage modulus = 2.2 GPa at 25 °C, a tensile strength of 38 MPa, and an elongation at break of 6.8%. Crucially, these thermomechanical properties showed no significant degradation after three cycles of grinding and hot-pressing reprocessing (180 °C, 20 MPa), confirming excellent recyclability and damage tolerance. For functional composites, incorporating polypyrrole-decorated multiwalled carbon nanotubes (PPy/MWCNTs) into BDSER yielded conductive composites (CNT/PPy/Vitrimer composite [CPV]). At just 1 wt% filler loading, the CPV composite achieved a low surface resistance of $4.9 \times 10^9\ \Omega$, meeting antistatic requirements. The composite retained its electrical functionality after reprocessing, with minimal resistance change (from $4.9 \times 10^9\ \Omega$ to $3.3 \times 10^9\ \Omega$). Furthermore, the BDSER matrix allowed the full recovery of conductive fillers via thiol-mediated disulfide reduction degradation in dithiothreitol/DMF, highlighting its environmental sustainability. Overall, the dual-disulfide design imparted BDSER and its composites with superior mechanical integrity, efficient self-healing/reprocessability, tunable electrical conductivity, and complete recyclability (Figure 5.6).

5.3.2.2 Dynamic Ester Bond

In the pursuit of high-strength self-healing materials, **dynamic ester bond** systems have transcended traditional limitations. Pei et al. [39] introduced dynamic ester

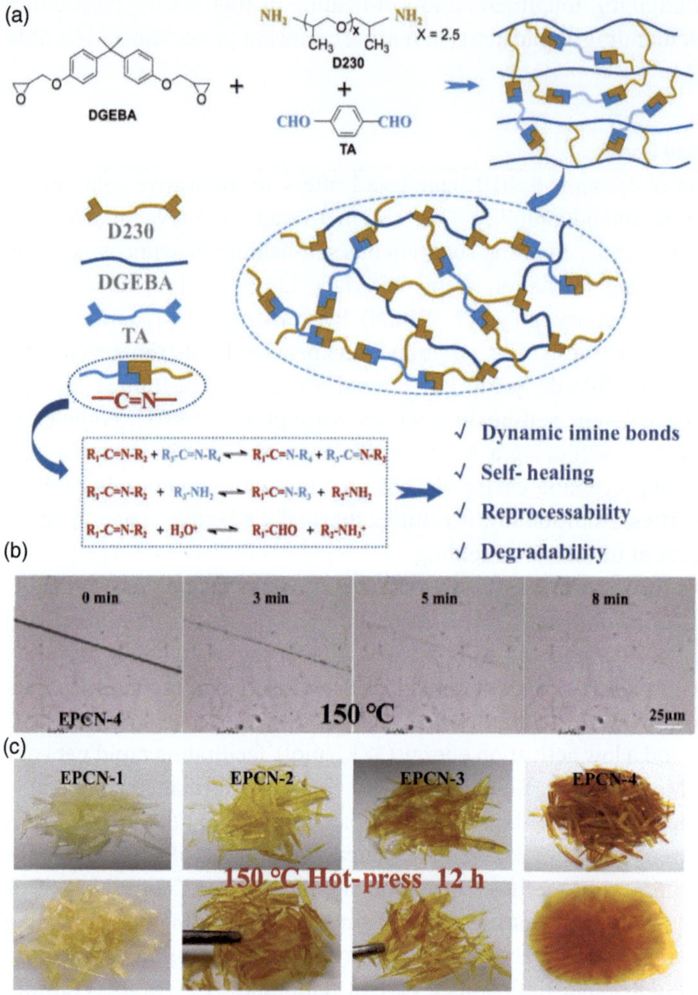

Figure 5.6 Structural design and performance characterization of **Schiff** dynamic resins. (a) Molecular structure design; (b) scratch repair characterization; (c) mechanical recovery characterization. *Source:* [37] / with permission of Elsevier.

bonds into semi-interpenetrating polymer networks (SIPNs), successfully fabricating a recyclable degradable base epoxy resin (DAS-S) combining high strength (maximum tensile strength of 123 MPa) and toughness. Under tertiary amine catalysis, ester exchange reactions confer self-healing capability and controllable degradability. However, residual zinc-based catalysts in these systems raise issues such as increased dielectric loss, which require urgent resolution (Figure 5.7).

Recently, catalyst-free boronic ester bond systems have offered a promising solution to insulation challenges caused by zinc catalysts. Wang et al. [40] constructed dynamically cross-linked glassy networks by incorporating nitrogen-coordinated cyclic boronic ester bonds (NCB). This bonding structure enables efficient

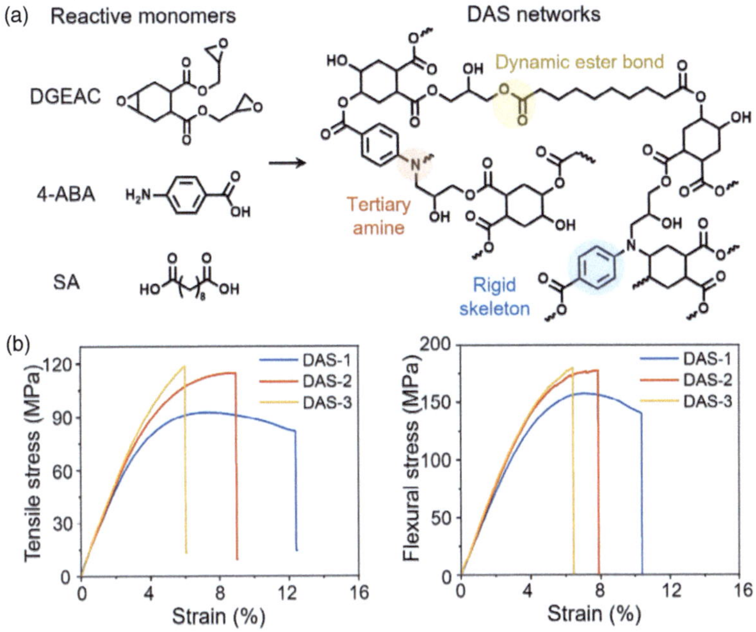

Figure 5.7 Structural design and performance characterization of **dynamic ester** resins. (a) Molecular structure design; (b) mechanical property characterization. *Source:* [40] / with permission of Elsevier.

thermal-triggered bond exchange without catalysts, allowing material recycling and multiple reprocessing cycles (over five cycles) while significantly enhancing thermal stability (T_g up to 122 °C) and mechanical properties (fracture toughness increased to 4.8 kJ/m^2). Critically, NCB bonds maintain polystyrene's inherently low dielectric properties and impart excellent high-frequency dielectric performance (dielectric constant of 3.89 and loss factor of 1.12×10^{-3} at 1 MHz), removing barriers for application in ultra-high voltage equipment.

5.3.3 Evaluation of Self-Healing Performance

The development of self-healing insulating materials, leveraging synergistic intrinsic and extrinsic mechanisms, offers innovative solutions for the long-term reliable operation of power equipment. Intrinsic self-healing systems repair damage via reversible cleavage and reformation of dynamic chemical bonds, such as reversible Diels-Alder bonds, hydrogen bonds, or ionic networks enabling adaptive responses to crack propagation and electrical tracking degradation. Extrinsic healing relies on microcapsules or three-dimensional networks embedding healing agents, which are released upon crack triggering to achieve efficient single-event filling but face limitations in cyclic stability. These dual mechanisms are driving self-healing materials toward multiscale damage responsiveness, with quantitative assessment of healing efficiency and in-depth understanding of dynamic mechanisms being critical for validating material engineering applicability.

Quantitative evaluation of self-healing performance requires a dual-dimensional metric system encompassing electromechanical properties. Mechanical recovery is assessed by comparing parameters such as tensile strength and fracture toughness before and after damage, reflecting the material's mechanical repair capability. Electrical insulation recovery focuses on the restoration of breakdown strength and volume resistivity: breakdown field testing simulates real operating conditions to evaluate dielectric repair effectiveness, while volume resistivity measured via high-resistance meters monitors the disruption of conductive pathways, effectively suppressing insulation failure caused by leakage currents. These quantitative indicators provide experimental bases for material selection and reliability grading. In-depth analysis of dynamic self-healing mechanisms demands advanced in situ characterization techniques. At the microscale, scanning electron microscopy (SEM) and atomic force microscopy (AFM) enable real-time observation of matrix morphology evolution during crack closure. Coupled with Raman spectroscopy, they facilitate tracking of dynamic bond breakage and reformation patterns, such as the reconstruction kinetics of hydrogen bond networks under thermal stimulation.

Current evaluation systems for self-healing materials have evolved from single-performance verification to multidimensional synergistic analysis, encompassing core parameters such as healing efficiency, dynamic response rate, and cyclic stability. By establishing cross-scale, multifield coupled composite assessment methods, the development of novel dynamic bond resins and microcapsule composites has been accelerated, facilitating their engineering applications in ultra-high voltage cable accessories, offshore wind power transformers, and other critical scenarios. This scientific evaluation paradigm lays a crucial technological foundation for the intelligent iteration and low-carbon transformation of the next-generation electrical insulating resins, significantly enhancing the full lifecycle sustainability of new energy power systems.

5.4 Recyclable/Degradable Insulating Resins

Driven by the dual imperatives of the carbon neutrality goals and circular economy, self-healing insulating resins can effectively extend the service life of next-generation electrical insulating materials. However, after long-term operation, end-of-life challenges remain inevitable. Therefore, it is imperative to further develop recyclable/degradable insulating resins to optimize the sustainability of future electrical insulation systems. As a frontier direction in polymer materials, the development of recyclable insulating resins requires coordinated resolution of the balance between functional performance and sustainability. This objective imposes stringent molecular design criteria in materials chemistry: maintaining the topological stability of cross-linked networks under service conditions while achieving controllable triggering of recycling processes and suppression of side reactions through precise arrangement of responsive functional groups. Current academic focus converges on two main technical pathways: network reconstruction driven by thermally activated dynamic covalent bonds (e.g. Diels-Alder bonds, ester exchange reactions) and reversible dissociation of physical/chemical bonds induced by solvents. The core lies in tuning the energy barrier thresholds and spatiotemporal responsiveness of dynamic bonds to optimize the performance degradation gradient over multiple recycling cycles. This section systematically analyzes the selection of dynamic bonds and network topology design, providing a theoretical framework for the molecular engineering and process optimization of recyclable insulating resins. It also outlines directions for their technical validation in green electrical equipment applications.

Recyclable insulating resins are primarily classified into two types based on recycling methods: mechanical recycling and solvent-based recycling. The core lies in designing dynamic reversibility or degradability of molecular chains. This chapter discusses both mechanically recyclable and solvent-recyclable insulating resins in detail.

5.4.1 Mechanically Recoverable Insulating Resins

Thermal recycling via mechanical recovery relies on heat-triggered reversible reactions of dynamic covalent bonds to enable material reprocessing. Its core is the reversible design of the cross-linked network, rather than high-temperature pyrolytic degradation. Heating serves two main functions: (1) shifting the equilibrium toward depolymerization and (2) accelerating bond breakage and reformation rates. These combined effects enhance flowability, thereby improving processability [41]. This enables such resins to combine the high performance of thermosets with the recyclability of thermoplastics. Consequently, most resins containing dynamic covalent bonds such as dynamic ester bonds, disulfide bonds, and imine bonds exhibit mechanical recyclability.

5.4.1.1 Dynamic Ester Bond

The core mechanism of dynamic ester bond networks relies on the thermally reversible behavior of β-hydroxy ester bonds. Under 120–180 °C, ester bonds undergo dynamic reconstruction via $Zn(OAc)_2$-catalyzed transesterification: hydroxyl and ester groups engage in nucleophilic substitution in a solvent-free environment, breaking original ester bonds and forming new ester networks [41]. The addition of $Zn(OAc)_2$ catalyzes this process by: (1) shifting the alcohol/alcoxide equilibrium toward the more nucleophilic alkoxide form; (2) activating the carbonyl bond via increased electrophilicity of the carbon atom; (3) bringing reactive centers into proximity [42]. For example, Albertini et al. [43] developed a fully bio-based epoxy vitrimer system from cashew phenol-derived epoxy monomer and itaconic acid, innovatively employing a dual catalyst system of 2-ethyl-4-methylimidazole and zinc triflate ($Zn(OTf)_2$). The zinc catalyst (5 mol%) facilitates epoxy-carboxylic esterification via a Lewis acid mechanism during cross-linking, enhancing network homogeneity (gel content 82–84%). During recycling, it efficiently catalyzes dynamic transesterification, endowing the material with excellent thermal reprocessability (hot pressing at 180 °C). Samples containing 5 mol% zinc maintained stable epoxy conversion (86%), gel content (85%), and cross-link density (3.10×10^{-4} mol/cm^3)

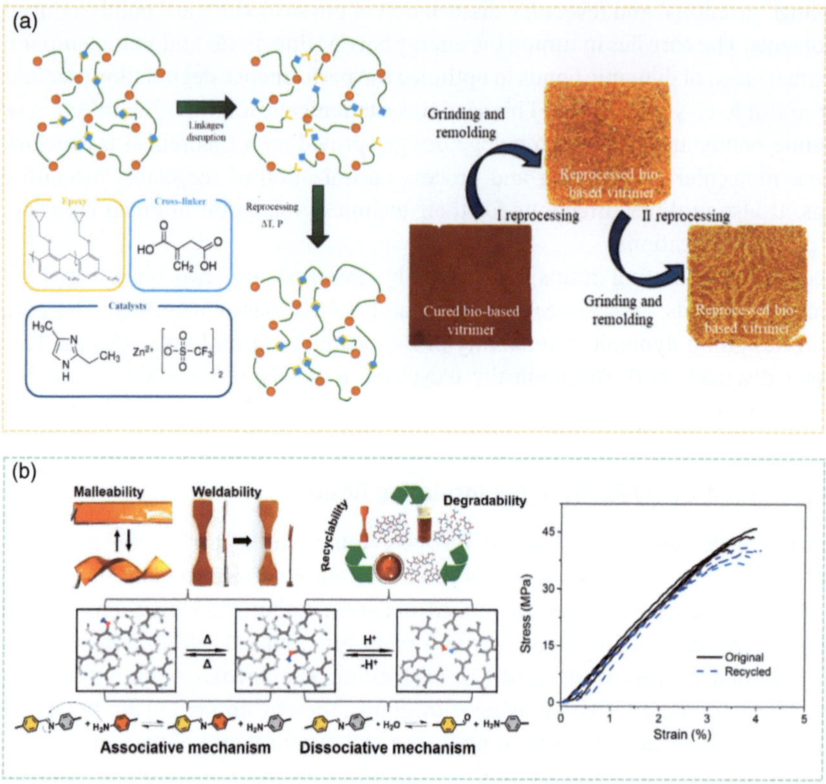

Figure 5.8 Design and characterization of recyclable dynamic resins for recycling performance. (a) **Dynamic ester bond.** *Source:* [43] / with permission of American Chemical Society. (b) **Dynamic imine bonds.** *Source:* [44] / with permission of American Chemical Society.

after two recycling cycles, whereas zinc-free samples failed secondary recycling, confirming zinc's critical role in multiple cycles (Figure 5.8).

However, the introduction of zinc-based catalysts poses significant challenges to the material's insulation properties. Thus, balancing dynamic performance and electrical insulation, and developing nonmetallic catalysts for transesterification, remain key challenges in the study of dynamic ester bonds. Hao et al. [45] developed a high-performance recyclable epoxy resin by incorporating vanillin-based hyperbranched epoxy (VEHBP) containing dual dynamic bonds – disulfide and imine – into a DGEBA/4-AFD network. The VEHBP's hyperbranched topology, combined with dual dynamic exchange mechanisms, simultaneously enhanced mechanical properties, creep resistance, and stress relaxation kinetics. The optimal formulation (5% VEHBP) achieved a T_g of 175 °C and a creep resistance temperature of 130 °C, significantly exceeding conventional dynamic epoxies. Mechanically, it exhibited a tensile strength of 96.5 MPa and toughness of 4.4 MJ/m^3, representing increases of 33.1% and 173.3%, respectively, over the control, attributed to intramolecular cavities improving energy dissipation and cross-linking density. The dual dynamic bonds enabled rapid stress relaxation (158 seconds at 140 °C) with a low activation energy (46.2 kJ/mol), facilitating efficient self-healing of scratches within 30 minutes at 180 °C. Reprocessability was demonstrated via hot-pressing (180 °C, 5 MPa), with reprocessed samples retaining 102.6% of original tensile strength after three cycles, supported by radical-mediated disulfide exchange confirmed via EPR and Raman spectroscopy. Degradation occurred within 30 minutes in DTT/DMF at 65 °C via thiol-disulfide exchange, enabling closed-loop recycling. This design overcomes the trade-off between reprocessability and high-temperature stability, positioning VEHBP-epoxy as a sustainable high-performance material.

5.4.1.2 Dynamic Imine Bonds

Schiff bases containing imine structures, formed by the condensation of aldehyde/ketone groups with amines under catalyst-free conditions, undergo thermally driven dynamic imine exchange upon heating [46]. This enables multiple reshaping cycles under mild, catalyst-free conditions, making them promising candidates for recyclable materials while maintaining high mechanical strength [47]. For instance, Abu-omar and coworkers [44] reported a recyclable and reprocessable epoxy thermoset based on aromatic imine (Schiff base) bonds. The aromatic imine bonds facilitate both dissociative mechanisms (acid/water-triggered hydrolysis-reformation cycles) and associative mechanisms (thermally activated imine exchange), endowing the material with controllable degradability, recyclability, and plasticity. The resin fully degrades into oligomers in mildly acidic aqueous solution (0.17 M HCl, 25 °C) or organic solvents, and the network can be reconstructed by drying and heating at 120 °C. After recycling, mechanical properties (tensile strength of 46 MPa, modulus of 2 GPa) and T_g (71 °C) remain comparable to the original resin, without the need for catalysts, additional monomers, or high pressure. Utilizing thermally driven imine exchange with an activation energy of 36.9 kJ/mol, the material can be reshaped into complex geometries (e.g. helices) or mechanically recycled at 80 °C, retaining 100% of its strength post-recycling (Figure 5.9).

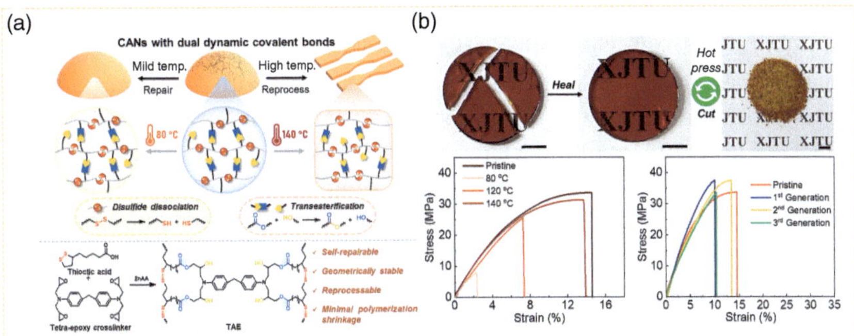

Figure 5.9 Structural design and performance characterization of **dynamic ester** resins. (a) Molecular structure design. (b) Characterization of mechanical recovery properties. *Source:* [48] / with permission of Elsevier.

5.4.1.3 Dynamic Disulfide Bonds

Disulfide bonds are weak dynamic covalent bonds with bond energies around 200 kJ·mol^{-1}, enabling mild recycling conditions but often resulting in compromised mechanical properties and poor chemical stability [49]. For example, Dong et al. [48] developed a bio-based dynamic disulfide epoxy vitrimer derived from 4-propyl guaiacol. The network, cross-linked solely by reversible disulfide bonds, leverages flexible propyl side chains to synergistically accelerate chain mobility. It demonstrates rapid stress relaxation with a relaxation time of only 15.8 seconds at 200 °C, significantly outperforming traditional epoxies. This dynamic nature grants excellent self-healing (complete scratch repair within 30 minutes) and enables up to four mechanical recycling cycles at 180 °C under 10 MPa, retaining over 60% of the original tensile strength after recycling. Additionally, Cui et al. [49] proposed a one-pot synthesis of dynamic polysulfide-containing resins from elemental sulfur and epoxides at room temperature. These polysulfide resins exhibit mild mechanical recycling conditions (60 °C, 20 minutes), yet their low bond energy leads to inferior mechanical performance (tensile strength 1.5 MPa, elongation at break 30%).

5.4.1.4 Hindered Urea Bonds

Urea bonds consist of two -NH_2 groups directly connected via a carbonyl group. Due to conjugation between the lone pair electrons on nitrogen and the π electrons of the carbonyl p orbital, the amide linkage in urea bonds exhibits significant stability, displaying dynamic behavior only under extreme conditions. However, increasing steric hindrance and electron-withdrawing effects on the urea bond hinder urea bonds (HUBs), reducing bond energy, thereby enhancing dynamic characteristics. Compared to other dynamic covalent bonds, HUBs offer controllable dynamics and mild recycling conditions. Moreover, Lyu et al. [50] developed a novel non-isocyanate synthesis route for covalent adaptable networks (CANs) using dynamic HUBs through sequential ring-opening of thiazolidin-2-one (TZ) with a sterically hindered diamine (1,3-di(piperidin-4-yl)propane [DPP]), followed by thiol-click reactions. The TZ-DPP intermediate, featuring terminal thiol groups, underwent thiol-Michael addition with trimethylolpropane triacrylate (TA) or thiol-epoxy coupling with trimethylolpropane triglycidyl ether (TE) to form CANs (TZ-DPP-TA and TZ-DPP-TE). Both networks exhibited dynamic behavior via HUB dissociation and reversible thiol-Michael adducts (for TA), with TZ-DPP-TA showing lower activation energy (59 kJ/mol above 110 °C) and faster stress relaxation than TZ-DPP-TE (172 kJ/mol). Mechanically, TZ-DPP-TE achieved higher tensile strength and modulus due to pendant hydroxyl groups enhancing hydrogen bonding, while TZ-DPP-TA demonstrated superior reprocessability, retaining structural integrity through five hot-pressing cycles (130 °C, 10 MPa) with minimal property degradation. Chemical recycling leveraged multiple dynamic pathways: HUB exchange with piperidine (40% conversion in 6 hours at 80 °C), retro-Michael addition catalyzed by 1,8-diazabicyclo[5.4.0]undec-7-ene (DBU) (full depolymerization in dimethylformamide/dimethyl sulfoxide [DMF/DMSO]), and transesterification with ethylene glycol/DBU. Repolymerization or reutilization enabled closed-loop recovery under solvent/catalyst-specific conditions, highlighting the methodology's versatility for sustainable thermoset design without toxic isocyanates.

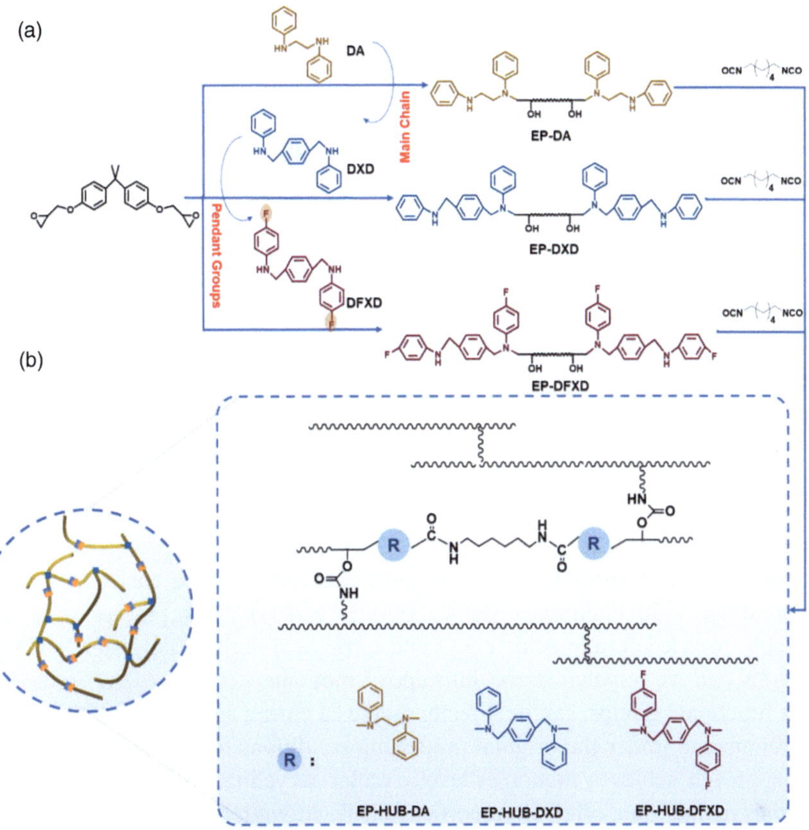

Figure 5.10 Structural design and performance characterization of **dynamic hindered urea bonds** resins. *Source:* [51] / Royal Society of Chemistry. (a) Molecular structure design; (b) Characterization of mechanical properties.

As shown in Figure 5.10, Zhang et al. [51] designed dynamic HUBs by introducing fluorinated aryl groups. The synergistic effect of steric hindrance and strong electron-withdrawing fluorine atoms significantly improved the epoxy resin self-healing and degradability. Theoretical calculations indicate that the strong electron-withdrawing nature of fluorine lowers the C—N bond energy of the urea bond to 330 $kJ{\cdot}mol^{-1}$ (a 17 $kJ{\cdot}mol^{-1}$ decrease compared to unmodified systems) and increases the bond length to 0.1409 nm, substantially reducing the activation energy for network rearrangement to 67 $kJ{\cdot}mol^{-1}$. This design endows the material with high electrical insulation strength (77 $MV{\cdot}m^{-1}$), mechanical strength (43 MPa), and a T_g of 113 °C, while enabling mechanical recycling at 180 °C.

5.4.2 Solvent Recoverable Insulating Resins

Solvent-based recycling triggers the dissociation of dynamic cross-linked networks through solvent molecule penetration, enabling low-temperature and low-energy regeneration. However, it demands higher control over solvent selectivity and toxicity. Common dynamic cross-linking systems amenable to solvent recycling include dynamic ester bonds, disulfide bonds, and imine bonds, whose characteristics and recycling mechanisms resemble those in mechanical recycling and were discussed in the previous section. This section focuses on closed-loop recyclable resins and degradable resins within solvent-based recycling.

Closed-loop recyclable resins represent an innovative approach aimed at sustainable material utilization, addressing the challenge of inefficient recycling in conventional epoxy resins [52]. The core concept involves chemically or physically depolymerizing used resin products back into raw materials suitable for production, enabling a closed-loop system of "monomer-polymer-monomer-repolymerization" and "manufacture-use-recycle-remanufacture." This strategy effectively reduces resource consumption and carbon emissions, eliminates pollution from landfill or incineration, and preserves the high-value recyclability of materials rather than resorting to downcycling [53]. Consequently, closed-loop recyclable resins have become a focal point in the development of next-generation electrical insulating resins.

Zhong et al. [54] developed a fully bio-based epoxy vitrimer using ferulic acid-based hyperbranched epoxy resin (FEHBP). The hydroxyl groups in FEHBP catalyze dynamic transesterification in situ, enabling selective cleavage of ester bonds under mild conditions (80 °C, 30 wt% NaOH). This process produces separable recycled ferulic acid (r-FA) and citric acid-derived polyester (r-CAP). The r-FA can be re-polymerized into epoxy monomers, while r-CAP, retaining active functional groups, can be directly used as a curing agent. Both components are recombined under the original processing conditions to form a new vitrimer. This approach achieves monomer-level circular recycling, with the regenerated materials maintaining original properties (tensile strength >88.3%, $T_g \approx 94\,°C$) and preserving cross-link density and mechanical performance throughout recycling cycles. As shown in Figure 5.11, Zhang et al. [16] developed a supramolecular thermoset material based on vanillin-derived hyperbranched epoxy resin (VanEHBP). Under mild conditions (room temperature, 0.1 M HCl/DMF solution), the imine bonds selectively cleave to generate aldehyde and amine monomers. After solvent removal, the depolymerization products can be directly reassembled into regenerated materials. The hyperbranched topology with intramolecular cavities facilitates chain mobility, synergistically accelerating depolymerization kinetics through dynamic transesterification and hydrogen bonding networks, achieving complete degradation within 180 minutes. The closed-loop recycling process reaches nearly 100% efficiency, with regenerated materials retaining the original mechanical properties (tensile strength of 104.5 MPa, toughness of $3.58\,MJ\cdot m^{-3}$) and thermal stability. Furthermore, after three reprocessing cycles, no loss in strength is observed. This strategy marks the first achievement of efficient room-temperature closed-loop recycling of thermosetting resins, providing a new paradigm for designing next-generation electrical insulating materials. In addition, the use of tertiary amine groups in the structure of anhydride-cured epoxy resins to facilitate dynamic ester exchange reactions also enables epoxy resins to undergo closed-loop recycling in aqueous solutions. For example, Hao et al. [55] reported a novel recyclable epoxy/anhydride thermosetting system for carbon fiber reinforced plastics (CFRPs),

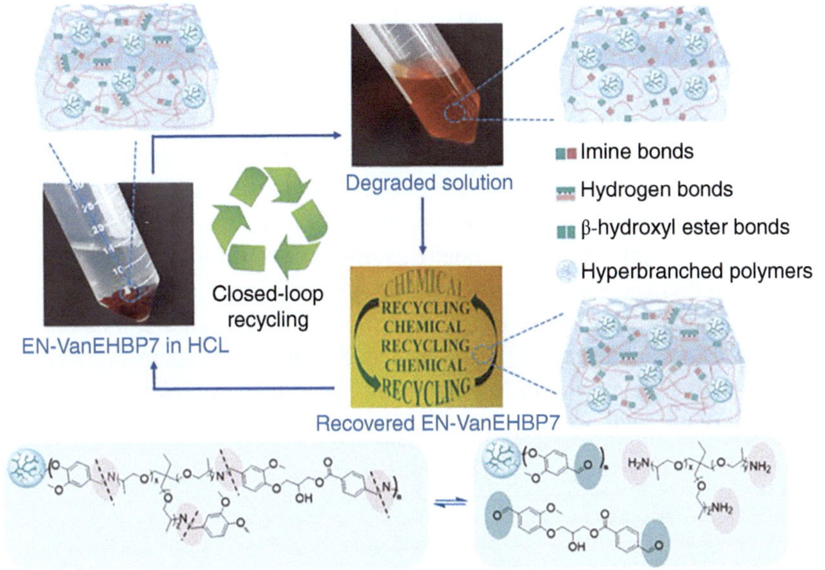

Figure 5.11 Structural design of a closed-loop recyclable resin and schematic diagram of closed-loop recycling. *Source:* [16] / Springer Nature / CC BY 4.0.

featuring strategically incorporated dynamic covalent bonds enabling closed-loop recycling without sacrificing high-performance properties. The matrix, formulated from tetraglycidyl methylenedianiline (TGDDM), nadic methyl anhydride (NMA), and triethanolamine (TEA) as a catalytic co-curing agent, possesses abundant ester linkages, hydroxyl groups, and inherent tertiary amines within its densely cross-linked network. This specific molecular design facilitates thermally activated dynamic transesterification reactions (DTR) and catalyzes ester bond hydrolysis under mild hydrothermal conditions. Consequently, the cured epoxy exhibits an exceptionally high T_g (up to 224 °C) and robust mechanical properties, including a tensile strength of 77.3 MPa and a high storage modulus (~2.7 GPa at 50 °C). CFRPs fabricated with this matrix demonstrate a tensile strength exceeding 500 MPa. Crucially, the material achieves full hydrothermal recyclability in pure water at 200 °C within 5 hours, without requiring added catalysts. This mild process efficiently decomposes the matrix into reusable oligomers and degraded matrix polymer (DMP), while causing minimal damage to the embedded carbon fiber. The recycled carbon fiber (rCF) retains 94% of its original tensile strength (2609 MPa vs. 2790 MPa). Furthermore, closed-loop recycling is demonstrated: the DMP is successfully epoxidized and blended with virgin resin to form new recyclable matrices, and the rCF is directly reused to fabricate new CFRPs exhibiting substantial tensile strength (300 MPa for a composite incorporating both rCF and modified DMP). This work establishes a practical pathway for producing high-Tg CFRPs amenable to efficient closed-loop recycling.

Degradable resins refer to the process of promoting thermochemical conversion of resins under catalytic conditions. At specific temperatures, catalysts such as acids, bases, or metal compounds selectively activate and cleave the three-dimensional cross-linked network, yielding low-molecular-weight depolymerization products [56]. This is primarily achieved via transition metal-catalyzed pathways [57], aminolysis under solvothermal conditions [58], or other depolymerization mechanisms (e.g. Lewis acid catalysis and alkoxy radical intermediates), which selectively break weak bonds in waste resins (e.g. C—O and C—N bonds). Subsequently, valuable molecular fragments are recovered for reuse. The process significantly lowers activation energy, optimizes energy input, and suppresses side reactions, enabling efficient recovery of monomers, high-value compounds, or carbon-based fuel precursors [59]. This provides a chemically controlled route for targeted resource recovery of resin waste while reducing fossil resource consumption and environmental pollution. Recently, degradation of resins has emerged as a

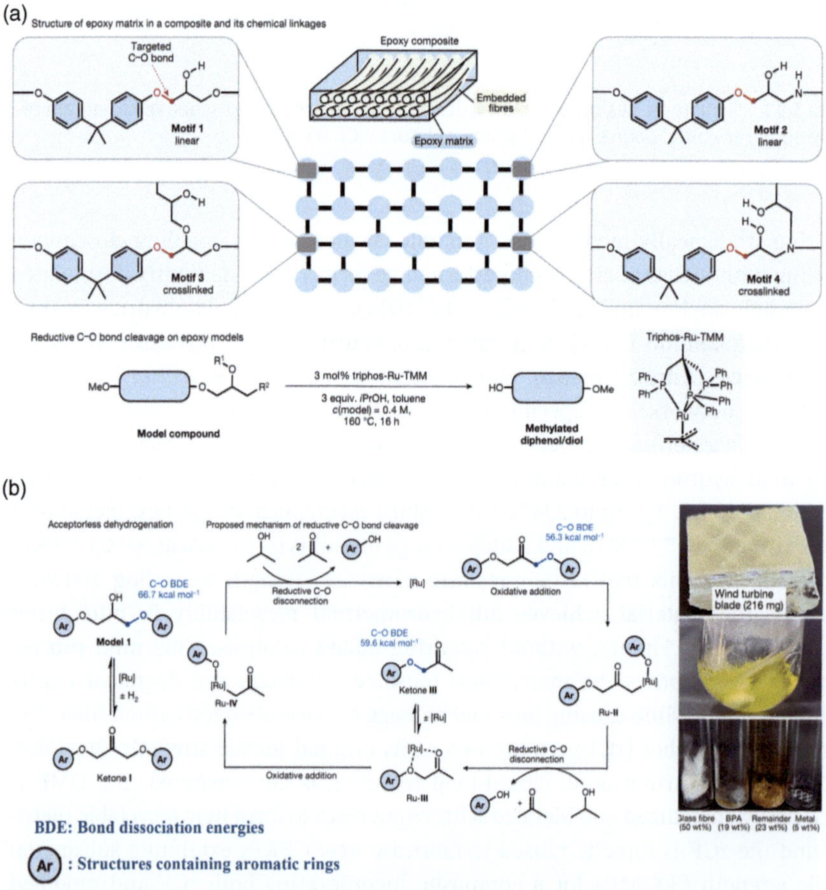

Figure 5.12 Molecular design of degradable epoxy resin based on "amine scissors" and schematic diagram of degradation efficiency. *Source:* [1] / with permission of Elsevier.

key research focus, driven by the demands of next-generation electrical insulation materials.

For example, Xie et al. [1] developed a synergistic "amine scissors" system based on the reaction between DCM and diethylenetriamine (DETA). This system generates a mixture of amines with varying nucleophilicity and basicity in situ via Hofmann alkylation, which selectively cleaves ester bonds (C—O) in resins through cooperative action. Under mild conditions (130–150 °C), it efficiently degrades eight types of cured epoxy resins, including two commercial products. Experimental results and theoretical calculations confirm that the carbonyl carbon of the ester bond is the nucleophilic attack site, undergoing aminolysis to form amides and hydroxyl groups while preserving the main polymer backbone. The system achieves nearly 100% degradation efficiency with a rate up to 119.99%·h^{-1}. Notably, bulky

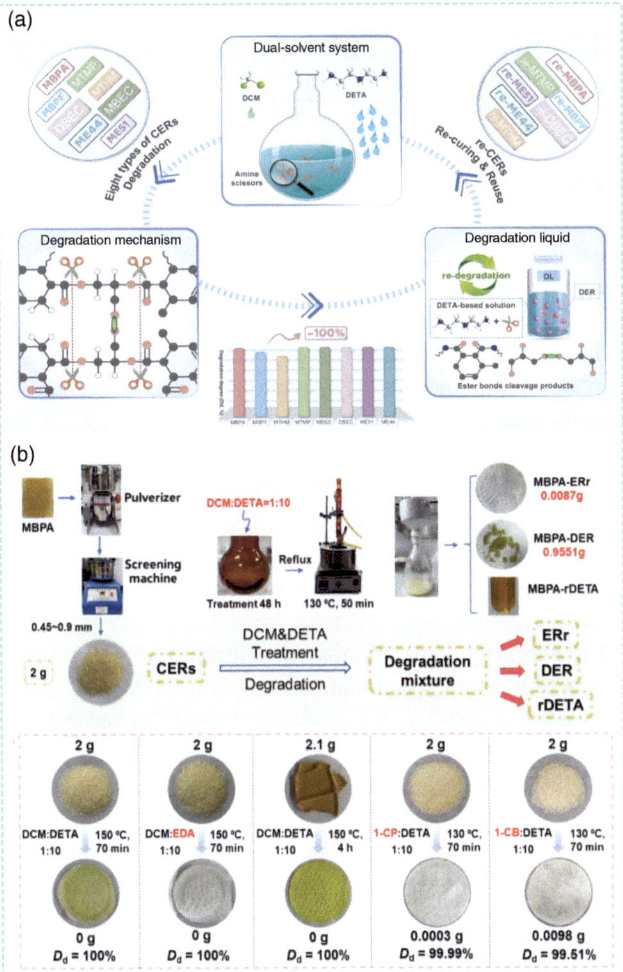

Figure 5.13 Structural design, catalytic degradation mechanism, and degradation schematic of degradable resin. *Source:* [57] / Springer Nature / CC BY 4.0.

resin samples can be completely degraded without pulverization, outperforming existing catalytic methods. The degradation products can be directly reused as monomers and curing agents to regenerate high-performance resins or subjected to further recycling, advancing closed-loop recycling in the resin industry (Figure 5.12).

Additionally, Ahrens et al. [57] proposed an innovative ruthenium-catalyzed method for the selective cleavage of C—O (alkyl) bonds in epoxy resins, enabling chemical recycling of thermoset epoxy composites. The catalytic system operates via a cascade of dehydrogenation, bond cleavage, and reduction: initial dehydrogenation of alcohol groups forms ketone intermediates that weaken the C—O bond (theoretical calculations show a bond dissociation energy reduction of 10.7 $kcal\cdot mol^{-1}$), followed by oxidative addition of the ruthenium complex to cleave the bond. This approach was successfully applied to commercial amine-cured epoxy resins and composites, including wind turbine blades, achieving high-purity BPA recovery at 56–81% yield. Simultaneously, glass and carbon fibers were recovered intact with 50–57% yield, maintaining mechanical properties comparable to the original fibers (Figure 5.13).

5.5 Conclusions and Outlook

The investigation of next-generation electrical insulating resins represents a significant endeavor in power equipment materials to address environmental challenges and technological advancement. Conventional petroleum-based insulating materials, constrained by their dependence on fossil resources and exhibiting recycling difficulties, are becoming increasingly apparent in their limitations. The lightweight design and extended service life requirements of high-voltage electrical equipment impose heightened demands on material sustainability, self-repairing capability, and recyclable performance. Although the exploration of bio-based resins, self-healing resins, and recyclable resins has infused new vitality into this field, their development still confronts multiple scientific and technological barriers, which necessitate breakthroughs through interdisciplinary collaboration and systematic innovation.

Current studies have preliminarily demonstrated the feasibility of substituting petroleum-based monomers with bio-based feedstocks (e.g. vegetable oils, lignin derivatives). Certain bio-based resins exhibit dielectric strength and thermal stability comparable to or even surpassing those of conventional materials in core performance metrics. Nevertheless, challenges such as the intricate purification processes of biomass feedstocks and insufficient functionalization efficiency of monomers hinder scalable production. Moreover, the long-term aging resistance behavior of bio-based resins, particularly molecular chain degradation under hygrothermal environments and its impact on dielectric properties has yet to undergo systematic evaluation.

Self-healing insulating resins achieve active repair of microcracks through dynamic chemical bond design. However, the introduction of dynamic bonds (e.g. hydrogen bond networks) may alter dielectric response characteristics.

For instance, network reconstruction could induce localized charge accumulation, thereby compromising long-term insulation reliability. Furthermore, existing self-healing mechanisms predominantly rely on thermal or chemical stimuli, whereas their compatibility with multifield coupling conditions (electric fields, mechanical stress, temperature fluctuations) in real-world power equipment operation remains unvalidated.

Recyclable resins incorporating reversible cross-linked networks offer potential for closed-loop resource utilization. However, the degradation of microstructures during pyrolysis or solvent-based recycling processes remains unresolved. Regenerated materials often exhibit significant deterioration in mechanical performance and dielectric strength, failing to meet the reuse requirements for high-voltage equipment.

Therefore, the development of bio-based resins should focus on precise directional conversion of feedstocks, such as enhancing lignin depolymerization efficiency through enzymatic catalysis or novel solvent systems to reduce performance interference from byproducts. Concurrently, establishing scientific evaluation methodologies for full life cycle environmental impacts is essential to mitigate potential "greenwashing" risks. Molecular design of self-healing resins must balance the reversibility of dynamic bonds with structural stability. This requires exploring low dielectric loss dynamic network construction strategies (e.g. steric hindrance-modulated dynamic covalent bonds) and advancing in situ characterization techniques to elucidate evolution mechanisms of self-healing behaviors under coupled electric-thermal fields. Innovation in recyclable resins requires breakthroughs beyond single dynamic bond limitations. Directional design should emphasize topology-adaptive multiple dynamic networks enabling precise control of recycling conditions (e.g. temperature gradient-responsive staged depolymerization), combined with interface engineering to optimize the performance homogeneity of regenerated materials.

Concurrently, breakthrough directions for next-generation electrical insulation materials have transcended the mere enhancement of individual performance metrics, instead embracing sustainability, intelligence, and circularity as inseparable, integrated objectives. By establishing a technical chain encompassing "bio-based monomer screening, dynamic network design, closed-loop recycling validation," this paradigm achieves low-carbon sourcing through bio-based feedstocks, enhances service life via self-healing functionality, and realizes resource circularity through recyclability. Synergistic integration of these three dimensions drives transformative advancements in next-generation electrical materials.

References

1 Xie, F., Chen, R., Deng, C. et al. (2025). Synergetic "amine scissors" break down cured epoxy resins with nearly 100% degradation efficiency. *Chemical Engineering Journal* 509: 161175.

2 Niu, P., Zhao, Z., Zhu, J. et al. (2025). Epoxy-based multifunctional re-bondable polymer with self-healing, shape memory and superb bonding properties. *Journal of Colloid and Interface Science* 678: 30–39.
3 Wen, Y., Chen, C., Ye, Y. et al. (2022). Advances on thermally conductive epoxy-based composites as electronic packaging underfill materials—a review. *Advanced Materials* 34 (52): 2201023.
4 Zhao, X., Long, Y., Xu, S. et al. (2023). Recovery of epoxy thermosets and their composites. *Materials Today* 64: 72–97.
5 Liu, Y., Lu, F., Wan, J. et al. (2022). Recyclable tough thermosets with an imide-hexahydrotriazine structure. *Polymer Chemistry* 13 (21): 3063–3075.
6 Guo, Y. and Miao, X.-Y. (2024). A wind of change in sustainability. *Nature Energy* 9 (12): 1455–1456.
7 Xu, Z., Liang, Y., Ma, X. et al. (2019). Recyclable thermoset hyperbranched polymers containing reversible hexahydro-S-triazine. *Nature Sustainability* 3 (1): 29–34.
8 Liu, X., Wu, H., Xu, W. et al. (2023). Ultrastrong and high-tough thermoset epoxy resins from hyperbranched topological structure and subnanoscaled free volume. *Advanced Materials* 36 (9): 2308434.
9 Wei, F., Zhang, J., Wu, C. et al. (2023). Closed-loop recycling of tough and flame-retardant epoxy resins. *Macromolecules* 56 (14): 5290–5305.
10 Cywar, R.M., Rorrer, N.A., Hoyt, C.B. et al. (2021). Bio-based polymers with performance-advantaged properties. *Nature Reviews Materials* 7 (2): 83–103.
11 Toto, E., Laurenzi, S., Pellegrini, R.C. et al. (2024). Eco-friendly synthesis of high-performance polyimide materials using bio-based greener solvents: towards sustainable technologies in space environment. *Materials Today Sustainability* 25: 100657.
12 Chen, X., Chen, S., Xu, Z. et al. (2020). Degradable and recyclable bio-based thermoset epoxy resins. *Green Chemistry* 22 (13): 4187–4198.
13 Zhao, L., Sun, J., and Fang, Q. (2024). A bio-based low dielectric polymer at high frequency derived from paeonol. *Progress in Organic Coating* 196: 108724.
14 Lu, Z., Hauschild, M., Ottosen, L.M. et al. (2024). Climate mitigation potential of biobased insulation materials: a comprehensive review and categorization. *Journal of Cleaner Production* 470: 143356.
15 Kuhire, S.S., Ichake, A.B., Grau, E. et al. (2018). Synthesis and characterization of partially bio-based polyimides based on biphenylene-containing diisocyanate derived from vanillic acid. *European Polymer Journal* 109: 257–264.
16 Zhang, J., Jiang, C., Deng, G. et al. (2024). Closed-loop recycling of tough epoxy supramolecular thermosets constructed with hyperbranched topological structure. *Nature Communications* 15 (1): 4869.
17 Yang, X., Ke, Y., Chen, Q. et al. (2022). Efficient transformation of renewable vanillin into reprocessable, acid-degradable and flame retardant polyimide vitrimers. *Journal of Cleaner Production* 333: 130043.

18 Xu, X., Ma, X., Cui, M. et al. (2024). Fully biomass-derived polyurethane based on dynamic imine with self-healing, rapid degradability, and editable shape memory capabilities. *Chemical Engineering Journal* 479: 147823.

19 Lyu, M., Yang, X., Li, J. et al. (2024). High toughness plant oil-based epoxy resins simultaneous with low dielectric properties. *Industrial Crops and Products* 207: 117731.

20 Zhang, X., Li, P., Zeng, J. et al. (2023). Dynamic covalent bond enabled strong bio-based polyimide materials with thermally-driven adaptivity, healability and recycling. *Chemical Engineering Journal* 465: 143017.

21 Danov, S.M., Kazantsev, O.A., Esipovich, A.L. et al. (2017). Recent advances in the field of selective epoxidation of vegetable oils and their derivatives: a review and perspective. *Catalysis Science & Technology* 7 (17): 3659–3675.

22 Wai, P.T., Jiang, P., Shen, Y. et al. (2019). Catalytic developments in the epoxidation of vegetable oils and the analysis methods of epoxidized products. *RSC Advances* 9 (65): 38119–38136.

23 Gao, H., Wu, X., Fan, J., and Wang, Z. (2024). Mechanically strong and tough plant oil-based poly(ester amide) vitrimers with closed-loop recyclability. *ACS Sustainable Chemistry & Engineering* 12 (7): 2668–2677.

24 Song, X., Deng, Z.-P., You, C.-W. et al. (2022). High-performance and fire-resistant epoxy thermosets derived from plant-derived ferulic acid. *Industrial Crops and Products* 187: 115445.

25 Zhou, S., Li, C., Gao, G. et al. (2023). Bio-based resins with tannin and hydroxymethylfurfural derived high-yield carbon for Zn-ion hybrid supercapacitors. *Journal of Cleaner Production* 389: 136067.

26 Wu, X., Hartmann, P., Berne, D. et al. (2024). Closed-loop recyclability of a biomass-derived epoxy-amine thermoset by methanolysis. *Science* 384 (6692): eadj9989.

27 Jiang, Y., Yun, J., and Pan, X. (2022). Renewable furan-based epoxy resins derived from 5-hydroxymethylfurfural and furfural. *ACS Sustainable Chemistry & Engineering* 10 (50): 16555–16562.

28 Zhang, C., Liang, H., Liang, D. et al. (2020). Renewable castor-oil-based waterborne polyurethane networks: simultaneously showing high strength, self-healing, processability and tunable multishape memory. *Angewandte Chemie International Edition* 60 (8): 4289–4299.

29 Zhu, X., Hao, Y., Huang, L.F. et al. (2025). Engineering of multi-dynamic bonds toward room-temperature self-healing epoxy/MXene adaptable network with record-high toughness. *Advanced Science* 12 (33): e08780.

30 Li, B., Cao, P.-F., Saito, T., and Sokolov, A.P. (2022). Intrinsically self-healing polymers: from mechanistic insight to current challenges. *Chemical Reviews* 123 (2): 701–735.

31 Diesendruck, C.E., Sottos, N.R., Moore, J.S., and White, S.R. (2015). Biomimetic self-healing. *Angewandte Chemie International Edition* 54 (36): 10428–10447.

32 Blaiszik, B.J., Kramer, S.L.B., Olugebefola, S.C. et al. (2010). Self-healing polymers and composites. *Annual Review of Materials Research* 40 (1): 179–211.

33 Heo, Y. and Sodano, H.A. (2014). Self-healing polyurethanes with shape recovery. *Advanced Functional Materials* 24 (33): 5261–5268.

34 Tian, W., Guo, Z., Wang, S. et al. (2024). Hydrogen and DA bond-based self-healing epoxy-modified polyurea composite coating with anti-cavitation, anticorrosion, antifouling, and strong adhesion properties. *Journal of Materials Science and Technology* 187: 1–14.

35 Rekondo, A., Martin, R., Ruiz de Luzuriaga, A. et al. (2014). Catalyst-free room-temperature self-healing elastomers based on aromatic disulfide metathesis. *Materials Horizons* 1 (2): 237–240.

36 Zhang, X., Lin, H., Hu, Y. et al. (2023). Electric tree intrinsic self-healing epoxy insulating materials based on disulfide bond. *Polymer Degradation and Stability* 218: 110567.

37 Liu, X., Zhang, E., Liu, J. et al. (2023). Self-healing, reprocessable, degradable, thermadapt shape memory multifunctional polymers based on dynamic imine bonds and their application in nondestructively recyclable carbon fiber composites. *Chemical Engineering Journal* 454: 139992.

38 Zhou, F., Guo, Z., Wang, W. et al. (2018). Preparation of self-healing, recyclable epoxy resins and low-electrical resistance composites based on double-disulfide bond exchange. *Composites Science and Technology* 167: 79–85.

39 Pei, D., Wang, P., Hao, J. et al. (2025). A strong and tough thermosetting epoxy resin for recyclable high-performance composites. *Angewandte Chemie International Edition* 137 (33): e202505526.

40 Wang, S., Wang, B., Zhang, X. et al. (2021). Fully recyclable and reprocessable polystyrene-based vitrimers with improved thermal stability and mechanical properties through nitrogen-coordinating cyclic boronic ester bonds. *Applied Surface Science* 570: 151157.

41 Montarnal, D., Capelot, M., Tournilhac, F., and Leibler, L. (2011). Silica-like malleable materials from permanent organic networks. *Science* 334 (18): 965–968.

42 Demongeot, A., Mougnier, S.J., Okada, S. et al. (2016). Coordination and catalysis of Zn^{2+} in epoxy-based vitrimers. *Polymer Chemistry* 7 (27): 4486–4493.

43 Albertini, E., Dalle Vacche, S., Bongiovanni, R. et al. (2025). Biobased and reprocessable vitrimers based on cardanol-derived epoxy for more sustainable thermosets. *ACS Sustainable Chemistry & Engineering* 13 (5): 2120–2131.

44 Zhao, S. and Abu-Omar, M.M. (2018). Recyclable and malleable epoxy thermoset bearing aromatic imine bonds. *Macromolecules* 51 (23): 9816–9824.

45 Hao, Y., Zhong, L., Li, T. et al. (2023). High-performance recyclable and malleable epoxy resin with vanillin-based hyperbranched epoxy resin containing dual dynamic bonds. *ACS Sustainable Chemistry & Engineering* 11 (30): 11077–11087.

46 Meyer, C.D., Joiner, C.S., and Stoddart, J.F. (2007). Template-directed synthesis employing reversible imine bond formation. *Chemical Society Reviews* 36 (11): 1705–1723.

47 Belowich, M.E. and Stoddart, J.F. (2012). Dynamic imine chemistry. *Chemical Society Reviews* 41 (6): 2003–2024.

48 Dong, K., Zhao, D., Pang, Y. et al. (2025). Multiple-reprocessable guaiacol-derived epoxy vitrimer with disulfide crosslinks and closed-loop recycling of carbon fiber-reinforced composites. *Chemical Engineering Journal* 508: 160754.

49 Cui, C., Wang, F., Chen, X. et al. (2024). Covalent adaptable networks with dual dynamic covalent bonds for self-repairing infrared transmitting materials. *Advanced Functional Materials* 34 (24): 2315469.

50 Lyu, J., Lee, S., Bae, H.E. et al. (2024). Non-isocyanate synthesis of covalent adaptable networks based on dynamic hindered urea bonds: sequential polymerization and chemical recycling. *Angewandte Chemie International Edition* 63 (43): e202411397.

51 Zhang, L., Sun, W., Guo, Z. et al. (2024). Enhancing the healability and degradability of epoxy via synergetic steric and electron-withdrawing effects for green electrical packaging. *Journal of Materials Chemistry A* 12 (25): 14991–15005.

52 Ye, G., Huo, S., Wang, C. et al. (2024). Strong yet tough catalyst-free transesterification vitrimer with excellent fire-retardancy, durability, and closed-loop recyclability. *Small* 20 (45): 2404634.

53 Chen, L. and Wang, Y.-Z. (2024). Recyclable flame-retardant thermosets and their fiber-reinforced composites. *Macromolecules* 57 (20): 9498–9517.

54 Zhong, L., Hao, Y., Zhang, J. et al. (2022). Closed-loop recyclable fully bio-based epoxy vitrimers from ferulic acid-derived hyperbranched epoxy resin. *Macromolecules* 55 (2): 595–607.

55 Hao, C., Liu, T., Liu, W. et al. (2022). Recyclable CFRPs with extremely high Tg: hydrothermal recyclability in pure water and upcycling of the recyclates for new composite preparation. *Journal of Materials Chemistry A* 10 (29): 15623–15633.

56 Nguyen, S.T., Fries, L.R., Cox, J.H. et al. (2023). Chemical recycling of thiol epoxy thermosets via light-driven C—C bond cleavage. *Journal of the American Chemical Society* 145 (20): 11151–11160.

57 Ahrens, A., Bonde, A., Sun, H. et al. (2023). Catalytic disconnection of C—O bonds in epoxy resins and composites. *Nature* 617 (7962): 730–737.

58 Zhao, X., Wang, X.-L., Tian, F. et al. (2019). A fast and mild closed-loop recycling of anhydride-cured epoxy through microwave-assisted catalytic degradation by trifunctional amine and subsequent reuse without separation. *Green Chemistry* 21 (9): 2487–2493.

59 Feng, Y., Zhang, Z., Yue, D. et al. (2024). Recent progress in degradation and recycling of epoxy resin. *Journal of Materials Research and Technology* 32: 2891–2912.

6

Environmental Degradation and Novel Development of Electrical Silicone Rubber Materials

Wang Xilin[1], Shen Zikui[2], Jia Zhidong[1], and Wang Liming[1]

[1] *Tsinghua University, International Graduate School, Taoyuan Street, Shenzhen, 518055, China*
[2] *South China University of Technology, School of Electric Power Engineering, Wushan Street, Guangzhou 510641, China*

6.1 Aging and Failure Issues of Silicone Rubber Materials Under Harsh Extreme Conditions

6.1.1 Degradation Caused by Alcohol and Its Volatile Emissions

In recent years, a rapid loss of hydrophobicity in insulators coated with room temperature vulcanized (RTV) silicone coatings has been observed shortly after commissioning at a converter station situated in western China. Since 2013, this facility has consistently applied RTV silicone coatings to insulators, bushings, and related electrical equipment. Initial hydrophobicity assessments conducted after coating consistently yielded HC1 to HC2 grades, indicative of excellent hydrophobic performance and high surface reliability. However, observations in subsequent years highlighted significant issues. By 2017, the hydrophobic properties of RTV-coated equipment and composite insulators at the station had notably deteriorated, with measured hydrophobicity levels decreasing dramatically to HC6 or even lower. Such results clearly indicated that these RTV coatings failed to meet the threshold criteria stipulated by the DL/T 627-2018 standard. In response to these deteriorations, the station implemented corrective recoating measures following the completion of the annual inspection. Initial hydrophobicity assessments performed immediately after recoating revealed a complete restoration of hydrophobicity to HC1 levels across all relevant devices. However, this improvement proved transient. During the subsequent 2018 annual inspection, the hydrophobicity was once again found to have deteriorated severely, reverting to HC6. Recognizing the risk of potential pollution flashover incidents associated with severely degraded hydrophobicity, the station proactively implemented contamination cleaning procedures to mitigate these hazards. Follow-up hydrophobicity assessments conducted after these cleaning interventions demonstrated a partial recovery, with recorded hydrophobicity values returning to HC3 to HC4. Nevertheless, the underlying

Next-Generation Electrical Insulating Materials: Low-Carbon, Eco-Friendly, and Sustainable Solutions, First Edition. Edited by Jun-Wei Zha.

Figure 6.1 Hydrophobicity changes of RTV coating in a converter station before and after 6 months. (a) Hydrophobicity of new samples; (b) hydrophobicity after 6 months.

issue persisted. A further assessment conducted one year after cleaning once more revealed a renewed decline to HC6 levels. The dynamic changes in hydrophobicity of equipment subjected to these severe environmental conditions over a typical six-month period are vividly illustrated in Figure 6.1. The repeated cycles of degradation, partial restoration through cleaning, and subsequent degradation again underscore the complexity and severity of environmental stressors influencing RTV silicone coatings in the studied region.

To uncover the root cause of such rapid and repetitive hydrophobicity loss, a comprehensive field investigation was conducted around the converter station. This investigation quickly revealed a significant contributing factor: a pronounced alcoholic odor permeated the ambient air, originating from a large domestic distillery situated roughly five kilometers from the station. Due to atmospheric migration and diffusion processes, the concentration of volatile organic pollutants emitted by the distillery gradually decreased with increasing distance from the emission source, resulting in varied deposition rates across different areas within and surrounding the station. On a macroscopic scale, RTV silicone specimens located in close proximity to the distillery exhibited significantly greater degradation in hydrophobicity, as well as a larger reduction in flashover voltage, relative to specimens located further away but subjected to equivalent aging durations. Importantly, RTV-coated specimens that had already undergone significant hydrophobicity degradation did not demonstrate any spontaneous recovery in hydrophobic performance even after removal from the polluted gaseous environment. Specifically, RTV specimens removed from the contaminated environment without any remedial cleaning measures continued to exhibit persistent hydrophobicity grades at HC6 even after five weeks of exposure to clean air. However, the deliberate mechanical removal of surface contaminants was found to yield partial restoration of hydrophobicity, with values returning to approximately HC3 within one week after cleaning.

These findings provided strong evidence that the organic contaminants responsible for hydrophobicity loss were not readily volatile or easily removed by natural weathering processes. Instead, they tended to form stable surface deposits, mechanically removable, yet their presence still facilitated intrinsic aging of the RTV coating itself. Further analytical research identified that the principal substances causing this rapid hydrophobicity deterioration are macromolecular organic compounds, specifically sucrose-6-acetate and di-nonyl phthalate. Sucrose-6-acetate is particularly problematic due to its high density of hydroxyl groups, making RTV-coated surfaces exceptionally prone to abrupt hydrophobicity loss upon surface accumulation. On the other hand, although phthalate-based compounds exhibit relatively limited water solubility, their excellent miscibility with polymer matrices enables penetration and diffusion deep within the RTV coating. Such penetration significantly restricts the recovery of hydrophobic performance even after mechanical contaminant removal. Infrared spectroscopic analysis further confirmed the presence of these contaminants through distinct characteristic absorption peaks appearing near 3300 and 1600 cm^{-1} [1], indicating specific chemical bonds associated with these organic pollutants.

To mitigate and ultimately prevent rapid hydrophobicity loss under these extreme conditions, innovative strategies such as precontamination migration techniques and advanced surface treatments were explored and successfully implemented. These methods demonstrated enhanced hydrophobic stability when exposed to atmospheric conditions characteristic of distillery environments. Accelerated aging tests showed that relevant organic contaminants generally did not accumulate directly on the outermost hydrophobic pollution layer. Instead, they predominantly resided in the interfacial region beneath the pollution layer but above the RTV specimen surface. This shift in contaminant location significantly extended the projected service life of the hydrophobicity of RTV coatings. Additionally, from a broader environmental management perspective, atmospheric emissions of contaminants from distilleries could be substantially curtailed through technical interventions. These included optimization of sealing performance of distillation equipment, reducing temperature and humidity levels during liquor extraction processes, and ensuring that distillation and distiller's grain airing occur within controlled, sealed environments.

6.1.2 Silicone Rubber Aging Caused by Algal Contamination

Under natural pollution accumulation conditions, free low-molecular-weight siloxanes readily adsorb onto the surfaces of contaminant layers, such that when the silicone rubber becomes moistened, water remains as discrete, isolated microdroplets on the silicone rubber surface rather than forming a continuous water film. This phenomenon endows the insulator with strong resistance to contamination-induced flashover [2]. However, in practical service environments, insulators are subjected to highly complex and dynamic ambient conditions, leading to variation in both the types and characteristics of contaminant deposits on their surfaces. In regions subject to conventional pollution, inorganic contaminants that deposit on insulator surfaces include soluble ionic species such as NH^{4+}, Ca^{2+}, K^+, Na^+, Mg^{2+}, SO_4^{2-}, and Cl^-, as well as insoluble salts and oxides like Fe_2O_3, MnO_2,

Figure 6.2 Phenomenon of algae growth on the surface of the insulator. (a) Yun Nan province; (b) Sichuan province; (c) Guangxi province; (c) Hainan Province.

Al_2O_3, and $CaSO_4$. In areas adjacent to sugarcane fields, contaminants on the insulator surface further include sugars such as glucose and sucrose [3]. In warm and humid regions, and on insulators near forested mountainous areas, biological contamination-predominantly algae, accompanied by fungal and bacterial growth has also been observed [4]. To date, microbial colonization of insulator surfaces has been documented in a variety of countries. This includes the proliferation of green algae in regions such as Sri Lanka [5], Tanzania, Sweden [6], Papua New Guinea [7], and the United States [8], as well as extensive lichen growth observed on transmission line insulators in Mexico, Paraguay, and New Zealand. In China, significant green algae colonization has been reported on insulator surfaces operating in the provinces of Yunnan, Sichuan, Guangxi, and Hainan [8]. Representative examples of such biological fouling are shown in Figure 6.2.

In contrast to the deposition and accumulation processes characteristic of abiotic contaminants, such as inorganic particulates and sugars, the accumulation of biological fouling, such as algae, on insulator surfaces involves not only the settlement of spores but also the subsequent growth, reproduction, death, and dormancy of algal cells. Under suitable environmental conditions, algae are capable of proliferating rapidly and, in conjunction with fungi and bacteria that colonize the insulator surface, can form biofilms exhibiting both structural organization and functional complexity. The physiological activities of algal cells encompass the uptake and release of inorganic salts, as well as the excretion of carbohydrates and extracellular enzymes, thereby increasing the compositional complexity of surface contaminants on the insulator. Furthermore, microbial growth may infiltrate the silicone rubber matrix itself, depleting its constituents and thereby expediting the aging process of the silicone rubber.

The presence of algal biofilms can substantially compromise the pollution flashover withstand capability of insulators. Both algae and their extracellular secretions exhibit pronounced hydrophilicity, which promotes the development of continuous water films on the insulator surface when combined with precipitation, thereby diminishing the operational reliability of the insulator [9]. Previous studies have demonstrated that in regions where algae cover the surface of insulators, hydrophobicity decreases to levels classified as HC5 to HC6, and the wet flashover voltage of algae-covered insulators is reduced by approximately 20% to 30%. The impact of algal contamination on the hydrophobicity of silicone rubber is greater than that of kaolin; under salt-fog conditions, algal fouling results in higher leakage current and the formation of dry-band discharges. Moreover, the extent to which algal contamination affects the pollution flashover voltage of insulators is strongly correlated with the coverage: when the coverage surpasses 20%, an increase in the coverage proportion further lowers the pollution flashover voltage [10].

The research team at Tsinghua Shenzhen International Graduate School developed a new RTV silicone coating incorporating the antimicrobial agent 4,5-dichloro-2-octyl-isothiazolone (DCOIT) mixed with silicon dioxide to inhibit algae growth on high-voltage insulators in humid regions [11]. The DCOIT powder was first heated above its melting point to obtain a liquid phase, after which finely powdered silicon dioxide (SiO_2), a material already commonly used in RTV formulations, was added to the molten DCOIT. The mixture was then heated and stirred for 20 minutes to ensure thorough combination and uniform dispersion of the antimicrobial agent with the silica. Following heating, the mixture was cooled and milled to produce a stable powder containing well-dispersed DCOIT crystals, as confirmed by micromorphology analysis, which showed that the mixed powder exhibited smaller particle sizes than pure DCOIT. Laboratory tests using *Apatococcus lobatus* algae demonstrated that adding 0.1 wt% of the DCOIT-silica mixture reduced photosynthetic activity to 6% of untreated samples and maintained surface hydrophobicity, with contact angles remaining above 80°. Surface and volume conductivity measurements confirmed that although algae growth increased conductivity by several orders of magnitude, electrical properties recovered after cleaning, and the antimicrobial addition did not degrade insulation performance. ATR-FTIR analysis indicated that physicochemical changes caused by algae were reversible upon removal, while tensile testing showed mechanical strengths over 3.6 MPa and strain rates above 285%, exceeding RTV-I standards. The environmentally friendly and long-acting algae-inhibitory RTV was demonstrated in substation-converter stations and lines. After three years of operation (2020 ~ 2023), there were almost no algae attached to the surface, and the operation was safe and reliable, as shown in Figure 6.3.

The extent of algal growth directly determines its impact on the performance of silicone rubber. Under optimal environmental conditions, algae may proliferate

Figure 6.3 Long service effect of anti-algae RTV coatings. (a) Effect after 3 years of outdoor operation; (b) local magnification.

rapidly, resulting in widespread contamination of silicone rubber surfaces and substantially undermining the pollution flashover withstand capability of insulators. A comprehensive investigation of algal growth characteristics on energized insulator surfaces, including analysis of natural spatial distribution patterns and influencing factors, as well as examination of the impact of algae with varying metabolic activity on insulator performance, can provide an important scientific basis for formulating operation and maintenance strategies for insulators in regions susceptible to algal colonization, which is of considerable engineering significance.

6.1.3 The Water Absorption and Permeability of Silicone Rubber

In southern regions of China, characterized by high humidity, composite insulators used in power transmission lines occasionally experience abnormal temperature rises, as shown in Figure 6.4a [12]. Currently, the reasons for such abnormal heating phenomena in composite insulators can be classified into three main categories: Firstly, increased polarization losses occur due to water absorbed by silicone rubber materials of sheds and sheaths under strong electric fields. Secondly, increased conductive losses result from surface contamination and decreased electrical resistivity caused by aging of silicone rubber materials and interfaces, as shown in Figure 6.4b. Thirdly, localized partial discharges occur at defects within the sheath–core rod interface. Among these identified mechanisms, relatively limited research has addressed the comprehensive moisture absorption characteristics of entire composite insulators in hot and humid environments, especially regarding variations in water absorption behavior at different locations of silicone rubber sheds.

Research into moisture absorption and diffusion in silicone rubber dates back to the 1960s, although studies specifically targeting silicone rubber used for high-voltage outdoor insulation began somewhat later. Lots of studies focused on high-temperature vulcanized (HTV) silicone rubber immersed in solutions of varying salinity. The following conclusion has been widely accepted. Higher temperatures increased water diffusion coefficients. Conversely, lower salinity resulted in higher saturated moisture absorption, with samples losing weight upon returning to dry air conditions. The dielectric constant of HTV silicone rubber increased during aging due to moisture absorption and decreased upon drying due to evaporation.

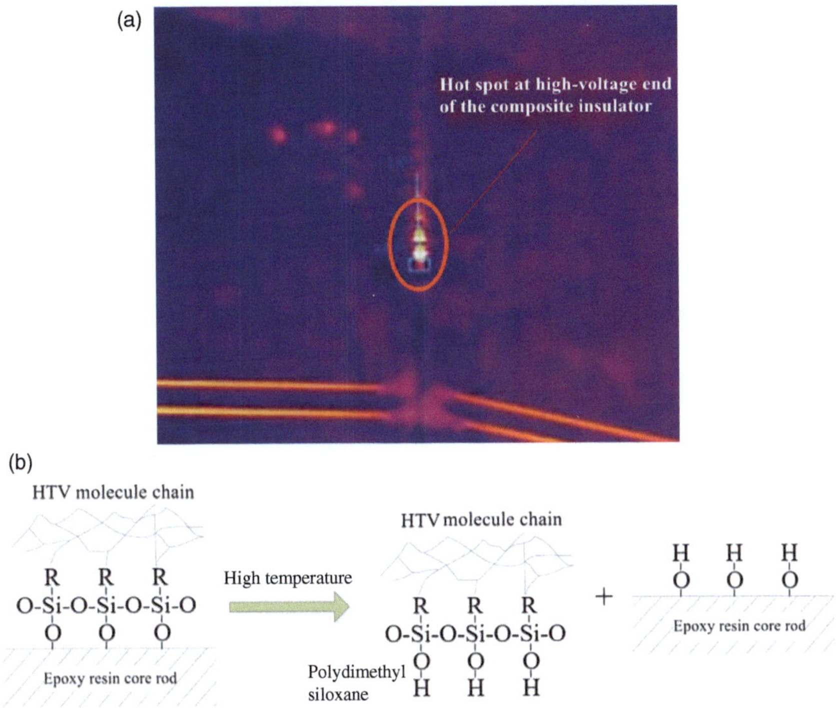

Figure 6.4 Abnormally overheating due to moisture absorption of the composite insulator. (a) Hot spot of the operating composite insulator; (b) hydrolysis mechanism of the interface coupling agent.

The research team from Tsinghua Shenzhen International Graduate School systematically examined the impact of fillers, environmental conditions, and mechanical stresses on the water and moisture absorption behavior of HTV silicone rubber, as well as the changes in electrical properties caused by moisture, as shown in Figure 6.5 [12]. They investigated the dependency of various electrical characteristics – including dielectric constant – on moisture content. Additionally, they conducted drying tests, confirming that HTV silicone rubber, once saturated with moisture, loses part of this absorbed water when returned to a dry environment.

Despite comprehensive existing studies into the factors influencing moisture absorption in newly produced HTV silicone rubber, most research overlooks the practical scenario of composite insulators during actual operation. The sheds and sheaths of in-service composite insulators experience complex environmental influences that significantly alter their physical and chemical properties, thus modifying the original moisture absorption behavior. Furthermore, composite insulators are exposed to varying humidity conditions in the field. Previous studies have shown that HTV silicone rubber exhibits a "breathing effect," absorbing moisture in humid environments and subsequently releasing moisture when moved to a drier environment. However, research into the environmental factors influencing this breathing effect for operational composite insulators remains limited. To address this gap, this

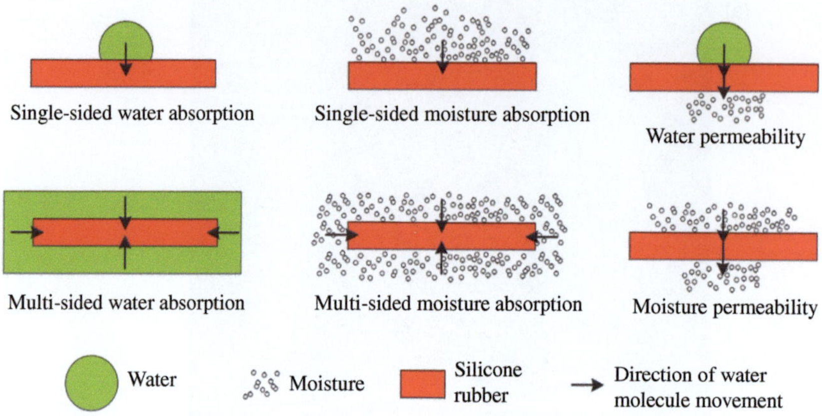

Figure 6.5 Schematics of water/moisture-absorbing, water/moisture-permeable properties of silicone rubber.

research team investigated the influence of moisture exposure duration, ambient temperature, and electric fields on the water absorption behavior of silicone rubber sheds specifically for composite insulators operating in humid climates [13]. They discovered that the quantity of water absorbed by sheds increased proportionally with exposure duration in highly humid conditions. Higher ambient temperatures accelerated both moisture absorption and subsequent drying. Additionally, applied electric fields, especially DC fields, significantly inhibited water absorption by silicone rubber and promoted moisture desorption. The Langmuir diffusion model effectively described moisture ingress and egress driven by concentration gradients in silicone rubber.

Building upon silicone rubber shed moisture absorption tests, further immersion and boiling-water tests were conducted on the entire composite insulators. Results demonstrated that only boiling-water treatments caused abnormal temperature rises in otherwise normally operating insulators, confirming thermal stress as a critical factor promoting moisture infiltration into the sheath–core rod interface. Simultaneously, thermal stress accelerated the hydrolysis of silane coupling agents, further weakening interfacial properties and intensifying temperature increases with prolonged boiling-water exposure.

To comprehensively evaluate water absorption characteristics across the entire composite insulator, moisture absorption tests were performed on silicone rubber sheds from various positions. It was found that the difference in moisture absorption among sheds at different positions was minimal. However, within individual sheds, water absorption capability consistently decreased from the external surfaces inward and from the shed edges toward the shed roots. Using micro-analytical techniques such as scanning electron microscopy (SEM), Fourier transform infrared spectroscopy (FTIR), and thermogravimetric analysis (TGA), it was determined that the silicone rubber at shed edges exhibited greater decomposition, increased surface cracks and voids, and more severe aging.

6.1.4 Chalking of Silicone Rubber in Hot and Humid Environments

The coastal regions of southern China are characterized by a typical hot and humid climate, under which composite insulators exhibit notably different operational behaviors compared with those in dry inland regions. Phenomena such as accelerated degradation of insulator sheds, abnormal heating, and brittle fracture of fiberglass cores frequently occur. In some cases, large-scale chalking of insulator sheds has been reported after 5 to 10 years of operational service. Once chalking occurs, the water absorption capacity of the sheds significantly increases, thereby compromising their protective ability for the internal core rod. According to field investigations, chalking predominantly occurs in coastal humid areas and in high-altitude regions subjected to strong ultraviolet radiation; occurrences have also been documented in certain mountainous areas within Guangdong, Fujian, Henan, and Zhejiang provinces. Inspection results from China Southern Power Grid from 2012 to 2019 indicated that approximately 40% of inspected composite insulators exhibited chalking phenomena.

At present, most studies primarily involve characterization analyses of chalked insulators, attempting to deduce degradation mechanisms via reverse inference, with relatively limited direct research into the fundamental chalking mechanisms. Furthermore, reproducing the chalking phenomenon in laboratory environments remains challenging. Xu C et al. [14] successfully replicated chalked layers on insulators by utilizing alkaline-catalyzed degradation methods involving reagents such as diethylamine and methylcyclohexane, hypothesizing that the chain-scission reactions predominantly occurred at the filler–matrix interfaces. Jiang T et al. [15] achieved secondary chalking on previously chalked insulator sheds through repeated cycles of soaking and drying. When this method was applied to fresh silicone rubber samples, surface discoloration occurred, accompanied by the emergence of white particulates. Based on the characterization of chalked insulators, Lu Ming concluded that the scission of hydrophilic Si–O–Si backbone chains was the principal cause of chalking, whereas decomposition of aluminum trihydroxide (ATH) served as a secondary contributing factor. Fang Su et al. [16] conducted accelerated aging experiments on in-service insulator sheds, simulating high-temperature, high-humidity, and combined hygrothermal conditions. They proposed that elevated temperatures primarily facilitated oxidative cross-linking and hydroxyl condensation reactions, while high humidity conditions promoted water ingress into silicone rubber, leading to hydrolytic reactions. Under combined hygrothermal conditions, simultaneous hydrolysis and cross-linking reactions occurred. Based on current research, moisture ingress appears to play a critical role in the chalking process of silicone rubber. Kaneko T. [17] reported experimental findings indicating that silicone rubber for cable applications exposed to 155 °C steam predominantly underwent hydrolysis reactions without evident cross-linking. However, cross-linking reactions gradually intensified with increased temperature, ultimately causing the silicone rubber to transition from flexible to brittle.

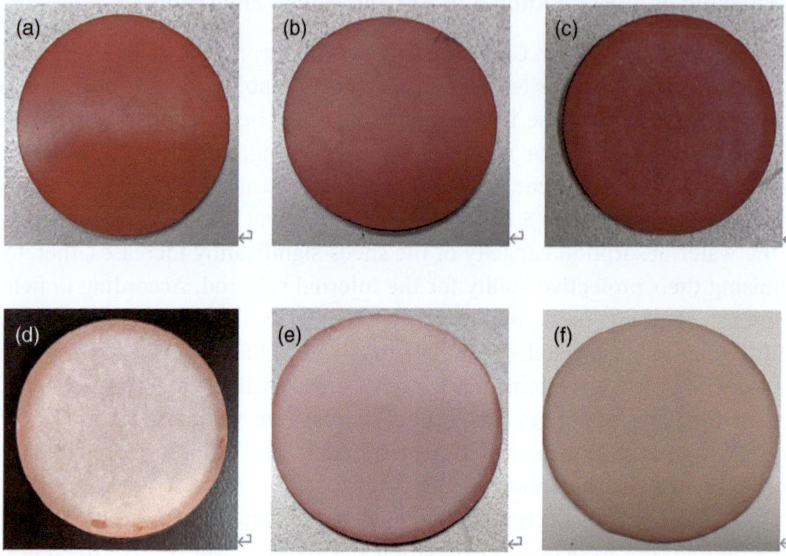

Figure 6.6 Results of the combined thermo-oxidation and steam-aging. (a) Initial sample; (b) thermo-oxidative 20d; (c) steam aging 1d; (d) steam aging 2d; (e) steam aging 3d; (f) steam aging 4d.

Although the studies mentioned above did not successfully replicate the chalking phenomenon solely by simulating hygrothermal conditions, they nonetheless demonstrated that moisture significantly promotes the migration of fillers. Based on this finding, the research group from Tsinghua Shenzhen International Graduate School proposed a combined thermo-oxidation and steam-aging approach, which specifically consists of two stages: thermo-oxidative aging at 150 °C for 20 days, followed by steam aging at 100 °C for 4 days. Using this method, they successfully observed chalking phenomena that closely resembled natural aging conditions, as shown in Figure 6.6 [18].

Both physical tests and the distribution of elements and functional groups can confirm the above equivalence, as shown in Figure 6.7. After thermo-oxidative aging, the Si–C bonds in the side chains of HTV silicone rubber exhibited slight weakening, whereas other functional groups showed negligible changes, with the content of Si–O–Si groups at 1008 cm^{-1} remaining largely unchanged. However, significant scission of the polymer backbone occurred, leading to a substantial reduction of organic groups. Additionally, there was a marked increase in hydroxyl (–OH) content, ATH migrated toward the rubber surface, and the hydrolysis of siloxane resulted in the formation of silanol (Si–OH) and related products. The thermo-oxidative pretreatment notably enhanced the subsequent hydrolytic reactions during steam aging. During the steam-induced accelerated degradation stage, Si–O bonds were consumed at a higher rate compared to Si–C bonds, with hydrolysis reactions becoming dominant, as shown in Table 6.1.

Infrared spectral analysis comparing artificially and naturally chalked powders revealed that the absorption peaks of both powders exhibited similar shapes and

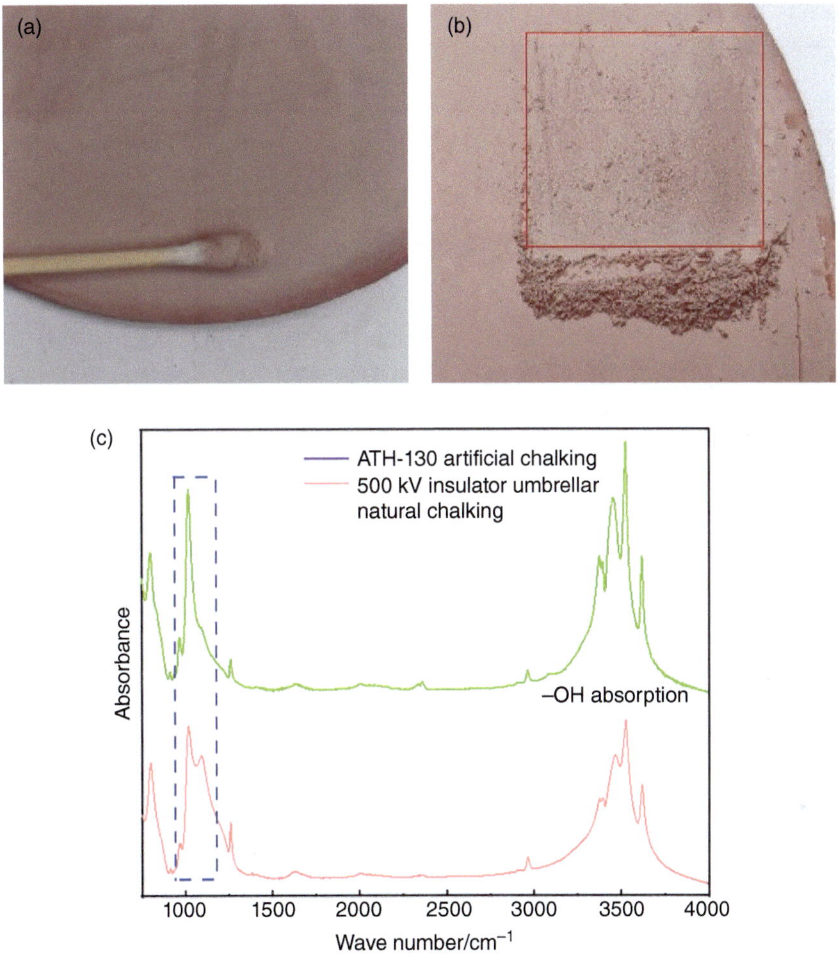

Figure 6.7 Characterization of the artificially chalking layer obtained in the laboratory. (a) Cotton swab wiping; (b) knife scraping; (c) Fourier infrared spectroscopy.

relative intensities, except for a weaker absorption at 1090 cm^{-1} observed in the artificially chalked samples. This peak is associated with long-chain siloxane networks or silica (SiO_2). Two possible reasons were proposed to explain this difference: firstly, artificial chalking induced more complete hydrolysis and cleavage of long-chain siloxanes; secondly, artificial chalking resulted in less silica (white carbon black) migration to the surface.

The research group proposed that the high equivalence to natural chalking observed in their approach was due to the limited penetration ability of water vapor into silicone rubber with an intact matrix and fully encapsulated fillers. However, after thermo-oxidative aging consumed methyl ($-CH_3$) side groups and weakened their protective effect, water molecules more aggressively attacked polar siloxane bonds, significantly intensifying hydrolytic reactions. Additionally, macro- and micro-scale cracks and pores formed on the silicone rubber surface

Table 6.1 Composition (parts per hundred) and preparation process of different samples.

		Artificial	Natural
Color		L = 71.5, a = 16.08	L = 61.23, a = 25.88
Hydrophobicity		HC1, 132.1°	HC1, 136.5°
Roughness		5.568 μm	7.641 μm
Silicone rubber content		7.90%	5.55%
ATH content		71.58%	73.38%
Functional group variation	$Si–CH_3$	−51.35%	−72.79%
	Si–O–Si	−49.55%	−68.81%
Elemental composition	C	7.55%	16.52%
	Si	3.27%	8.79%

facilitated water vapor ingress, thus providing vulnerable points and pathways for deeper penetration. Finally, the research group recommended that for composite insulators operating in hot and humid environments, it is advisable to reduce ATH filler content while still meeting the TMA 4.5 flame retardancy standard, rationally decrease the shed-design margin based on inclined-plane tests, and consequently enhance chalking resistance.

6.2 Research and Development of Advanced Silicone Rubber Materials

In order to enhance the resilience of silicone rubber materials against diverse operational environments, prolong their service lifespan, and endow them with targeted functional attributes, researchers have undertaken extensive investigations across multiple research directions.

6.2.1 Self-Healing Silicone Rubber Materials

In 2001, White et al. introduced the concept of self-healing materials in *Nature*, referring to materials that are capable of autonomously repairing damage, thereby restoring their mechanical properties and prolonging their operational longevity [19]. The degradation of silicone rubber composites is commonly accompanied by the nucleation and propagation of microcracks, which are challenging to detect and virtually impossible to repair during the operational lifespan of composite insulators. The advancement of these microcracks may precipitate phenomena such as shed cracking and loss of mechanical integrity, ultimately culminating in insulation failure. Therefore, the incorporation of self-healing mechanisms into silicone rubber insulating materials presents considerable potential for the autonomous remediation of internal microcracks within silicone rubber composites, markedly retarding the degradation process and substantially extending the service lifetime of composite insulator sheds.

In 2001, White et al., upon introducing the microcapsule-based self-healing technology, utilized computational simulation to investigate the propagation dynamics of microcracks. They pointed out that when the elastic modulus of the matrix is less than one-third of the elastic modulus of the microcapsules, microcracks preferentially propagate around the microcapsules, thereby making it difficult to trigger the self-healing process. Owing to the fact that the elastic modulus of silicone rubber is substantially lower than that of epoxy resin insulating materials, extrinsic self-healing systems have predominantly adopted epoxy resins as matrix materials, whereas polysiloxane matrices have been infrequently implemented. To date, only international researchers such as Keller and White have developed microcapsule-based self-healing silicone rubber systems employing a PDMS/Pt catalyst combination [20], while Allahdini et al. have established analogous materials utilizing a PDMS/organotin catalyst system [21].

The research team at Tsinghua Shenzhen International Graduate School employed low-viscosity methylhydrogen polysiloxane with exceptionally low hydrogen content as a prototypical PDMS healing agent, and systematically investigated the preparation methods for urea-formaldehyde (PUF) microcapsules encapsulating the PDMS healing agent. The wall material of the microcapsules governs critical physicochemical attributes, including stability, permeability, and service life. The polymeric wall materials intended for microcapsules should meet the following criteria: (1) The wall material should possess excellent physicochemical properties, enabling effective protection of the core material after curing. (2) The wall material should not react chemically with either the matrix or the core material. (3) The wall material should possess appropriate mechanical properties to ensure that the prepared microcapsules exhibit suitable mechanical robustness, thus preventing premature rupture during processing of the self-healing system, while allowing rupture to occur upon the formation of microcracks in the matrix, thereby releasing the healing agent and activating the self-healing function.

The most widely utilized method for the preparation of PUF microcapsules is in situ polymerization. The fundamental principle of this technique is depicted in Figure 6.8.

Initially, the precursors for the wall-forming material are dissolved in the continuous phase (commonly water), whereas the core material remains insoluble within this medium; consequently, through the combined action of emulsifiers and

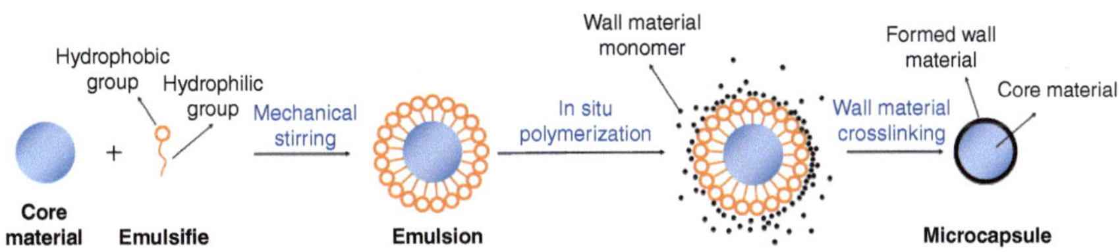

Figure 6.8 Schematic diagram of the principle of in situ polymerization.

mechanical agitation, the core material forms the dispersed phase. Upon achieving a homogeneous emulsion, the cross-linking reaction of the shell-forming precursors is initiated. The intermediate products generated during the formation of the shell are insoluble in the continuous phase and, consequently, precipitate from the solution, becoming adsorbed at the interface between the dispersed and continuous phases (i.e. on the surface of the core material). As the reaction progresses, these intermediate species undergo further reactions, ultimately generating a continuous and compact membrane on the surface of the dispersed phase, thereby ensuring complete encapsulation of the core material. Finally, the microcapsules are isolated by sequential processes of filtration, washing, and drying. Through experimental investigation, the optimal formulation for preparing PUF microcapsules encapsulating polydimethylsiloxane (PDMS) healing agents via in situ polymerization was established: a combined emulsifier system of EMA and PVA, with concentrations of 0.13 wt.% EMA and 0.05 wt.% PVA, an initial pH of 4.0, a core-to-shell ratio of 1.5 : 1, and a sodium chloride concentration of 2 wt.%. The microcapsules (Figure 6.9), silicone rubber matrix, and modifiers are thoroughly blended according to the specified formulation, after which a curing agent is introduced and further mixed.

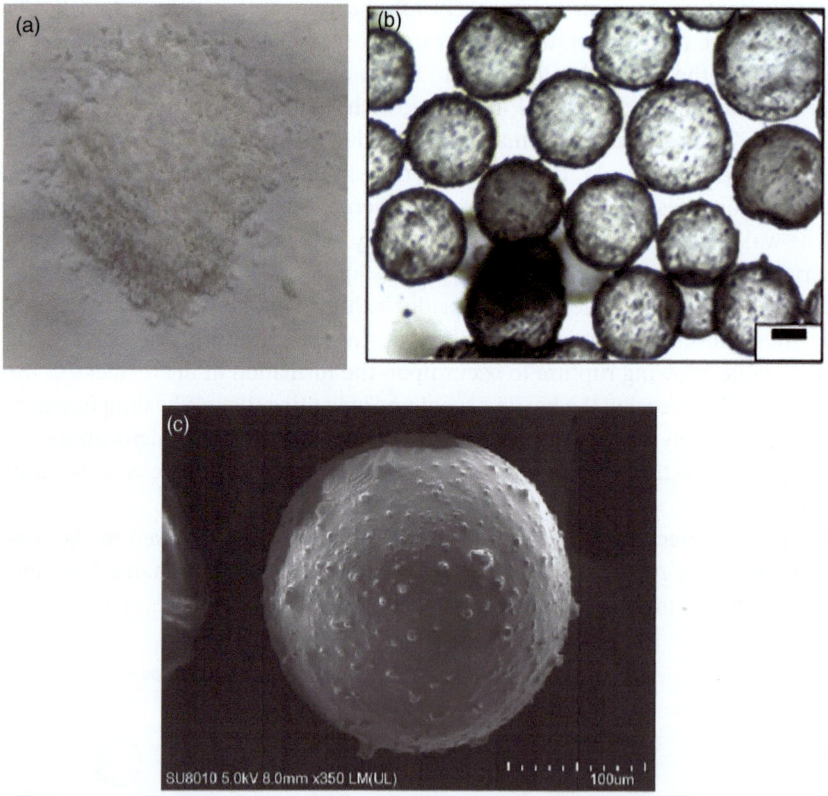

Figure 6.9 Morphology and microstructure of microcapsules. (a) Appearance of microcapsules; (b) morphology of microcapsules (optical microscopy); (c) microstructure of microcapsules (SEM).

Following degassing and curing, the material is demolded to obtain self-healing silicone rubber specimens [22].

The underlying self-healing mechanism of microcapsule-based systems can be systematically described as follows: (1) Microcapsule rupture: Within the matrix, microcapsules are subjected to mechanical and thermal stresses, which may arise from the propagation of microcracks or local heating at the matrix surface. When the stress imposed upon the microcapsule wall exceeds a critical threshold, the microcapsule ruptures, thereby releasing the encapsulated healing agent. (2) Healing agent cross-linking: Following release, the healing agent migrates into the microcrack region by means of capillary action. In single-component systems, the healing agent encounters a catalyst located at the fracture interface and subsequently undergoes curing initiated by catalytic activity during infiltration. Conversely, for dual-component systems, the two separate healing components converge and mix as they traverse the crack, resulting in in situ curing and effective restoration of the damaged domain.

The fractured silicone rubber specimens obtained from the tensile test were carefully realigned and joined together. The fracture interface was immobilized using polytetrafluoroethylene (PTFE) tape and further clamped between two glass slides to ensure optimal interfacial contact. Subsequently, the specimens were subjected to self-healing at ambient temperature for 24 hours, followed by a further 24-hour period at 60 °C, resulting in a total healing duration of 48 hours. The tensile strength of the samples was subsequently reassessed.

The self-healing efficiency is defined as the recovery efficiency of tensile strength, calculated as follows:

$$\eta_{\text{Heal}} = \frac{T_{\text{Healed}}}{T_{\text{Virgin}}} \tag{6.1}$$

In this equation, T_{Virgin} corresponds to the original tensile strength, whereas T_{Healed} refers to the tensile strength measured posthealing.

The qualitative evaluation of self-healing performance for the specimens is presented in Figure 6.10. Under the specified curing and healing conditions of this experiment, the self-healing efficiency of the pristine silicone rubber sample was less than 2%, indicating that it possessed negligible self-healing capability. By contrast, the microcapsule/silicone rubber composite materials exhibited pronounced self-healing functionality, as evidenced by effective restoration of their tensile properties. Therefore, the introduction of microcapsules was demonstrated to endow the composite system with an intrinsic self-healing capability in terms of its mechanical performance.

Building upon the aforementioned research, the project team proposed a theoretical equation for calculating the self-healing efficiency of microcapsule-based systems. The model indicates that the self-healing efficiency is positively correlated with several critical parameters, such as the interfacial strength between the healing agent and the matrix, the rupture probability of microcapsules at microcrack sites, the mean diameter of the microcapsules, the loading content of microcapsules, and the core fraction within each capsule. Conversely, the self-healing efficiency exhibits a negative correlation with crack width. Accordingly, these theoretical

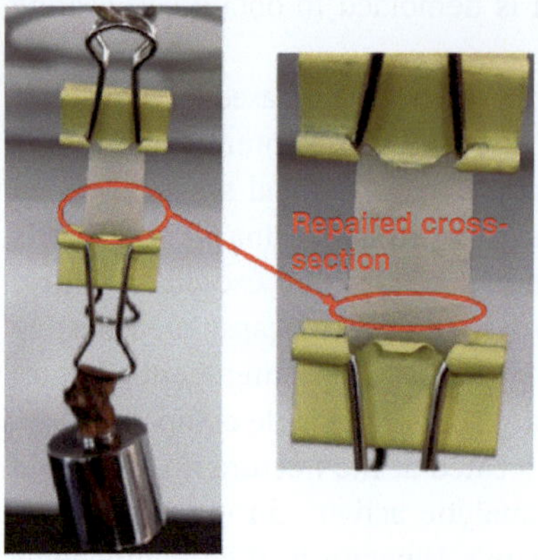

Figure 6.10 Qualitative test results of the self-healing performance of samples.

considerations offer a scientific foundation for the rational design and optimization of microcapsule-based self-healing systems in silicone rubber matrices [22, 23].

6.2.2 Nonlinear Conductive Silicone Rubber Materials

The electrical conductivity of nonlinear conductive composite materials depends on the externally applied electric field; as the magnitude of the applied field increases, the conductivity correspondingly rises, which can, in turn, facilitate optimization of the electric field distribution. The concept of dielectric functionally graded materials (FGMs) derives from FGMs in the field of materials science, specifically referring to composite materials in which the dielectric constant or electrical conductivity exhibits a spatial gradient distribution [24]. In AC systems, regions with higher electric field strength employ materials with greater dielectric constants, which can alleviate localized field concentration and enable flexible regulation of the electric field, while simultaneously avoiding unnecessary complexity in the insulation structure [25, 26]. The voltage equalization approach based on electric field–adaptive functional materials offers incomparable advantages over traditional structural methods, both in terms of application scenarios and the effectiveness of voltage equalization. Moreover, electric field–adaptive functional materials are well aligned with the ongoing trends in electrical equipment development toward miniaturization, intelligent functionality, and environmental sustainability. The successful development of high-performance, widely applicable, and industrially manufacturable electric field–adaptive functional materials and components would cause a paradigm shift in insulation structure design and catalyze transformative advances in the domains of electrical insulation materials and electrical equipment manufacturing.

The research team from Tsinghua Shenzhen International Graduate School has proposed the novel concept of "Electrically modulating electric fields" [27], whereby the application of an AC electric field during the fabrication of composites serves to drive the movement, collision, overlap, and assembly of fillers. This strategy facilitates the formation of tailored interfacial contacts, conductive networks, and parallel alignment, thereby precisely directing the autonomous assembly of filler

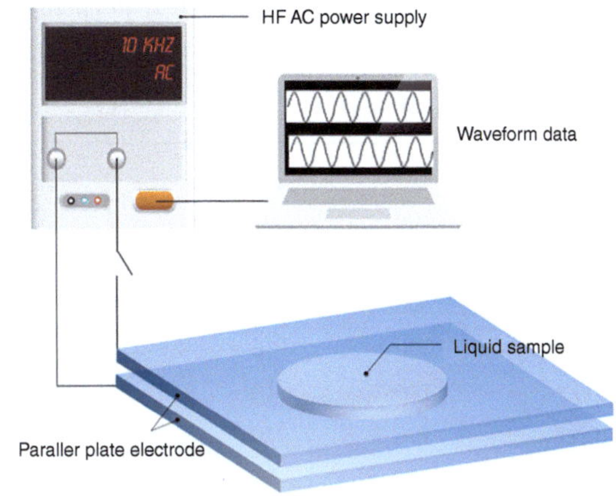

Figure 6.11 Schematic of the device for AC electric field-assisted preparation of composite samples.

particles and enabling the fabrication of dielectric gradients with high spatial fidelity [28, 29]. Ternary T-ZnOw/CF/LSR composites prepared via AC electric field-assisted methodology manifest pronounced nonlinear electrical conductivity, even under comparatively low filler concentrations. Initially, components A and B of liquid silicone rubber are mixed in a 1 : 1 weight ratio, followed by the incorporation of T-ZnOw, CF, and a defoaming agent, with subsequent mechanical agitation sustained for 30 minutes. The mixture is then transferred to a vacuum oven and subjected to degassing under ambient conditions for five minutes; this cycle is repeated thrice to ensure complete removal of entrapped gases. A schematic diagram of the apparatus for electric field-assisted composite preparation is presented in Figure 6.11: the liquid mixture is introduced between parallel-plate electrodes (dimensions: $130 \times 130 \times 12$ mm), with the electrode gap precisely regulated by PTFE spacers positioned at each corner [30].

At ambient temperature (25 °C), an AC voltage is applied across the electrodes for 30 minutes, after which the assembly is transferred to an oven for a programmed thermal ramp, elevating the temperature from room temperature to 125 °C over a period of 30 minutes, followed by an isothermal hold at 125 °C for 2 hours. The AC voltage is continuously maintained for a total duration of 90 minutes from the commencement of processing, thereby ensuring the presence of the electric field throughout the initial curing phase. Table 6.2 presents the compositions and detailed fabrication procedures for each specimen; specifically, the notation "10&2-E" denotes a formulation incorporating 10 phr T-ZnOw and 2 phr CF, prepared under electric field-assisted conditions.

Specimens fabricated in the absence of electric field assistance displayed pressure-sensitive conduction threshold field strengths greater than 2 kV/mm, as shown in Figure 6.12. For specimens produced under the assistance of an AC electric field, exemplified by 10&2-E and 10&5-E, nonlinear conduction phenomena were not detected within the investigated field intensity range. However, upon increasing the T-ZnOw filler concentration, as in 20&2-E, 20&5-E, and 30&5-E, distinct nonlinear conduction behavior became apparent within the measurement window. The threshold electric fields for pressure-sensitive conduction in samples

Table 6.2 Composition (parts per hundred) and preparation process of different samples.

Sample name	LSR	T-ZnOw	CF	Defoamer	Applied field strength(V/mm)
Pure LSR	100	0	0	0.5	0
0&5	100	0	5	0.5	0
0&5-E	100	0	5	0.5	200
10&2-E	100	10	2	0.5	200
10&5	100	10	5	0.5	0
10&5-E	100	10	5	0.5	200
20&2-E	100	20	2	0.5	200
20&5	100	20	5	0.5	0
20&5-E	100	20	5	0.5	200
30&2-E	100	30	2	0.5	200
30&5	100	30	5	0.5	0
30&5-E	100	30	5	0.5	200

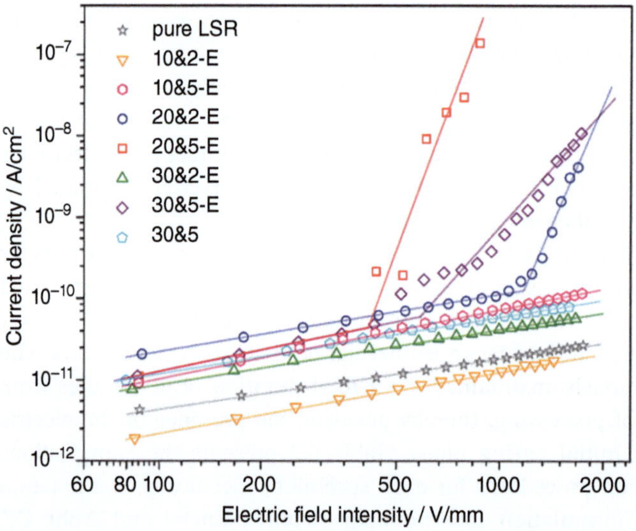

Figure 6.12 J-E characteristics of different silicone rubber composites at room temperature.

20&2-E, 20&5-E, and 30&5-E were 1200, 420, and 600 V/mm, respectively, with corresponding nonlinear coefficients of 10.7, 11.7, and 5.2, thereby demonstrating a broad tunability of nonlinear conduction parameters. Figure 6.13 compares these results with those reported in the literature, revealing that this methodology effectively fills the gap in low filler content, low threshold field, and high nonlinearity coefficient regime, thereby conferring a distinct competitive advantage.

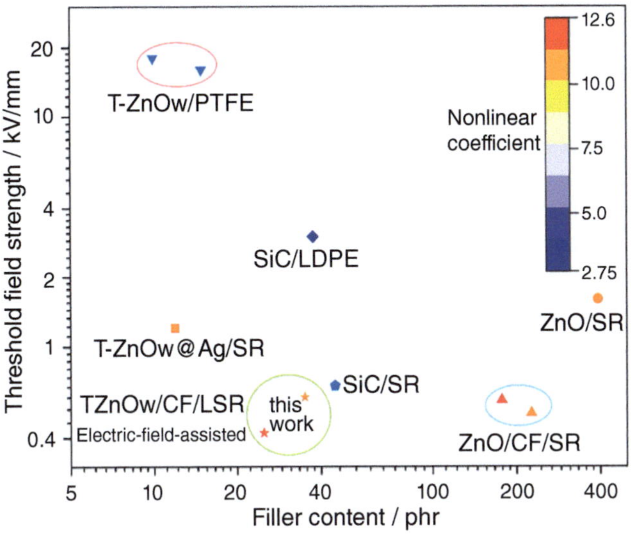

Figure 6.13 Comparison with literature reports. *Source:* [30] / Chinese Society of Electrical Engineering / CC BY-NC-ND 4.0.

Silicone rubber composites fabricated under the assistance of an AC electric field exhibit a substantial increase in shallow trap density, accompanied by a reduction in characteristic trap energy levels (from 0.52 to 0.34 eV). These changes are considered to positively influence both charge transport and the spatial distribution of space charge within the dielectric. The electric field–assisted fabrication strategy effectively mitigates surface charge accumulation and expedites the dissipation of superficial charges. Conductive pathways induced and constructed via electric field assistance simultaneously serve as thermal conduction channels, thereby markedly enhancing the thermal conductivity of the composites. The rotation or translational displacement of filler particles under an applied electric field serves to minimize their electrostatic potential energy. Consequently, in contrast to composites with randomly dispersed fillers, those synthesized under electric field–assisted conditions manifest reduced internal stress when exposed to operational electric fields.

6.2.3 Silicone Gel for Encapsulation

High-voltage, high-power semiconductor devices have been widely used in flexible DC transmission projects and are key core equipment for achieving power conversion. Moreover, they are being developed toward higher voltages and higher power densities, which impose stricter requirements on the insulation, mechanical, and thermal properties of encapsulation materials. Power electronic modules typically employ two primary insulation materials: ceramic substrates and silicone gel. The ceramic substrates serve as the principal insulation component, providing electrical isolation between active components and the grounded base plate. Meanwhile, silicone gel functions as an encapsulating medium that prevents air-gap discharges while safeguarding substrates, semiconductor devices, and interconnections from environmental factors including moisture, contaminants, and mechanical vibrations.

Silicone gels represent a unique category of encapsulating materials that solidify into highly flexible substances. The Si–O–Si bonds in these gels exhibit a broad angular variation ranging from 105° to 180°, enabling nearly free rotation around these bonds and resulting in highly flexible polymer network chains. This molecular structure creates substantial free volume within the material while minimizing intermolecular forces. As a result, silicone gels demonstrate characteristics of both solid and liquid phases. The cured material maintains a soft yet elastic consistency, combining the stress-relief and self-repair capabilities typical of liquids with the dimensional stability characteristic of cross-linked elastomers. This self-healing property enables the material to automatically recover from damage caused by mechanical stress or partial discharge activity associated with electrical treeing. Additionally, silicone gels retain their physical and electrical properties across an extensive temperature range spanning from −45 °C to 200 °C. Their inherently tacky surface facilitates adhesion to most common substrates without requiring primer treatments. These combined characteristics establish silicone gels as particularly effective encapsulation solutions for power electronic applications.

During operation over a wide temperature range, differences in the temperature sensitivity coefficients of a material's dielectric parameters can alter the electric field distribution and charge transport, while mismatches in thermal expansion coefficients can lead to thermal stresses at heterogeneous interfaces and high-cycle fatigue failures, subsequently inducing insulation degradation. Additionally, prolonged operation of silicone gel across a wide temperature range can induce reactive free radicals within the material, leading to secondary cross-linking and chemical structural rearrangement. This process results in the continuous densification of the cross-linked network and the depletion of flexible chain segments, manifesting as typical issues such as reduced penetration, increased modulus, and weakened interfacial adhesion. These effects significantly compromise its capabilities for flexible sealing, insulation protection, and long-term stable service performance. The Fraunhofer Institute in Germany investigated the thermo-electro-mechanical coupling characteristics at the interface between silicone gel and ceramic substrates, demonstrating that partial discharges and electrical treeing at the interface are the primary forms of insulation failure in devices under high-frequency square-wave excitation. As shown in Figure 6.14, the welded power semiconductor module is entirely encapsulated in soft silicone gel.

The research team from Hefei University of Technology highlights that analyzing methyl oxidation activation energy provides a quick method to predict long-term thermal aging behavior [31, 32]. Three encapsulants were compared in terms of thermal stability, molecular structure evolution, and mechanical properties under accelerated thermal oxidative aging at 250 °C. Furthermore, they investigate the coupling relationship between molecular structures at silica/silicone gel interfaces and ionic conductivity in polymer composites. The team prepared silicone gel composites with varying fumed silica loadings and characterized

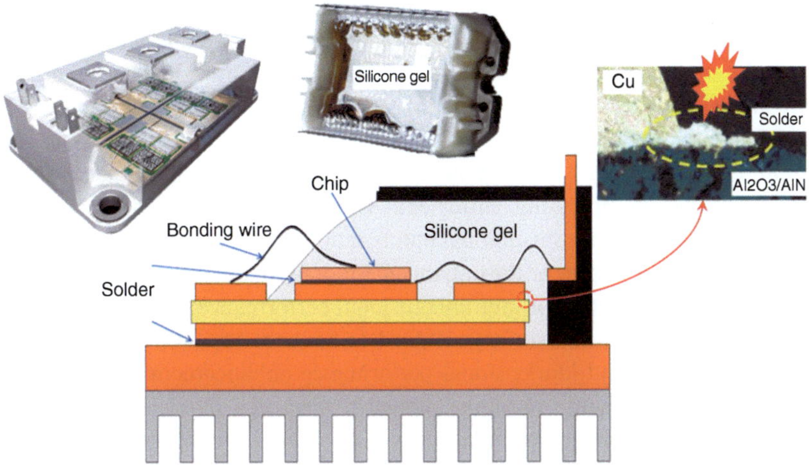

Figure 6.14 Cross-sectional structure of a soldered power module.

their morphology, dielectric properties, and molecular dynamics using broadband dielectric spectroscopy. The addition of silica significantly increased the concentration of mobile charges (by approximately four orders of magnitude) due to surface-adsorbed water molecules and interfacial double-layer effects, while also raising the ion dissociation energy barrier by about five times. Molecular chain analysis revealed that interfacial regions exhibited more flexible, short-length chain conformations (loop and tail types), enabling higher cooperative rearrangement and faster relaxation than bulk matrix chains, thus accelerating ion diffusion. Quantitative analysis showed negative decoupling exponents between interface molecular relaxations and ion diffusion coefficients, indicating that ions diffuse more slowly than interface segmental dynamics but faster than bulk matrix relaxations. Furthermore, the ionic conductivity exhibited a nonmonotonic dependence on silica content, increasing initially due to enhanced interfacial pathways and then decreasing at high loading due to filler aggregation and percolation effects. These findings provide a molecular-level understanding of how interface structures govern ionic conduction, offering guidelines for designing polymer composites with tunable electrical insulation performance.

The research team from Xi'an University of Architecture and Technology synthesized phenyl-modified silicone gels (PMSG) with varying phenyl contents and systematically compared with conventional pure silicone gel (PSG) in terms of thermal and dielectric properties [33]. Structural analyses using 1H-NMR and FT-IR confirmed successful phenyl incorporation, while TGA showed that PMSG exhibited significantly enhanced thermal stability, with an optimal composition (phenyl:methyl silicone oil = 1 : 1.2) achieving a 5% weight-loss temperature of 383 °C and 58.85% carbon residue at 800 °C. Electrical tests revealed a maximum room-temperature breakdown strength of 32.62 kV/mm, 17.42% higher than PSG, and only a 25.38% reduction at 150 °C, indicating superior high-temperature electrical resistance. Thermally stimulated depolarization current measurements showed deeper trap energy levels (0.77 eV vs. 0.68 eV), while dielectric spectroscopy demonstrated lower dielectric loss in PMSG across frequencies and temperatures. These results confirm that phenyl modification significantly improves both thermal stability and electrical insulation, providing a promising approach for advanced IGBT packaging materials.

Current research primarily focuses on strategies such as designing heat-resistant functional groups, incorporating antioxidant additives, and modifying dynamic reversible bonds. However, conventional optimization approaches often emphasize either single-molecule design or the enhancement of individual properties, lacking comprehensive consideration of multidimensional performance improvement and operational stability. There is an urgent need to advance research on molecular design for synergistic multidimensional performance enhancement and the construction of multicomponent cross-linked network structures. This will facilitate the development of a silicone gel formulation system tailored for high-voltage power device encapsulation, enabling coordinated improvements in multidimensional performance and supporting scalable production of silicone gel materials under wide-temperature-range and long-term service conditions.

6.2.4 Electroluminescent Silicone Rubber

The electroluminescent phenomenon, electroluminescence (EL), refers to a solid-state emission process where the radiative recombination of electrons and holes occurs under the influence of AC or DC electric fields, thus facilitating the direct transduction of electrical energy into radiant energy. When subjected to a high electric field, electroluminescent materials emit light as a result of deep-level electrons being accelerated by the electric field to become hot electrons; upon colliding with luminescent centers, these hot electrons facilitate an electronic transition in the luminescent center from the ground state energy level E_1 to an excited state energy level E_2. Subsequently, as the electron returns from the excited state E_2 to the ground state E_1, energy is released in the form of photon emission, thereby giving rise to observable luminescence. The energy transition involved in this process can be expressed as $E_2 - E_1 = \mathrm{h}\nu$, where h denotes Planck's constant and ν signifies the frequency of the emitted photon which determines the emission color.

Traditionally, electroluminescent materials utilize pure compounds as the host matrix, into which trace amounts of specific impurities, referred to as "activators" are introduced to endow the matrix with luminescent properties. Through this approach, host matrices that inherently exhibit negligible or suboptimal luminescence can attain significantly improved electroluminescent efficiency. Electroluminescent materials exhibit considerable diversity; following their initial discovery in 1936, research efforts in the scientific community predominantly concentrated on sulfide-based electroluminescent systems. In subsequent decades, substantial progress has also been achieved in the development of organic electroluminescent materials. Currently, the spectrum of available electroluminescent materials comprises ZnS-based semiconductor compounds, SiC systems, rare-earth organic complexes, europium-complex phosphors, and related materials. This broad array facilitates the selection of optimally suited electroluminescent materials for a wide range of application environments.

Professor Ji Sheng Chang's research group at Xi'an Jiaotong University has conducted preliminary investigations into the luminescent properties of ZnS: Cu-based electroluminescent materials [34]. Specifically, a layer of coating containing ZnS:Cu was applied to the surfaces of both disc-type and pin-type insulators; after pressurizing the insulators in an SF_6-filled chamber, optical images were captured, and a grayscale analysis method was employed to evaluate the luminescent behavior of the coated insulators under high voltage conditions, as well as to assess the correlation between luminescence intensity and parameters such as voltage, frequency, and electric field strength. Employing the electroluminescent material-based detection method, the mean measurement error for electric field distribution on disc-type insulators was less than 3%, whereas that for bushing-type insulators was below 4%. However, the team has not yet conducted investigations into the integration of electroluminescent materials with actual power grid infrastructure.

The research team at Tsinghua Shenzhen International Graduate School conducted systematic investigations into the luminescence mechanisms of ZnS:Cu composite materials, as well as strategies for the modulation of their luminescent performance [35]. Four sets of ZnS:Cu composite samples were fabricated for comparative experiments, with each set differing in phosphor content and thickness, and two specimens were prepared for each group. The preparation procedure was as follows: (1) The ZnS:Cu phosphor powder and RTV matrix were weighed in accordance with the designated mass ratios for each experimental group. (2) After mixing, the blend was stirred at a speed of 5 r/s for 10 minutes. (3) The homogeneous solution was poured into circular molds with a diameter of 3 cm. (4) The molds were placed in a vacuum drying oven at a constant temperature of 50 °C and thermally cured for 5 hours. The synthesized ZnS:Cu powder was subsequently formulated into a coating and uniformly applied onto the surfaces of insulators, in order to evaluate the voltage-responsive luminescent display characteristics of the ZnS:Cu coating. The applied voltage was gradually increased up to 18 kV, and the luminescence images of the upper and lower surfaces of the glass insulator were obtained as shown below (Figure 6.15).

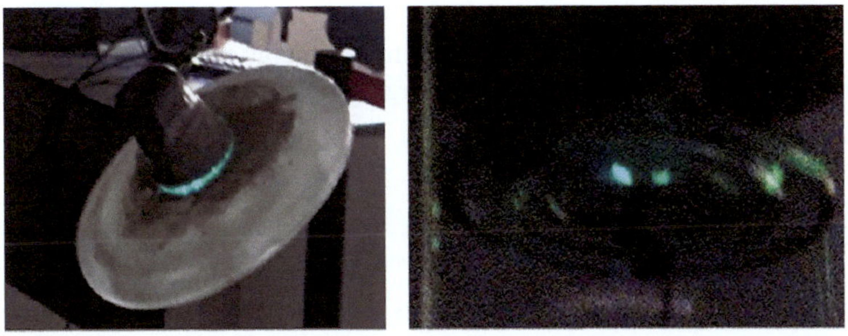

Figure 6.15 Luminescence of a single insulator under 18 kV applied voltage.

ZnS: Cu electroluminescent materials are characterized as energy-injection-type systems, wherein the emission process necessitates the acquisition of adequate external energy, ensuring that the energy delivered to the luminescent centers exceeds their respective activation energy barriers. The generation of high-energy electrons is the principal driving force for EL, rendering ZnS:Cu composites prototypical electron-driven luminescent materials. ZnS:Cu composites intrinsically exhibit luminescent properties. In the absence of external electron injection, ZnS:Cu composites utilizing an RTV matrix exhibit observable luminescence when exposed to an electric field intensity of approximately 5.12 kV/mm.

6.3 Conclusion and Outlook

Recent research avenues concerning silicone rubber materials for electrical insulation are predominantly focused on high-performance composite design, self-healing capability and durability enhancement, intelligent functionality, enhanced environmental adaptability, and sustainable manufacturing and recycling. (1) High-performance composite design: Through the incorporation of innovative fillers or advanced polymer modifications, material properties such as thermal stability, mechanical strength, and thermal conductivity are significantly improved, thereby meeting insulation requirements in extreme service conditions. (2) Self-healing capability and durability enhancement: by introducing dynamic chemical bonds (such as hindered urea bonds) or polymerizable nanoparticles, materials acquire autonomous self-healing functionalities, which substantially increase operational longevity and lower maintenance expenditures. (3) Intelligent functionality: the adoption of three-dimensional printing techniques allows for the fabrication of oriented thermally conductive networks, while the integration of fiber-optic sensors enables real-time monitoring and early detection of insulation states. (4) Enhanced environmental adaptability: material performance in harsh conditions – such as elevated temperatures, high humidity, and saline atmospheres – is optimized by means of surface fluorination or rare-earth oxide doping, thereby improving corona resistance and antiaging properties. (5) Green manufacturing and recycling: the development of bio-based, environmentally compatible fillers and recyclable processing technologies contributes to lower carbon emissions and facilitates intelligent life-cycle management. Collectively, these research directions are poised to advance silicone rubber materials toward greater safety, extended service lifetimes, and minimized energy consumption, thus supporting the sustainable development of emerging fields such as new energy systems and smart grids.

References

1 Yang, C., Ren, S., Yang, C. et al. (2023). Analysis on rapid failure causes of anti-pollution flashover coating hydrophobicity in winery environment. *Insulating Materials* 56 (05): 109–113.

2 Lu, Z. and Yang, X. (2001). Influence of contaminant with sugar on outer insulation performances of insulator in substation. *Proceedings of the CSEE* 21 (5): 51–55+73.

3 Mei, H., Mao, Y., Wang, L., and Guan, Z. (2012). Influence of contaminant with sugar on outer insulation performances of insulator in substation. *High Voltage Engineering* 38 (1): 66–74.

4 Zhang, X., Li, Y., Bai, H. et al. (2016). Natural Pollution Component For RTV-coated insulator in some substationof Sichuan province and its effect on flashover characteristics. *High Voltage Engineering* 42 (9): 2922–2928.

5 Gorur, R.S., Montesinos, J., Roberson, R. et al. (2003). Mold growth on nonceramic insulators and its impact on electrical performance. *IEEE Transactions on Power Delivery* 18 (2): 559–563.

6 Fernando, M.A. and Gubanski, S.M. (2010). Ageing of silicone rubber insulators in coastal and inland tropical environment. *IEEE Transactions on Dielectrics and Electrical Insulation* 17 (2): 326–333.

7 Wallstrom, S. and Karlsson, S. (2004). Biofilms on silicone rubber insulators; microbial composition and diagnostics of removal by use of ESEM/EDS: composition of biofilms infecting silicone rubber insulators. *Polymer Degradation and Stability* 85 (2): 841–846.

8 Jia, Z., Wang, Y., Ouyang, X. et al. (2017). Influence of green algae on hydrophobic property of HTV silicon rubber. *High Voltage Engineering* 43 (3): 885–890.

9 Yang, S. (2020). *Researches on Hydrophobic Properties of Algae-Contaminated Silicone Rubber and Its Inhibition Strategy*. Beijing: Tsinghua University.

10 Ouyang, X., Jia, Z., Yang, S. et al. (2018). Influence of algae growth on the external insulation performance of HVDC insulators. *IEEE Transactions on Dielectrics and Electrical Insulation* 25 (1): 263–271.

11 Wang, S. (2020). *Research on the Prevention Method of Algae Growing on RTV Coating Surface under Warm and Humid Environment*. Beijing: Tsinghua University.

12 Wang, Z. (2016). *Researches on Water Absorptivity and Permeability of Composite Insulator Silicone Rubber and Its Influence Mechanisms*. Beijing: Tsinghua University.

13 Zhang, J. (2024). *Study on the Moisture Infiltration Mechanism of Composite insulator's Shed and Sheath in Hot and Humid Environment*. Beijing: Tsinghua University.

14 Xu, C., Jia, Z., and Guan, R. (2023). Research on controllable artificial chalking method for silicone rubber. *IEEE Transactions on Dielectrics and Electrical Insulation* 30 (3): 1345–1352.

15 Zhou, Y., Wang, T., Jiang, T. et al. (2023). Effect of water ingress into on-site aged silicone rubber on its properties. *Insulating Materials* 56 (12): 78–84.

16 Fang, S., Xiao, W., Zhang, F. et al. (2020). Study on the aging factors and mechanisms for HTV in composite insulators in hot and humid environments. *2020 IEEE International Conference on HighVoltage Engineering and Application (ICHVE)*, 1–4. Beijing: IEEE.

17 Kaneko, T., Ito, S., Minakawa, T. et al. (2019). Degradation mechanisms of silicone rubber under differentaging conditions. *Polymer Degradation and Stability* 168: 108936.

18 Wang, X. (2024). *Research on Chalking Mechanism of High Temperature Vulcanized Siliconerubber in Hydrothermal Environment*. Beijing: Tsinghua University.

19 White, S.R., Sottos, N.R., Geubelle, P.H. et al. (2001). Autonomic healing of polymer composites. *Nature* 409 (6822): 794–797.

20 Keller, M.W., White, S.R., and Sottos, N.R. (2007). A self-healing poly(dimethyl siloxane) elastomer. *Advanced Functional Materials* 17 (14): 2399–2404.

21 Allahdini, A., Jafari, R., and Momen, G. (2022). Development of a dual capsule self-healing silicone composite using silicone chemistry and poly(melamine-urea-formaldehyde) shells. *Journal of Applied Polymer Science* 139 (8): 51670.

22 Li, M., Cao, B., Liu, Y., and Wang, L. (2022). Novel micro-capsule-doped insulating material with hydrophobicity transfer capability triggered by creepage discharge. *High Voltage* 7 (5): 950–959.

23 Li, M., Cao, B., Shang, R. et al. (2022). Synthesis of poly(urea-formaldehyde) microcapsules for the self-healing system of silicone rubber insulating material. *Journal of Applied Polymer Science* 139 (45): e53021.

24 Naebe, M. and Shirvanimo, K. (2016). Functionally graded materials: a review of fabrication and properties. *Applied Materials Today* 5: 223–245.

25 Wang, C., Li, W., Yang, X., and Zhan, M. (2020). A comprehensive strategy for electric field regulation of GIS/GIL spacer by using structure and dielectric distribution optimization. *Proceedings of the CSEE* 40 (22): 7166–7174.

26 Du, B., Wang, Z., Li, J. et al. (2020). Optimization of permittivity distribution for surface functionally graded insulator. *High Voltage Engineering* 46 (5): 1594–1601.

27 Shen, Z. (2021). *Research on Mechanism of AC Electric Field Assisted Preparation of Electric Field Grading Functional Materials*. Beijing: Tsinghua University.

28 Shen, Z., Wang, X., Zhang, T., and Jia, Z. (2020). In situ electric field driven assembly to construct adaptive graded permittivity $BaTiO_3$/epoxy resin composites for improved insulation performance. *Applied Materials Today* 20: 100647.

29 Shen, Z., Wang, X., Xin, Z. et al. (2021). Analytical model for the spatiotemporal permittivity of uncured-composite devices in an AC electric field. *Journal of Physics D: Applied Physics* 54: 155302.

30 Shen, Z., Xu, C., Wang, X. et al. (2025). The impact of electric field treatment on nonlinear conductivity of tetra-needle-shaped ZnO whisker/carbon fiber/liquid silicone rubber composites. *CSEE Journal of Power and Energy Systems* 11 (2): 900–908.

31 Lin, Y., Dong, L., Dang, M. et al. (2025). Silicone encapsulants with high thermal stability for SiC power devices under high temperatures. *Journal of Materials Science: Materials in Electronics* 36 (1): 18.

32 Lin, Y., Dong, L., He, C. et al. (2024). Coupling relationship between molecule structure at silica/silicone gel interfaces and ionic conductivity in polymer composites. *Composites Science and Technology* 249: 110507.

33 Wang, Z., Luo, M., Wang, R. et al. (2024). Research on heat resistance and dielectric properties of phenyl modified silicone gels for IGBT packaging. *Proceedings of the CSEE* 44 (14): 5833–5844.

34 Yang, X., Jia, Y., Gao, L. et al. (2021). Method for measuring surface electric field intensity of insulators based on electroluminescence effect. *High Voltage Engineering* 47 (4): 1411–1419.

35 Rao, W. (2024). *Research on Luminescence Theory and Regulation of Luminescence Characteristics of ZnS:Cu Composite Material*. Beijing: Tsinghua University.

7

The Evolution of Insulating Paper: Bio-Based and Circular Solutions for High-Performance Electrical Insulation

Junwen Ren, Jiacheng Zhang, Zi Wang, Junxue Chen, and Lin Zhang

Sichuan University, College of Electrical Engineering, 1st Ring Road, Chengdu, 610065, China

7.1 The Enduring Legacy of Conventional Insulating Papers

Paper, a material of such profound simplicity, is itself a chronicle of civilization spanning millennia. In 105 AD, Cai Lun, an official of the Chinese Han Dynasty, first perfected the art of modern papermaking by blending mulberry bark, hemp fibers, and old rags. This innovation provided an unprecedentedly inexpensive and efficient medium for the documentation and dissemination of knowledge [1]. From that moment, paper carried scriptures, histories, and art across continents, connecting disparate cultures and epochs. Yet, few could have foreseen that this ancient material would experience a second renaissance nearly 2000 years later, at the dawn of the electrical age. As scientists and engineers of the late 19th century wrestled with the challenge of harnessing high-voltage electricity, they were in desperate need of a reliable medium to confine the flow of current. After experimenting with early insulators like glass and rubber, they turned their attention to paper. The porous network of interlaced cellulose fibers, originally conceived to absorb ink, proved to be an unexpectedly ideal scaffold for insulating oil. Through impregnation, the microscopic voids once filled with air were saturated with oil of high dielectric strength, transforming this venerable material into an indispensable pillar of modern high-voltage apparatus and imbuing it with newfound vitality [2].

Through decades of intensive development, conventional insulating paper has evolved from a single material into a sophisticated family encompassing everything from basic natural cellulose to complex chemically modified and fully synthetic polymers. This section will delve into the two foundational pillars of this family: cellulose-based insulating papers and synthetic aramid papers. By meticulously elucidating their manufacturing principles, performance characteristics, and the international standards that govern their application, we will establish a solid and comparative benchmark for the next-generation low-carbon, eco-friendly, and sustainable insulating paper materials explored in subsequent chapters.

Next-Generation Electrical Insulating Materials: Low-Carbon, Eco-Friendly, and Sustainable Solutions, First Edition. Edited by Jun-Wei Zha.

7.1.1 Cellulose-Based Insulation: From High-Purity Pulp to Thermally Stabilized Systems

For over a century, cellulosic insulating paper has formed the bedrock of insulation systems in everything from large power transformers to various electric motors, owing to its exceptional cost-effectiveness and reliable performance within established engineering frameworks. Its success is no accident; it is built upon a deep and evolving understanding of its material chemistry, microstructure, and service behavior under complex electro-thermal stresses. Broadly, the performance of cellulosic insulating paper is rooted in the purity of its raw materials and the microstructure created by the papermaking process. This performance is further enhanced through chemical modifications, which have expanded its application boundaries and led to a comprehensive product line ranging from standard Kraft paper to Thermally Upgraded Kraft (TUK) paper.

7.1.1.1 Electrical-Grade Kraft Pulp: The Foundation of Performance

The production of electrical insulating paper begins with a highly specialized chemical pulp known as electrical-grade Kraft pulp. Its manufacture relies on the Kraft process (or sulfate process), a chemical method designed to isolate high-purity cellulose from wood. The core chemical principle of this process involves treating wood chips with a strongly alkaline aqueous solution of sodium hydroxide (NaOH) and sodium sulfide (Na_2S), known as "white liquor," under high temperature (typically 170–176 °C) and pressure. This harsh chemical environment effectively cleaves the ether bonds linking lignin, hemicellulose, and cellulose, degrading the lignin and most of the hemicellulose into smaller, alkali-soluble fragments. This allows them to be stripped away from the cellulose backbone and removed.

The ultimate objective of this process is to achieve the highest possible content of α-cellulose, which is paramount for the dielectric integrity of the insulating paper; typically, the cellulose content is no less than 90%. The value of a high α-cellulose content is multifaceted and extends beyond a simple notion of "purity." First, it signifies the maximum removal of impurities like lignin and hemicellulose. These non-cellulosic components contain numerous polar functional groups and have a more amorphous structure, making them more susceptible to polarization and moisture absorption than crystalline cellulose. This directly leads to an increase in the dielectric loss factor ($\tan \delta$) and a decrease in insulation resistivity. Second, the production of electrical-grade pulp involves an extremely rigorous washing stage, often using deionized water, to eliminate residual metallic cations (e.g. iron, copper, calcium) that may have been introduced during pulping. These ions are mobile charge carriers under an electric field, and their presence significantly increases the material's conductivity, representing a key factor in the deterioration of dielectric properties. Finally, by removing the more hydrophilic hemicellulose, paper made from high-α-cellulose pulp exhibits lower intrinsic moisture absorption. Given that water is one of the primary agents accelerating the aging of cellulose insulation and reducing its dielectric strength, controlling moisture content is critical for ensuring the long-term reliability of the insulation system. Therefore, the pursuit of high α-cellulose content is the fundamental prerequisite for ensuring the final insulating paper possesses low dielectric loss, high insulation resistance, and excellent aging resistance.

7.1.1.2 From Pulp to Sheet: The Papermaking Process

Obtaining high-purity electrical-grade pulp is only the first step; the subsequent challenge lies in transforming these fibers, suspended in water, into a solid, structurally uniform, and reliable paper sheet. In industrial settings, this transformation is primarily accomplished using a papermaking machine. On a classic Fourdrinier machine, the process can be divided into several key stages. First, at the "wet end" of the machine, the refined pulp is further diluted to form a slurry with over 99% water content. This slurry is then evenly sprayed onto a wide, rapidly moving wire mesh screen. Under the combined forces of gravity and vacuum suction, most of the water drains through the mesh, while the cellulose fibers interlace and deposit on the screen, forming a continuous, wet, initial paper web. For electrical-grade insulating paper, the purity of the water used in this process is of utmost importance; deionized water is typically used to prevent the introduction of any conductive ionic impurities. Subsequently, this fragile wet web is transferred to the "press section." Here, it passes through a series of massive press rolls, which mechanically squeeze out a significant portion of the remaining water. Next is the "dryer section," where the paper sheet serpentines through dozens of steam-heated cylinders. Through a precisely controlled heating process, the remaining water is evaporated, allowing strong hydrogen bonds to form between the fibers, which imparts the paper with its initial mechanical strength. The final, and for insulating paper, critically important step is "calendering." The dried paper sheet passes through a set of hard, smooth, heated metal rollers. Under the effects of high temperature and pressure, the paper is further compacted, resulting in a more uniform thickness, a smoother surface, and a significant increase in density. This process yields the finished cellulosic insulating paper, as shown in Figure 7.1. This process not only improves the paper's mechanical properties but, more importantly, reduces the size and number of internal micropores, thereby enhancing the material's dielectric strength and overall insulation performance. During drying, the fibers undergo a process known as "hornification," which involves irreversible structural changes that can reduce the fiber's water retention capacity and flexibility but increase tensile strength. For thicker materials like pressboard (also known as transformer board), the wet web can be wound layer by layer on a forming cylinder until the desired thickness is achieved, after which it is pressed and dried to form a dense, rigid board.

7.1.1.3 The Paper-Oil Composite: A Symbiotic System

In mainstream liquid-immersed high-voltage equipment, such as power transformers, cellulosic insulating paper is never used alone. Instead, it is wound onto components like windings and coils to serve as the solid framework within the paper-oil composite insulation system, as illustrated in Figure 7.2. The core physical characteristic of Kraft paper is its porous network, formed by highly purified cellulose fibers

Figure 7.1 Cellulosic insulating paper.

Figure 7.2 Cellulosic insulating paper wrapped around a transformer winding (in factory).

interconnected by hydrogen bonds. This porosity provides the channels for impregnation with an insulating liquid, traditionally mineral oil. During the final stages of transformer manufacturing, a vacuum drying process is employed to thoroughly remove moisture and air from the paper. Subsequently, degassed insulating oil is introduced, allowing it to fully penetrate and fill the microscopic voids between the fibers [3].

This impregnation process is the key to achieving high dielectric strength. Its fundamental principle lies in replacing the air between the paper fibers, which has a very low dielectric strength (approximately 3 kV/mm under standard conditions), with insulating oil, which has a much higher dielectric strength (typically >20 kV/mm for AC breakdown). However, this combination also introduces new complexities. While the impregnation process is macroscopically homogeneous, at the microscopic level it creates a complex system composed of two materials with different dielectric constants (ϵ_r). According to standards from the International Electrotechnical Commission (IEC) and the Institute of Electrical and Electronics Engineers (IEEE), as well as industry practice, the relative permittivity of cellulosic paper is approximately 4.4, whereas that of commonly used mineral oil is about 2.2. When an electric field is applied to this composite system, the mismatch in permittivity causes the field to redistribute non-uniformly, with the field strength becoming disproportionately concentrated in the material with the lower permittivity – the insulating oil.

Furthermore, under sustained direct current (DC) voltage or low-frequency AC voltage, the different electrical conductivities of the paper and oil cause charge to accumulate at their interface. This phenomenon is known as Maxwell-Wagner interfacial polarization. This charge accumulation further distorts the local electric field distribution, potentially causing the actual field strength in certain areas to far exceed the design value. This concentration of the electric field along the paper-oil interface is a primary cause of "surface flashover" or "creepage discharge," a common mode of insulation failure inside transformers and a core challenge in insulation design. Therefore, while the paper-oil composite system greatly enhances the overall insulation level, its performance limits and weak points are ultimately determined by this microscopic liquid–solid interface.

7.1.1.4 Thermally Upgraded Kraft (TUK): Engineering for Durability

The long-term operating temperature of standard Kraft paper is limited, and its rate of thermal aging is a key determinant of a transformer's service life. To meet the demands for higher operating temperatures or longer design lifetimes, the industry developed Thermally Upgraded Kraft (TUK) paper. TUK is essentially standard electrical-grade Kraft paper whose thermal stability has been enhanced through chemical means [4]. This modification is typically carried out during the final stages of papermaking by adding nitrogen-containing chemical stabilizers to the pulp. The most used stabilizer is dicyandiamide (DICY), though others like melamine and polyacrylamide (PAM) are also employed.

The mechanism of these stabilizers is multifaceted and "sacrificial" in nature. The primary degradation pathways for cellulose at high temperatures are hydrolysis and pyrolysis, both of which are catalyzed by acidic substances. Since water and acids are byproducts of the aging process of both the insulating oil and the paper itself, a self-accelerating vicious cycle is established. The nitrogen-containing stabilizers added in TUK are typically weak alkaline compounds. Their primary function is to act as acid neutralizers, reacting with the acidic degradation products of aging and thereby breaking the acid-catalyzed degradation chain. More importantly, these additives are themselves susceptible to hydrolysis, and their reactivity is higher than that of the β-1,4-glycosidic bonds in the cellulose backbone. Consequently, they preferentially react with any moisture present in the system, acting as "moisture scavengers" and thus protecting the cellulose from hydrolytic chain scission. This protective action is sacrificial; experimental studies have confirmed that the concentration of stabilizers like DICY gradually depletes over time during thermal aging. Through this sacrificial chemical action, the rate of decline in the Degree of Polymerization (DP) of TUK paper is significantly slower than that of unmodified Kraft paper, effectively extending its thermal life. Research has shown that paper modified with DICY can exhibit a DP value 58% higher than that of an unstabilized blank sample after extended thermal aging, clearly demonstrating its anti-aging effect.

7.1.1.5 Standardization and Classification of Cellulosic Papers

The production and application of cellulosic insulating papers are governed by a mature system of international standards. The IEC 60641 series of standards, published by the International Electrotechnical Commission, is a foundational set of documents in this field. It provides detailed definitions for pressboard and presspaper for electrical use, along with general requirements (Part 1), methods of test (Part 2), and technical specifications for various individual materials (Part 3) [5–8]. For instance, IEC 60641-1 explicitly states that electrical-grade pulp should be made from 100% virgin plant fibers to ensure its chemical purity and its electrical and mechanical properties.

When evaluating the thermal endurance of insulating materials, the "Thermal Class" defined in the IEC 60085 standard is the common language of the industry. It classifies insulation systems into different grades based on the maximum operating temperature they can withstand for a long duration without premature aging. Standard Kraft paper impregnated with mineral oil is a typical Class A insulation material, with a maximum allowable hot-spot temperature of 105 °C.

For improved materials like TUK, a higher thermal rating (e.g. 120–140 °C) is not self-proclaimed but must be validated through rigorous system-level testing. The IEEE C57.100 standard provides the authoritative test procedure for this purpose [9]. This standard mandates accelerated thermal aging tests on complete insulation system models or actual equipment, which include the paper, oil, and conductors. By monitoring the degradation of key performance indicators, such as dielectric and mechanical strength, a thermal life curve and a Temperature Index (TI) are determined for the insulation system, allowing for the assignment of an appropriate thermal class. The existence of these two distinct but complementary sets of standards – one for the material (IEC 60641) and one for the system (IEEE C57.100) – reveals a fundamental principle of high-voltage engineering. It demonstrates that the performance of an insulation system is an emergent property that is more than the sum of its parts. Material specifications are necessary but insufficient for guaranteeing reliability. True performance can only be validated by testing the complete system, which accounts for the complex physical and chemical interactions at material interfaces. This comprehensive standards chain, from material specification to system validation, ensures the high degree of reliability and predictability of traditional cellulosic insulation in power equipment.

7.1.2 The High-Performance Benchmark: Synthetic Aramid Paper

The success of cellulose-based insulating paper is a testament to maximizing the potential of a natural polymer. However, as a material derived from nature, its inherent chemical structure also sets an insurmountable ceiling on its performance. The β-1,4-glycosidic bonds within the cellulose molecular chain are relatively fragile in hot and humid environments, which means that even with advanced chemical modifications like those in TUK paper, the long-term operating temperature can rarely exceed a barrier of 140 °C. As mid-20th-century advancements in aerospace, military technology, and high-power-density motors accelerated, they placed stringent demands on insulation materials to maintain long-term stability under extreme temperatures. Engineers gradually recognized that merely “patching up” existing natural materials was no longer sufficient to meet the needs of these cutting-edge applications. A paradigm shift from “harnessing nature” to “creating molecules” was imperative. The challenge was no longer about how to better purify or modify cellulose, but about designing, from first principles, a completely new polymer whose molecular backbone was inherently engineered for high temperatures and high electric fields. It was in this context, when application requirements for thermal resistance, mechanical toughness, and chemical stability surpassed the limits of cellulose-based materials, that meta-aramid paper, exemplified by DuPont’s Nomex® brand, emerged as a benchmark for high-performance insulation.

7.1.2.1 The Manufacturing of Aramid Paper: A New Paradigm in Material Design

The advent of aramid paper marked a fundamental shift in the design of paper-based insulation, moving from a reliance on purified natural products to a fully synthetic, molecularly engineered approach. Its chemical foundation is poly(*m*-phenylene isophthalamide) (MPIA), a polymer synthesized through the polycondensation of *m*-phenylenediamine and isophthaloyl chloride. The alternating rigid benzene rings and amide linkages in its molecular backbone endow this polymer with extraordinary intrinsic thermal stability and chemical inertness [10].

The manufacturing process of aramid paper is itself an exercise in microstructural engineering, fundamentally distinct from traditional papermaking. The core of the process begins with converting the single poly(*m*-phenylene isophthalamide) resin into two components with different forms and functions as shown in Figure 7.3. The first component consists of short-cut fibers, which are typically straight, smooth-surfaced staple fibers created by chopping spun aramid filaments. These fibers serve as the structural skeleton, providing the paper’s primary mechanical strength. The second component is a fibrid, a particle with a large surface area that acts as a hot-melt adhesive. Fibrids are tiny, film-like particles with numerous curly branches, produced by precipitating the aramid resin into a coagulation bath under specific processing conditions. Their role is to melt and bond the short-cut fibers during subsequent hot pressing.

These two components are then blended in specific proportions in water to form a slurry, which is then formed into an initial, low-strength paper web via a wet-laid

Figure 7.3 Meta-aramid short-cut fibers (left) and fibrids (right).

process. The critical step that determines the final properties is hot calendering. The web is passed through a series of rollers at high temperature (typically above the glass transition temperature of the fibrids) and high pressure. During this process, the fibrids soften and melt, acting like an adhesive to securely bond the high-strength short-cut fibers together, ultimately forming a dense, tough, and extremely low-porosity sheet, as illustrated in Figure 7.4.

This structure, held together by strong, fused polymer bonds rather than the weaker hydrogen bonds of cellulose paper, combined with the outstanding intrinsic properties of meta-aramid itself, is the fundamental source of aramid paper's exceptional mechanical properties, such as high tear strength and abrasion resistance.

7.1.2.2 The Profile of Excellence: Aramid's Superior Properties

Aramid paper's comprehensive performance profile has set an extremely high standard for high-performance insulation, with its key characteristics well-documented in numerous technical specifications and data sheets.

Its most prominent advantage is its unparalleled thermal stability. As certified by Underwriters Laboratories (UL), Nomex® aramid insulation paper is designated as a Class 220 (or Class R) material, suitable for continuous operation at 220 °C. Its useful properties are maintained for at least 10 years of continuous exposure at this temperature. This means it can operate reliably for long periods at temperatures far exceeding the limits of cellulosic materials.

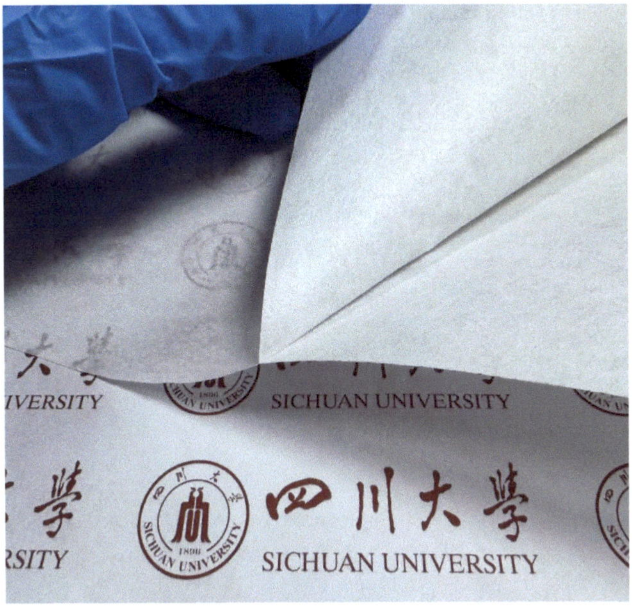

Figure 7.4 The effect of hot-pressing on meta-aramid paper.

In terms of electrical properties, aramid paper exhibits a high inherent dielectric strength. Depending on the grade and thickness, its AC breakdown strength typically ranges from 18 to 34 kV/mm, which is often more than 30% higher than that of Kraft paper. Concurrently, it possesses excellent chemical compatibility and stability, allowing it to be used with virtually all types of transformer oils (mineral, synthetic ester, silicone), varnishes, and adhesives without adverse chemical reactions.

This combination of superior properties makes it an indispensable material in critical sectors such as aerospace, military, rail transportation, and special-duty transformers that require high overload capacity. However, this high performance comes at a steep price, with a cost index far exceeding that of Kraft paper. Consequently, it is not a universal substitute for Kraft paper but rather a high-end specialty material used to solve insulation challenges in extreme operating conditions. The tremendous success of aramid paper has also established a clear target for the development of new bio-based insulating materials: in the pursuit of sustainability, they must strive to match, or even surpass, the high performance standards set by this synthetic benchmark. This "designed for persistence" philosophy, which created a molecule with exceptional in-service durability, also inadvertently created a material with extreme environmental persistence. The very chemical stability that makes it an excellent insulator also makes it non-biodegradable, a point that will be explored further in the context of its sustainability deficit.

Table 7.1 summarizes the key characteristics of the conventional insulating papers discussed in this section, providing a clear illustration of the trade-offs among performance, cost, and environmental impact.

Table 7.1 Comparative analysis of conventional electrical insulating papers.

Parameter	Kraft paper	Thermally upgraded Kraft (TUK)	Aramid paper (e.g. Nomex®)
Raw material source	Virgin wood pulp	Virgin wood pulp, nitrogenous stabilizers	Petrochemical feedstocks (e.g. xylene)
Key mfg. process	Kraft (sulfate) pulping, papermaking	Kraft pulping, chemical stabilization	Polymer synthesis, wet-laid forming, hot calendering
Thermal class (°C)	105 (Class A)	120–140	220 (Class R)
Dielectric strength (kV/mm)	12–20 (oil-impregnated)	12–20 (oil-impregnated)	18–34 (dry, varies by grade)
Dielectric loss ($\tan\delta$ @ 50/60 Hz)	~0.005 (oil-impregnated)	~0.005 (oil-impregnated)	~0.01 (dry, varies by grade)
Thermal conductivity ($Wm^{-1}K^{-1}$)	~0.1	~0.1	~0.15
Relative cost index	1	1.2–1.5	10–20
Primary environmental burden	High water/energy use, pulping chemicals	High water/energy use, pulping chemicals	Fossil fuel dependency, non-biodegradable

7.1.3 The Sustainability Deficit of Conventional Insulating Papers

Despite their immense commercial success and indispensable role in electrical equipment for over a century, conventional insulating papers are facing increasingly stringent scrutiny regarding their environmental footprint and reliance on finite resources. This sustainability deficit is not a single issue but a systemic challenge that spans the entire material lifecycle, from raw material acquisition and energy- and pollution-intensive manufacturing processes to the intractable problem of end-of-life disposal. A thorough analysis of this deficit not only reveals the inherent limitations of existing materials but also establishes a clear and urgent rationale for the next-generation sustainable solutions that will be explored in the subsequent chapters of this book.

7.1.3.1 The Life-Cycle Burden of the Kraft Process: Resource Intensity and Chemical Pollution

The core of electrical-grade Kraft paper production relies on the Kraft process, a method that is inherently resource-intensive. Its manufacturing is characterized by the massive consumption of two key resources: water and energy. Quantitative analysis shows that producing one ton of pulp typically requires over 60 cubic meters of water. In terms of energy, the pulping process, particularly the steam required for cooking wood chips and evaporating water during sheet drying, demands enormous thermal input. This makes the pulp and paper industry one of the most energy-intensive industrial sectors globally, with energy consumption per unit of product comparable to that of steel production. A detailed energy audit of a typical U.S. Kraft pulp mill indicates process steam and electricity demands of approximately 20.4 GJ and 687 kWh/ton of pulp, respectively [11].

This immense energy demand reveals a profound contradiction: although the raw material for Kraft paper (wood) is renewable, its manufacturing process is far from "zero-carbon." The global pulp and paper industry remains heavily dependent on fossil fuels to meet its energy needs. In the United States, for example, natural gas and coal together account for over 30% of the industry's fuel mix. This means that every sheet of seemingly natural, "bio-based" cellulosic insulating paper carries a significant carbon footprint from the combustion of fossil fuels. This fact forcefully challenges the simplistic notion that "bio-based" is synonymous with "low-carbon" and underscores the absolute necessity of adopting a cradle-to-gate Life Cycle Assessment (LCA) perspective when evaluating a material's sustainability.

Beyond resource consumption, the Kraft process leaves a complex legacy of chemical pollution. Although modern pulp mills widely employ Elemental Chlorine Free (ECF) bleaching processes, which use chlorine dioxide instead of elemental chlorine and thus significantly reduce the formation of highly toxic chlorinated organic compounds, this does not signify the end of pollution. The wastewater effluent from ECF processes still contains a range of strictly regulated toxic and persistent pollutants. According to regulatory bodies like the U.S. Environmental Protection Agency (EPA), these pollutants include Adsorbable Organic Halides (AOX), chloroform, and substances of high concern such as the carcinogen and endocrine disruptor 2,3,7,8-tetrachlorodibenzo-*p*-dioxin (TCDD). Furthermore, sulfides from the pulping process and sulfur dioxide and nitrogen oxides from combustion are major air pollutants.

A deeper examination reveals systemic barriers to truly clean production. The technologically superior Totally Chlorine Free (TCF) bleaching process, which uses agents like ozone and hydrogen peroxide, can eliminate the generation of chlorinated organics at the source. However, TCF has not achieved market dominance, primarily due to economics. TCF processes typically involve higher chemical costs, making their operating expenses greater than those of ECF processes. There have even been cases of pulp mills reverting from TCF back to ECF, clearly reflecting a market failure: existing regulations and market mechanisms have failed to internalize the full environmental externality costs of chlorine chemistry. This situation has led the industry to widely adopt a "good enough" rather than "best available" technology, leaving the production of traditional insulating paper burdened with a chemical pollution legacy that is avoidable but entrenched by economic factors.

7.1.3.2 The Fossil Fuel Dependency of Synthetic Materials

If the sustainability deficit of Kraft paper is manifested in the resource and environmental impacts of its production process, then the deficit of high-performance synthetic insulating papers, such as meta-aramid paper (e.g. Nomex®), is rooted in its most fundamental material source: a complete dependence on non-renewable fossil resources. The exceptional properties of aramid paper – its ultra-high thermal rating, excellent dielectric strength, and chemical stability – are derived from its unique polymer backbone: poly(*m*-phenylene isophthalamide).

The production of this high-performance polymer is inextricably linked to the petrochemical value chain. Its synthesis begins with two key monomers: *m*-phenylenediamine (MPD) and isophthaloyl chloride (IPC). These monomers do not exist in nature; they are derived through complex, multistep chemical reactions from basic chemicals, such as xylene, which are separated during the refining of crude oil. This unbroken petrochemical chain, from the oil well to the insulating paper, means that the entire lifecycle of aramid paper is tied to the carbon footprint and environmental externalities of the fossil fuel industry, including greenhouse gas emissions, environmental risks from extraction and transportation, and the depletion of finite resources.

This profound dependency reveals the core dilemma facing aramid paper. The very success of its material design – the creation of a rigid, aromatic polymer molecule with extreme chemical and thermal stability – is the root cause of its sustainability failure. The same strong aromatic rings and amide bonds that grant it superior performance necessitate enormous energy inputs during its production and lead to its extreme environmental persistence at its end of life. In other words, the "high performance" of aramid paper and its "high environmental legacy" are two sides of the same chemical coin. This design philosophy, which prioritizes ultimate in-service performance while neglecting the full lifecycle impact of the material, is characteristic of traditional materials engineering. It is also precisely this philosophy that sets it at odds with modern principles of sustainable development and provides a clear target for the innovation of next-generation bio-based materials: to break free from this reliance on fossil resources and the resulting permanent environmental liabilities, without sacrificing – and perhaps even exceeding – its performance.

7.1.3.3 The End-of-Life Dilemma

The sustainability deficit of traditional insulating paper materials manifests most starkly at the end of their lifecycle. Whether it is oil-impregnated cellulose paper or synthetic aramid paper, both face a severe disposal predicament upon decommissioning from equipment, preventing them from being integrated into a closed-loop circular economy and ultimately turning them into long-term environmental burdens.

1) The Recycling Impasse of Oil-Impregnated Cellulose
   In theory, cellulose paper is recyclable, much like common household paper products. In practice, however, recycling the oil-impregnated insulating paper removed from decommissioned transformers is nearly impossible. The core obstacle is the deep impregnation of insulating oil. In conventional repulping processes, which rely on an aqueous medium and the reformation of hydrogen bonds between fibers, the hydrophobic mineral oil severely contaminates the

pulp. It prevents the effective dispersion and rebonding of fibers, resulting in recycled paper products of inferior quality with mechanical strength and electrical properties that are far below standard, rendering them economically worthless. Furthermore, after decades of operation, the cellulose chains in the insulating paper have been severely degraded. The Degree of Polymerization (DP) drops from an initial value of over 1000 to 250 or even lower, making the fibers brittle and fragile and stripping them of their value as a quality recycling feedstock.

This technical challenge is further compounded by a legacy of chemical contamination, creating a "pollution cascade" that elevates the issue from a technical barrier to a legal mandate. Many older transformers used insulating oil that contained or was cross-contaminated with Polychlorinated Biphenyls (PCBs). PCBs are highly toxic, persistent organic pollutants that are strictly regulated by global conventions. In countries like the United States, any material containing PCBs above a specific concentration threshold (e.g. 50 ppm) is legally defined as regulated hazardous waste. Consequently, insulating paper contaminated with such oil is legally reclassified from "difficult-to-recycle industrial waste" to "hazardous waste." This transformation completely closes the door to standard recycling pathways and mandates its disposal through specific, costly hazardous waste management procedures, typically high-temperature incineration or permanent disposal in a secure chemical waste landfill. Documents from the International Council on Large Electric Systems (CIGRE) also implicitly acknowledge that during transformer refurbishment, the old insulation system is typically discarded outright. This progression of barriers – from technical and material to legal – collectively dictates that the lifecycle of oil-impregnated cellulose insulation is a one-way street from cradle to grave, in complete opposition to the principles of a circular economy.

2) The Environmental Persistence of Aramid Polymers

If the end-of-life fate of cellulose is a failure to be recycled, the fate of aramid is its "eternal" persistence. Aramid polymers were engineered to withstand harsh environments, and the result is a material that is virtually non-biodegradable in nature. Its product safety data sheets explicitly state that aramid fiber is not biodegradable. Its highly stable chemical structure resists microbial decomposition, allowing it to persist in the environment for hundreds of years.

This extreme persistence poses a severe disposal challenge. Because its chemical structure contains benzene rings, simple incineration is not an ideal solution, as it can produce toxic gases under incomplete combustion. Currently, the primary disposal method is landfilling. However, this means burying a high-value, high-embodied-energy synthetic material as permanent waste, which not only represents a tremendous waste of resources but also places a long-term burden on land resources. There are reports that some landfills have begun to express concern about or even refuse to accept aramid waste due to its non-biodegradable nature and the potential presence of chemical additives.

This situation profoundly reveals the fundamental conflict between traditional materials engineering and modern principles of sustainable design. The "designed for persistence" philosophy of aramid, which guarantees its success during its service life, becomes its fatal flaw at its end of life. This design paradigm, which focuses solely on in-service performance while ignoring full lifecycle impacts, is no longer tenable in the face of the demands of a circular economy. A circular economy requires that materials be designed from their inception with their future fate in mind – easy to disassemble, convenient to recycle, or able to return safely to the biosphere. The predicament of aramid proves that without a revolutionary change at the design source, high-performance materials will inevitably become the environmental liabilities of the future.

In summary, whether derived from biomass like Kraft paper or from petrochemicals like aramid paper, conventional insulating materials exhibit deep, systemic sustainability deficits. The production of Kraft paper is resource-intensive and accompanied by chemical pollution; aramid paper is entirely dependent on non-renewable fossil resources. At their end of life, one is rendered unrecyclable by contamination and degradation, while the other persists indefinitely due to its excessive stability. These fundamental flaws collectively form a powerful "problem statement," clearly indicating the necessity of a paradigm shift in insulating paper technology and laying a solid argumentative foundation for the low-carbon, eco-friendly, and sustainable solutions explored in the subsequent chapters of this book.

7.1.4 Emerging Performance Challenges in Modern Electrical Systems

In addition to mounting sustainability pressures, traditional insulating papers are beginning to encounter performance bottlenecks at a technical level, struggling to meet the stringent demands of modern electrical equipment that is evolving toward higher power densities, higher voltage levels, and higher operating frequencies. This performance gap, coupled with the environmental deficit, constitutes a dual driving force for technological change in insulating paper, providing clear technical targets for material innovation. These challenges are no longer issues of a single physical field but are systemic problems arising from the coupled interaction of electrical, thermal, and material physics.

7.1.4.1 Thermal Management Bottlenecks and the "Hot Spot" Effect

A fundamental physical limitation of traditional cellulosic insulating paper is its extremely low thermal conductivity. In its oil-impregnated state, its thermal conductivity is typically only about $0.1\,W\,m^{-1}\,K^{-1}$. In older, bulkier electrical equipment with lower power densities, this flaw was not a prominent issue. However, in modern power transformers, traction motors, and power electronic modules that pursue ultimate compactness and lightweight design, this low thermal conductivity has become a severe thermal management bottleneck [12]. The heat generated in the windings cannot be efficiently conducted away, leading to the formation of "hot spots" – localized areas where the temperature is significantly higher than the average temperature of the winding.

The existence of these hot spots poses a fatal threat to the reliability and lifespan of the equipment. The life of an insulating material is intimately related to its operating temperature; an increase in temperature dramatically accelerates the chemical degradation processes of cellulose (such as hydrolysis and pyrolysis), leading to a decrease in its Degree of Polymerization (DP) and a loss of mechanical strength. According to a widely accepted rule of thumb in industry standards from IEEE and IEC, for every 6–7 °C increase in temperature, the aging rate of insulating paper doubles, and its effective life is halved. Consequently, the insulation in hot spot regions will become brittle far more quickly than in other parts of the equipment. It can ultimately fail mechanically under the stress of a short-circuit fault or even normal operating vibrations, leading to an inter-turn short circuit and a catastrophic failure. It can be said that the low thermal conductivity of traditional insulating paper directly limits the potential for increasing the power density of modern electrical equipment and has become the "Achilles' heel" of its long-term operational reliability.

7.1.4.2 The Space Charge Challenge in High-Voltage Direct Current (HVDC) Applications

High-Voltage Direct Current (HVDC) transmission technology, particularly in converter transformers used to connect renewable energy sources and build transnational power grids, presents a set of challenges for insulation systems that are fundamentally different from those in traditional alternating current (AC) systems. Under an AC electric field, the field distribution is primarily determined by the dielectric constant of the materials and is relatively stable and predictable. However, under a sustained high-voltage DC electric field, the electrical conductivity of the insulating materials becomes the dominant factor determining the field distribution.

The most critical problem for oil-impregnated paper insulation systems under DC voltage is the accumulation of space charge. Due to the significant difference in electrical conductivity between insulating oil and cellulose paper, and the fact that the conductivities of both materials are highly sensitive to temperature and electric field strength, charge carriers (ions) migrate and accumulate at the oil-paper interface and within the bulk of the insulator under a DC field. This accumulation of space charge severely distorts the original electric field distribution within the insulation system, potentially causing the local field strength in certain areas to far exceed the design value. An even more dangerous situation arises during transient conditions such as voltage polarity reversal (a common occurrence in Line-Commutated Converter HVDC systems) or sudden load changes. The accumulated space charge cannot dissipate instantaneously, and the electric field it generates can superimpose on the externally applied field. This can produce a transient overvoltage in localized areas that is several times the normal operating field strength, making the insulation highly susceptible to premature breakdown. This fundamental physical phenomenon is the primary technical bottleneck limiting the application of traditional oil-paper insulation in higher-voltage-level HVDC converter transformers and is a long-standing research focus for international organizations like IEEE and CIGRE.

7.1.4.3 The Threat of Partial Discharge Under High-Frequency, High-dV/dt Conditions

With the rapid development of power electronics centered on wide-bandgap semiconductor devices like silicon carbide (SiC), a new generation of variable frequency drives, solid-state transformers, and switching power supplies is advancing toward higher switching frequencies and faster switching speeds (high dV/dt). This subjects the insulation materials within them, especially the winding insulation of high-frequency transformers and motors, to unprecedented stress from high-frequency, high-voltage pulse-width modulated (PWM) waveforms.

This rapidly changing voltage stress dramatically increases the risk of Partial Discharge (PD). PD refers to weak electrical sparks that only partially bridge the insulation between conductors, typically occurring in microscopic gas-filled voids or defects within the insulation system. Under high-frequency and high-dV/dt pulsed voltage, several factors combine to exacerbate the occurrence and destructiveness of PD. First, a high dV/dt can create a non-uniform voltage distribution and transient overvoltages across the turns of a winding, potentially reaching twice the DC bus voltage. This can exceed the local Partial Discharge Inception Voltage (PDIV) of the insulation. Second, a high switching frequency means that PD events occur with extremely high repetition rates. Each tiny discharge inflicts irreversible chemical and physical damage on the organic molecular chains of the insulating material. The continuous bombardment by high-energy electrons and ions, ultraviolet radiation, and the generation of corrosive byproducts like ozone collectively erode the insulation, causing it to age rapidly, carbonize, and ultimately lead to dielectric breakdown. The anti-PD performance of traditional insulating paper has proven inadequate in the face of this new type of electrical stress, becoming a critical weak link that constrains the reliability and lifespan of high-frequency power electronic equipment.

In conclusion, from the physical limits of thermal management and the charge-related dilemmas of DC high voltage to the risk of dielectric breakdown under high-frequency pulses, traditional insulating paper is facing a series of profound performance challenges. These challenges are not isolated; they represent a multi-physics assault on a material that was historically optimized for a single function: providing dielectric strength under 50/60 Hz AC conditions. The thermal challenge is a heat transfer problem, the HVDC challenge is an electrostatic and charge transport problem, and the high-frequency challenge is a high-frequency dielectric breakdown problem. Modern systems now demand that a single material excel in all these domains simultaneously. These challenges clearly indicate that incremental improvements to existing materials are no longer sufficient to meet the needs of a future electrified society. A material revolution, oriented toward multifunctionality and high performance, is imperative. This sets the direction for the development of next-generation insulating materials: the creation of novel solutions capable of simultaneously withstanding the multi-faceted extreme conditions of high temperature, high field strength, and high frequency.

7.2 New-Generation Raw Materials: Advanced Bio-Based and Nanostructured Materials

As the core chapter of the "Solutions" section, this part will systematically and in-depth analyze the new material systems that form the foundation of next-generation insulating paper. By integrating multidisciplinary perspectives including materials science, chemical engineering, and dielectric physics, it will not only present the composition and structural characteristics of the materials, but also reveal the scientific principles behind their performance, providing a comprehensive cognitive framework from microscopic mechanisms to macroscopic applications.

7.2.1 Nanocellulose

Nanocellulose refers to a nanostructured material extracted or prepared from natural cellulose (sourced from plants, animals, and bacteria) via simple, scalable, and efficient separation technologies [13]. With a size range of 10–70 nm, it is composed of microfibrils with a diameter of 0.1–1.0 μm, and a single cellulose fiber consists of multiple bundles of such microfibrils [14]. It possesses intrinsic high insulation, large specific surface area, excellent mechanical strength, tunable dielectric properties, as well as good biocompatibility and degradability, thus demonstrating significant advantages in the field of insulating materials such as cable insulation layers, electronic packaging, and flexible insulating films. Its unique one-dimensional nanostructure and strong intermolecular forces enable it to meet diverse insulation requirements through compounding and modification, providing a green and sustainable solution for insulating materials and showing great potential in the field of green high-performance insulation.

Based on its main sources and properties, nanocellulose is generally classified into three categories: Cellulose Nanofiber (CNF), Cellulose Nanocrystal (CNC), and Bacterial Nanocellulose (BC) [15].

Cellulose Nanofibers (CNF) consist of crystalline regions and amorphous regions, and are typically prepared by a combination of chemical and mechanical methods. Raw materials such as wood, bamboo, rice straw, and potato tubers can all be used for its production. CNF has a large aspect ratio and a 3D network structure of irregularly entangled fibers, thus exhibiting good flexibility, along with the advantages of low cost and environmental friendliness, and has attracted much attention in the development of high-performance energy storage systems. Generally, CNF is prepared by subjecting pulp to high-intensity mechanical treatment (such as high-pressure homogenization or microfluidization), which is usually supplemented with enzymatic or chemical pretreatment to reduce energy consumption. This process decomposes micron-sized cellulose fibers into flexible filaments with a diameter ranging from several nanometers to tens of nanometers and a length up to several micrometers. These filaments with a high aspect ratio can form a highly entangled network structure during the papermaking process, thereby producing dense, low-porosity paper whose mechanical strength is far superior to that of traditional paper. In actual production, researchers also conduct surface modification on CNF to further optimize its performance. For example, treating the surface of CNF with a silane coupling agent can enhance its compatibility with other polymer matrices, which is particularly important in the preparation of composite insulating materials.

Cellulose Nanocrystals (CNC) present as rod-like or needle-like nanoparticles with highly crystalline properties, and have a wide range of extraction sources, which can be extracted from raw materials such as wood pulp, algal cellulose, microcrystalline cellulose, and cotton linter [16]. CNC exhibit a rod-like morphology, with a width ranging from 3 to 50 nm and a length varying from several tens of nanometers to several hundreds of nanometers; these characteristics differ depending on the source of cellulose [17]. The main properties of CNC include high crystallinity, excellent mechanical properties, thermal stability, surface chemical activity, as well as biocompatibility and degradability, and these properties endow it with the ability to significantly enhance the mechanical properties of composite materials. Compared with CNF, CNC show a lower aspect ratio and higher rigidity. CNC possess a large specific surface area, high crystallinity, and a chiral nematic structure. CNC are prepared by hydrolyzing natural cellulose raw materials with strong acids (such as sulfuric acid). The acid selectively digests the amorphous regions in cellulose, retaining the highly crystalline regions, and finally forms rigid rod-like nanoparticle CNC. CNC have extremely high crystallinity and stiffness, and can spontaneously form a chiral nematic liquid crystal structure at a specific concentration, which provides the possibility for the preparation of highly ordered functional films.

Bacterial Cellulose (BC) is a type of cellulose synthesized by specific microorganisms, such as those belonging to the genera Rhizobium, Acetobacter, Agrobacterium, and Sarcina. Bacterial cellulose is free of hemicellulose, lignin, and pectin; its ultra-fine network structure endows it with high crystallinity and degree of polymerization, good biocompatibility and degradability, as well as better elastic modulus and tensile strength than plant fibers. BC is a high-molecular polysaccharide secreted by bacteria, composed of glucose residues linked by β-1,4-glycosidic bonds, with a diameter generally ranging from 20 to 500 nm and a length between several hundred nanometers and several micrometers. It has high porosity, high purity, and high crystallinity; compared with plant cellulose, bacterial cellulose contains no lignin or hemicellulose, thus being easier to process. BC is a nano-scale cellulose produced through the metabolism of certain bacteria (e.g. *Acetobacter xylinum*) in liquid culture media. Unlike plant-sourced nanocellulose, the preparation process of BC does not require mechanical or chemical treatment – bacteria can directly synthesize a cellulose network with extremely high purity (free of impurities such as lignin and hemicellulose). It possesses extremely high hydrophilicity and air permeability, along with a unique three-dimensional nano-porous structure, which gives BC potential advantages in insulation scenarios that require high flexibility and good moisture management (Figure 7.5).

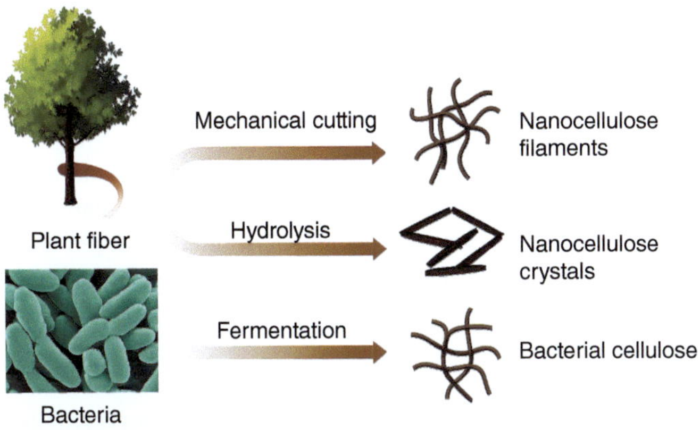

Figure 7.5 Flow chart for nanocellulose preparation.

At present, the preparation of nanocellulose mainly relies on various chemical treatment methods, including sulfuric acid hydrolysis, oxidation treatment, ionic liquids (ILs), and deep eutectic solvents (DESs). Table 7.2 summarizes the comparison of common chemical preparation methods for nanocellulose [18].

Table 7.2 Comparison of common preparation methods for nanocellulose.

Method	Reagent	Raw material	Advantages	Disadvantages
Acid hydrolysis	Sulfuric acid	Lemon seeds	Mature technology, small product size	Easy to corrode equipment, high cost
	Oxalic acid	Pineapple residue	Lower corrosivity	Low reaction efficiency
TEMPO oxidation	TEMPO oxidant $[Bmim][HSO_4]$	Water hyacinth Corn cob	Uniform product size, large aspect ratio	High cost of oxidant, long reaction time
Ionic liquid	$[Emim][HSO_4]$/ $[Bmim][HSO_4]$/ $[Hmmim][HSO_4]$ [Emim][OAc]/ [Bmim][OAc]/ [Bmim][Cl]	Micro-crystalline cellulose	Low toxicity, good solubility for cellulose	Easy to destroy fiber structure
Deep eutectic solvent	Choline chloride/ citric acid Choline chloride/ oxalic acid dihydrate	Kraft pulp Barley straw	Environ-mentally friendly	Mechanical assistance required

7.2.1.1 Excellent Properties of Nanocellulose

Nanocellulose is regarded as a superior dielectric component because of its unique physicochemical properties exhibited at the nanoscale.

1) Mechanical Properties
   The strength and rigidity of nanocellulose far exceed those of ordinary natural fibers. This is mainly due to its unique structural features: the fiber diameter is extremely small, there are few internal defects, the crystallinity is high, and the multi-dimensional hydrogen bond network structure strengthens the connection between molecular chains. The typical mechanical property parameters are shown in Table 7.3:
2) Dielectric Properties
   Paper composed of nanocellulose (commonly referred to as nanopaper) exhibits extremely high density and crystallinity. This structure significantly reduces the free volume and defects inside the material, thereby inhibiting the migration of charge carriers and the avalanche-like development of space charge, while also reducing leakage current and dielectric loss. As a result, nanopaper demonstrates extremely high dielectric breakdown strength (which can be several times that of traditional paper) and extremely low dielectric loss, making it an ideal candidate material for high-performance capacitors and high-field insulation.
   In high-frequency electric field environments, the dielectric properties of nanopaper remain stable. This is because nanocellulose has a regular molecular structure and a relatively low content of polar groups; under the action of a high-frequency electric field, its polarization response is rapid with no obvious hysteresis, and the dielectric constant and dielectric loss vary slightly with frequency. This endows it with broad application prospects in insulating components of high-frequency electrical equipment, such as radars and communication devices.
3) Thermal Stability
   Compared with traditional cellulose containing a large number of amorphous regions, nanocellulose (especially cellulose nanocrystals, CNC) has a regular crystal structure, with closely arranged molecular chains and strong intermolecular forces, resulting in a higher thermal decomposition temperature. The thermal decomposition temperature of natural cellulose is generally around 250 °C, while that of nanocellulose can be increased to above 300 °C. Chemical modification methods such as TEMPO oxidation and phosphorylation can further enhance the thermal stability of nanocellulose.

Table 7.3 Mechanical parameters of nanocellulose.

Performance index	Nanocellulose	High-strength steel	Polypropylene (PP)
Tensile strength (MPa)	200–300	400–600	30–40
Young's modulus (GPa)	80–100	200	1–2
Density (g/cm^3)	1.6	7.8	0.9

For example, TEMPO oxidation introduces carboxyl groups on the surface of nanocellulose, enhancing intermolecular interactions and increasing the thermal decomposition temperature; phosphorylation introduces phosphate ester groups onto the molecular chains, forming thermally stable cross-linked structures and delaying the thermal decomposition process. These modification methods enable nanocellulose to adapt to some thermoplastic processing technologies, such as injection molding and extrusion, broadening the application scope of bio-based insulating materials and allowing them to be used in the preparation of insulating products that require high-temperature processing.

4) Chemical Stability and Surface Modifiability
Nanocellulose has a specific surface area as high as 100–200 m^2/g, and its surface is rich in functional groups such as hydroxyl and carboxyl groups. A large number of hydroxyl groups make it have good hydrophilicity and can be well compatible with water and other polar solvents; at the same time, these functional groups also provide abundant reactive sites for chemical modification.
By grafting different functional groups such as alkyl groups, silane, or polymer chains, the surface energy, hydrophilicity and hydrophobicity, and interfacial compatibility with other materials of nanocellulose can be regulated. For example, grafting long-chain alkyl groups can reduce the surface energy of nanocellulose and improve its hydrophobicity, making it suitable for insulation in humid environments; grafting silane coupling agents can enhance the interfacial bonding force between nanocellulose and polymer matrices, improving the mechanical and dielectric properties of composite materials; grafting conductive polymers can endow nanocellulose with a certain degree of conductivity, expanding its application in conductive insulating composite materials. In addition, nanocellulose can also interact with metal ions, inorganic nanoparticles, etc., to form functionalized composite materials, further enriching its properties and application fields.

7.2.2 Biobased Synthetic Polymers and Hybrid Materials

In addition to nanonizing natural cellulose, developing new biobased polymers and functional hybrid materials is also an important research direction.

7.2.2.1 Biobased Alternatives to Aramid

As a high-performance insulating material, aramid paper is widely used in electrical, aerospace and other fields. However, the production of traditional aramid relies on fossil fuels and is difficult to biodegrade, which does not meet the requirements of green and sustainable development. To solve this problem, the research community is actively developing high-performance biobased polyamides (Bio-PAs) and other polymers derived from renewable resources (such as castor oil, biobased monomers).

For example, a polyamide polymerized from sebacic acid extracted from castor oil and biobased hexamethylenediamine can maintain high thermal stability comparable to some aramids while having a certain degree of biodegradability. It can be gradually decomposed by microorganisms in the natural environment, reducing environmental pollution. Through Life Cycle Assessment (LCA) analysis, its carbon footprint is reduced by about 40% compared with traditional aramid, greatly reducing greenhouse gas emissions during the production process and providing a more sustainable solution for high-temperature insulation. In practical applications, this biobased polyamide can be used to prepare motor insulation paper, transformer insulation films, etc., and can maintain stable insulation performance under high-temperature working conditions, meeting the long-term operation requirements of equipment.

In addition, research on biobased polymers using soybean protein as a raw material is also advancing. Soybean protein has a wide range of sources and low prices, but its own mechanical properties and water resistance are poor, which limits its application in the field of insulation. Through chemical cross-linking and other means, such as using cross-linking agents such as formaldehyde and glutaraldehyde to react with amino groups, hydroxyl groups, etc. in soybean protein molecules to form a cross-linked structure, its mechanical strength, water resistance and thermal stability can be significantly improved. The modified soybean protein-based polymer shows application potential in the field of insulating materials and can be used to prepare insulating gaskets, insulating sleeves, etc. for low-voltage electrical appliances, replacing traditional petroleum-based insulating materials to a certain extent (Figure 7.6).

1) For Flame Retardancy
   Flame retardancy is a key property ensuring the safe operation of electrical equipment. Introducing nanoscale flame retardants – such as montmorillonite, magnesium hydroxide nanoparticles, and nanoscale intumescent flame retardants – into cellulose paper can significantly enhance its flame retardancy. The mechanism of action of these nano-flame retardants is as follows: Montmorillonite forms a layered barrier structure at high temperatures to block the transmission of heat and oxygen; magnesium hydroxide nanoparticles decompose upon heating to release water vapor, which dilutes combustible gases and lowers the temperature; intumescent flame retardants form an expanded char layer at high temperatures to isolate the substrate from the fire source.
2) For Aging Resistance
   Insulating materials in electrical equipment are exposed to complex environments for a long time, making them prone to aging, which affects the service

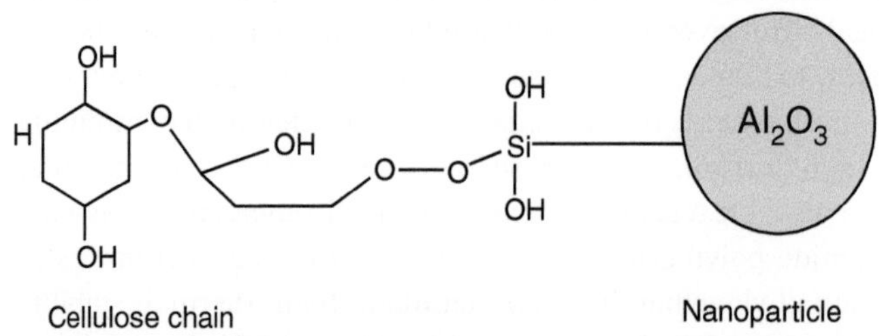

Figure 7.6 Interfacial connection structure of Al_2O_3 between cellulose chains.

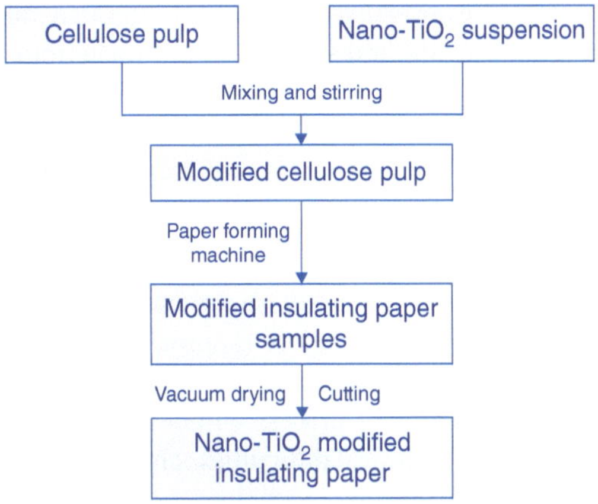

Figure 7.7 Preparation flow chart of cellulose insulation pressboard modified by nano-TiO_2.

life and safety of the equipment. By adding nanofillers with anti-oxidation and anti-ultraviolet functions – such as nano-zinc oxide (ZnO) and nano-titanium dioxide (TiO_2) – to nanocomposite paper, its aging resistance can be significantly improved. Nano-ZnO and TiO_2 have excellent ultraviolet shielding capabilities, which can absorb and scatter ultraviolet rays, reducing the damage of ultraviolet rays to the cellulose matrix. At the same time, they also have a certain catalytic activity, which can decompose free radicals generated during the aging process and delay the aging rate of the material.

For example, after adding nano-TiO_2 particles, the number of hydrogen bonds in the cellulose molecular model increases, the relative dielectric constant of the modified cellulose insulating paper decreases, and the electrical strength increases; there is a threshold for the improvement of cellulose modification effect by nano-TiO_2 content, and the modification effect on cellulose samples is optimal when the mass fraction of nano-TiO_2 is 5% (Figure 7.7).

7.2.3 Diversification of Resources

To achieve true sustainable development in the insulating material industry, the raw material sources for the next generation of insulating paper must break free from reliance on virgin forests and shift toward a more diversified and circular biomass resource system. This diversification of resources not only reduces the industry's pressure on the ecological environment but also endows insulating paper with more diverse functions through the differentiated characteristics of raw materials.

7.2.3.1 Agricultural Wastes as Raw Materials

Converting by-products generated during agricultural production – such as sugarcane bagasse, wheat straw, rice husks, corn cobs, cotton stalks, peanut shells, and rapeseed straw – into raw materials for insulating materials is a direct and efficient way to practice the concept of circular economy. The annual output of such non-food biomass resources is enormous; in China alone, the total amount of agricultural wastes exceeds 3 billion tons each year, which contains extremely rich cellulose resources. Converting these wastes into high-value-added insulating materials not only realizes the transformation of "waste into treasure" and builds a new green value chain but also effectively reduces the felling of forest resources and eases the pressure of ecological protection.

Due to differences in their own structures and compositions, different agricultural wastes exhibit unique advantages in the preparation of insulating materials:

1) Corn Cobs
   Corn cobs have a cellulose content as high as 45–55%, with a well-balanced proportion of hemicellulose (approximately 35%) and lignin (approximately 15%). This reasonable component ratio enables efficient separation of cellulose during the pretreatment process without complex technologies, reducing preparation costs and technical difficulties. Thus, corn cobs are an ideal raw material for producing low-cost nanocellulose.
   After mild alkali treatment (5% sodium hydroxide solution, 60 °C, 1 hour), hemicellulose and lignin in corn cobs can be effectively removed, with a cellulose extraction rate of over 80%. The prepared nanofibrils exhibit good dispersibility and mechanical properties; the tensile strength of paper formed by these fibrils can reach 80 MPa, which meets the mechanical requirements for insulation in medium and low-voltage electrical equipment. In addition, corn cobs are widely available and low-cost; large-scale use can further reduce the production cost of insulating paper and enhance the market competitiveness of the product.
2) Cotton Stalks
   Cotton stalks have tough fiber texture, with a cellulose content of approximately 40–45% and relatively long fiber length (2–3 mm on average). After cooking-bleaching pretreatment, microfibrils with excellent mechanical strength can be prepared. The tensile strength of insulating paper produced from these microfibrils is more than 30% higher than that of traditional wood pulp paper.
3) Sugarcane Bagasse
   As the main waste of the sugar industry, sugarcane bagasse has a cellulose content of approximately 40–50%, with concentrated sources and easy collection. After combined enzymolysis-mechanical treatment, sugarcane bagasse microfibrils can form a uniform composite system with other nanofillers; the dielectric constant stability of the prepared insulating paper is superior to that of wood pulp-based materials.
4) Wheat Straw
   Wheat straw has a cellulose content of approximately 35–40%, and its fiber structure contains a certain amount of siliceous components. Although this characteristic increases the difficulty of pretreatment, it also makes the insulating paper prepared from wheat straw perform better in arc resistance. After appropriate modification, the arc resistance index can reach more than 120 seconds.
   Of course, the use of agricultural wastes to prepare insulating materials also faces many technical challenges:
   Complexity of Composition: For example, the silica content in rice husks is as high as 15–20%. These siliceous components will reduce the purity of cellulose

and affect the subsequent nanonization treatment and insulating performance. To address this issue, in addition to the traditional alkali leaching method (10% sodium hydroxide solution, 80 °C, 2 hours, with a silica removal rate of over 85%), the emerging microwave-assisted alkali treatment technology can shorten the treatment time to 30 minutes and increase the silica removal rate to 90%. The biological pretreatment method, on the other hand, screens microorganisms with high silica enzyme production (such as Streptomyces strains) and cultivates them at 30 °C for 5 days, enabling the biodegradation of silica and avoiding pollution caused by chemical treatment.
Variations in Raw Material Uniformity: Agricultural wastes from different batches and regions exhibit fluctuations in composition and structure, which can lead to unstable product performance. By establishing a standardized raw material pretreatment process – such as near-infrared spectroscopy online detection technology – it is possible to analyze the raw material composition in real time and adjust pretreatment parameters, thereby controlling the deviation of cellulose extraction rate among different batches of raw materials within 5%.
High Energy Consumption in Pretreatment: The traditional high-temperature and high-pressure pretreatment process consumes a large amount of energy. The new ultrasonic-assisted pretreatment technology can loosen the fiber structure at room temperature, reducing energy consumption by 40% while increasing the cellulose retention rate to over 90%.

7.2.3.2 New Biomass Sources

Researchers are actively exploring more cutting-edge biomass sources, which not only further expand the raw material channels but also endow insulating materials with performance that traditional raw materials can hardly match through unique biosynthetic mechanisms.

1) Microalgal Cellulose
   As photosynthetic autotrophic organisms, microalgae have significant advantages such as short growth cycles (usually a growth cycle of 3–7 days), no occupation of arable land, and cultivation using seawater or industrial wastewater. The cellulose contained in their cell structure is mostly type I crystalline structure with high crystallinity (65–75%), which makes their potential dielectric properties more excellent. Currently, Chlorella and Spirulina, which have been extensively studied, can achieve a cellulose yield of 1.5–2.0 g/L of culture medium under optimized culture conditions (temperature: 25–30 °C, light intensity: 3000–5000 lux, CO_2 concentration: 5%). More importantly, microalgae can absorb large amounts of pollutants such as carbon dioxide, nitrogen, and phosphorus during growth, so the cultivation process also has environmental protection functions. For example, cultivating Chlorella using secondary effluent from municipal sewage treatment plants can not only reduce the content of nitrogen and phosphorus in water (removal rate over 80%) but also simultaneously produce high-purity cellulose, realizing the synergy of "wastewater treatment-resource production." However, large-scale extraction of microalgal

cellulose still faces challenges; currently, the freeze–thaw-enzymolysis combined method is mostly used in laboratories, and more efficient separation technologies need to be developed in the future to reduce costs.

2) Bacterial Cellulose
Bacterial cellulose (BC) produced by the metabolism of bacteria such as *Acetobacter xylinum* and *Acetobacter aceti* has distinct structural advantages compared with plant cellulose. Its fiber diameter is only 20–100 nm, and it has a three-dimensional network interwoven structure with a purity of over 99% (free of impurities such as lignin and hemicellulose), which makes the paper formed by it have extremely high density and uniformity, with a dielectric breakdown strength of over 0.8 MV/cm. During the cultivation process, the morphology of bacterial cellulose can be controlled by regulating the composition of the culture medium (e.g. adding citric acid can promote fiber refinement, and adding ethanol can increase fiber yield) and the cultivation method (static cultivation forms a membrane structure, and dynamic cultivation forms a spherical structure). At present, the production cost of bacterial cellulose is relatively high; by developing low-cost culture media (e.g. using food industry waste liquid) and continuous cultivation devices, it is expected to reduce its cost by more than 50%, laying a foundation for industrial application.

3) Fungal Cellulose
Some fungi (e.g. Trichoderma and Penicillium) secrete cellulose during growth. The fiber length can reach 5–10 μm, with a moderate aspect ratio and good biocompatibility. The uniqueness of fungal cellulose lies in that it can be modified through genetic engineering methods; for example, by introducing specific genes, the cellulose surface can have functional groups (such as hydroxyl and carboxyl groups) inherently, enabling good compounding with other nanofillers without subsequent chemical treatment. At present, the research on fungal cellulose is still in its initial stage, but its highly designable feature gives it great potential in the field of functional insulating materials.
The development of these new biomass sources not only enriches the raw material system of insulating materials but also, more importantly, provides a new idea for the performance optimization of insulating paper through the precise regulation of biosynthesis, promoting the development of insulating materials towards the direction of "green synthesis and functional customization."

7.2.4 Summary

The design of a new generation of insulating materials is gradually moving away from the traditional "experience accumulation" approach and shifting toward a "bottom-up" precise construction strategy. Its core lies in hierarchical regulation: First, in response to insulation requirements, engineering design of basic nanocomponents is carried out at the molecular scale – for example, achieving controllable performance by adjusting the crystallinity of cellulose nanocrystals (CNC) or optimizing the aspect ratio of cellulose nanofibrils (CNF). Subsequently, using technologies such as self-assembly and directional alignment, these nanocomponents are constructed into ordered structures at the mesoscopic scale, such as chiral nematic films formed by CNC and three-dimensional entangled networks composed of CNF. Finally, by introducing functional nano-components (e.g. boron nitride nanosheets (BNNS) for building thermal conductive networks, zinc oxide (ZnO) particles with anti-aging capabilities), a three-level hierarchical system of "matrix–structure–function" is formed inside the material. This hierarchical construction strategy not only ensures structural controllability but also breaks the limitations of traditional material performance, enabling the coordinated coexistence of "conflicting properties" such as electrical insulation and thermal management.

This paradigm shift directly addresses the comprehensive requirements of modern electrical equipment under multi-physical fields including electrical, thermal, mechanical, and environmental conditions. Taking large power transformers as an example, new nanocomposite insulating paper integrates partial discharge resistance and anti-aging functions into its material system, extending the equipment's major maintenance cycle from the traditional approximately 15–25 years, which significantly reduces operation and maintenance costs and shutdown risks. It can be seen that hierarchical design not only achieves an overall improvement in performance but also provides new material support for the safety and reliability of electrical equipment.

Notably, this "bottom-up" design paradigm benefits from the support of diversified biomass raw materials. From lignocellulosic waste to aquatic plants, various renewable resources provide abundant basic units for constructing nano-insulating materials. On this basis, the concept of "hierarchical nanostructure construction" maximizes the potential value of these raw materials, promoting insulating materials from "single-function satisfaction" to "multi-performance coordinated optimization." In the future, with the further development of green manufacturing and function integration technologies, such materials are expected to be more widely applied in fields such as new energy power generation, high-voltage power transmission, and power electronics, becoming an important basic material supporting the upgrading and iteration of electrical equipment.

7.3 Green Manufacturing and Circular Economy Framework: From Material Innovation to Systemic Sustainability

Part Two provided an in-depth analysis of the novel material building blocks that form the foundation of next-generation insulating papers, particularly nanocellulose and functional hybrid materials. However, the revolutionary potential of these advanced materials can only be fully realized when their entire life cycle is systematically engineered through the principles of green chemistry and the circular economy. This section expands the perspective from the materials themselves to their entire process of manufacturing, use, and end-of-life, aiming to answer a core question:

How can we ensure that the birth and demise of these high-performance materials are equally clean and efficient?

The central argument of this section is that the transition from traditional insulating paper to next-generation bio-based solutions represents a dual paradigm shift. The first shift occurs at the microstructural level of the material – evolving from macroscopic fiber networks to precisely engineered nanoscale architectures, as detailed in Part Two. The second, which this section will elaborate on, is a fundamental revolution in the philosophy of production and life cycle management. This new philosophy is built on three pillars: first, adopting low-carbon, low-environmental-impact manufacturing pathways to minimize the ecological footprint at the source; second, utilizing Life Cycle Assessment (LCA) as a rigorous quantitative tool for the scientific and transparent measurement and optimization of environmental benefits; and finally, pre-planning the material's regenerative pathway after its service life through the principles of "Design for Circularity," ensuring its value continues to flow within the ecosystem or industrial system. This section will systematically argue that true sustainable innovation is not an isolated technological breakthrough, but a comprehensive system engineering that integrates material science, process engineering, environmental science, and economic models.

7.3.1 Low-Carbon and Low-Impact Manufacturing Pathways

The first step toward achieving sustainability is to subvert the traditional high-energy, high-pollution models at the front end of production and manufacturing. This means fundamentally rethinking how cellulose is extracted from biomass raw materials and how these fibers are assembled into the final paper product [19].

7.3.1.1 A Paradigm Shift in Pulping: From Harsh Chemicals to Biocatalysis and Green Solvents

While the traditional Kraft process is highly efficient in producing high-purity electrical-grade pulp, its environmental cost is substantial: it is a resource-intensive process that consumes large amounts of water and energy, relies on harsh chemicals like sodium hydroxide (NaOH) and sodium sulfide (Na_2S), and is accompanied by malodorous sulfur-containing gas emissions. To overcome these drawbacks, a series of milder, more selective "green" pulping and fibrillation technologies have emerged [20].

1) Enzymatic Refining: Gentle Biocatalysis
   Enzymatic refining represents a highly promising biotechnological path that uses highly specific enzymes (such as cellulases and hemicellulases) to replace or assist the traditional high-energy mechanical beating process. Its core mechanism is that these enzymes can selectively act on the fiber surface, slightly "modifying" the fiber structure through hydrolysis to promote the separation of microfibrils and fiber swelling.
   This biocatalytic action can be carried out under mild pH and temperature conditions. Compared to mechanical beating, which relies purely on physical shear

forces, its advantages are obvious, including significant energy savings, reduced water consumption, and improved product performance. Studies have shown that using enzymatic treatment as a pretreatment step for mechanical beating can reduce subsequent energy consumption by up to 20%. At the same time, by improving the drainage properties of the fibers, it lowers the water circulation load in the entire papermaking process and can enhance the mechanical strength and uniformity of the final paper.

Although enzymatic refining is often insufficient to completely replace mechanical treatment, as a powerful auxiliary technology, it greatly reduces the environmental footprint of the critical step from pulp to paper.

2) Deep Eutectic Solvents (DES): Designable Green Solvents

   Deep Eutectic Solvents (DES) are a recent breakthrough in the field of green chemistry, considered an ideal alternative to traditional harsh chemicals and high-cost ionic liquids (ILs). DES are typically formed by mixing two or more inexpensive, safe, and biodegradable components (one as a hydrogen bond acceptor, HBA, such as choline chloride; the other as a hydrogen bond donor, HBD, such as organic acids like oxalic acid or lactic acid) at a specific molar ratio, resulting in a melting point far below that of the individual components.

   In pulping applications, the core mechanism of DES lies in their powerful hydrogen bond network, which can efficiently and selectively break the β-*O*-4 aryl ether bonds – the most dominant linkage in the lignin macromolecule – thereby dissolving the lignin with minimal impact on the cellulose structure. Acidic DES, in particular, are especially effective in depolymerizing lignin and disrupting the fibrous structure by providing a large number of activated protons (H^+), which promotes fiber swelling and increases reaction efficiency.

   Compared to traditional methods, DES exhibit multiple advantages, including environmental friendliness due to their typically low-toxicity, biodegradable components and simple, low-energy preparation process; high selectivity and efficiency in separating lignocellulose to obtain high-purity components; and a simpler recovery process compared to ionic liquids, further enhancing their economic viability and sustainability.

3) Organosolv: A Mature Sulfur-Free Pathway

   The Organosolv process is a more mature, sulfur-free pulping technology that uses a mixture of water and an organic solvent (such as ethanol or formic acid) at temperatures of 140–220 °C to extract lignin and hemicellulose. The core advantage of this process is that it fundamentally eliminates the malodorous sulfur-containing gases produced by the Kraft process and allows for the recovery of high-purity, sulfur-free lignin. This lignin, which has a structure close to that of native lignin, is a valuable raw material for producing advanced materials (such as carbon fiber and biobased resins), rather than just being used as fuel. Furthermore, the organic solvents used in the process can be efficiently recovered through methods like distillation, creating a closed-loop system and reducing water pollution.

   These emerging pulping technologies collectively point to a core paradigm shift: the evolution from a "pulp mill" to a "biorefinery" [21]. The economic model of the traditional Kraft process is built around a single high-value product – cellulose pulp – with other components like lignin and hemicellulose mostly treated as low-value byproducts, primarily burned for energy recovery.

In contrast, technologies like Organosolv and DES are essentially fractionation techniques. Their goal is not simply to remove lignin, but to finely separate the biomass raw material into three pure, high-value streams: cellulose, hemicellulose, and lignin. This model transforms the factory from a single-product producer into a "biorefinery" with a multiproduct portfolio, fundamentally enhancing the value-added utilization efficiency of renewable resources. Therefore, the driving force for adopting these green chemistry pathways stems not only from the pressure of environmental compliance but also from the immense economic potential and new business models they entail.

7.3.1.2 A Revolution in Resource Efficiency: Decoupling Papermaking from the Water-Energy Nexus

Another inherent "original sin" of the traditional wet-laid papermaking process is its immense dependence on water and energy. Fibers are dispersed in water to form a very low-concentration slurry, and then, after forming and pressing, a massive amount of thermal energy is consumed to evaporate the remaining water from the paper sheet [22]. Breaking this Water-Energy Nexus is key to achieving sustainable manufacturing.

1) Foam-Forming Technology: A Revolution of Replacing Water with Air
   Foam-forming technology is one of the most disruptive innovations in this field. Its working principle is to replace water with a stable aqueous foam (composed of a surfactant, water, and a large volume of air microbubbles) as the medium for transporting and dispersing fibers. The unique rheological properties of foam – high viscosity and a three-dimensional network structure – can effectively inhibit fiber movement and aggregation (i.e., flocculation), achieving an extremely uniform fiber distribution even at much higher fiber concentrations than in traditional water-based processes. This technology can reduce process water consumption by more than 90%. More importantly, due to the lower initial water content, the paper sheet has a higher dryness after pressing, and experimental data indicate that this can directly translate into drying energy savings of up to 20%. Additionally, foam-forming can create bulkier (high-caliper), more porous structures, which means that less fiber raw material is needed to achieve the same product thickness and stiffness, with a potential for product lightweighting of 15–25%. It is worth noting that foam-forming technology has moved beyond the laboratory stage, attracting investment from an international industrial consortium, including equipment giant Valmet, and is undergoing pilot-scale and commercialization trials, demonstrating its mature technological potential and broad application prospects (Figure 7.8).

   The importance of this technology extends far beyond improvements in process efficiency; it is a key enabling technology. Many of the high-performance components discussed in Part Two, such as long-fiber bio-based polymers or

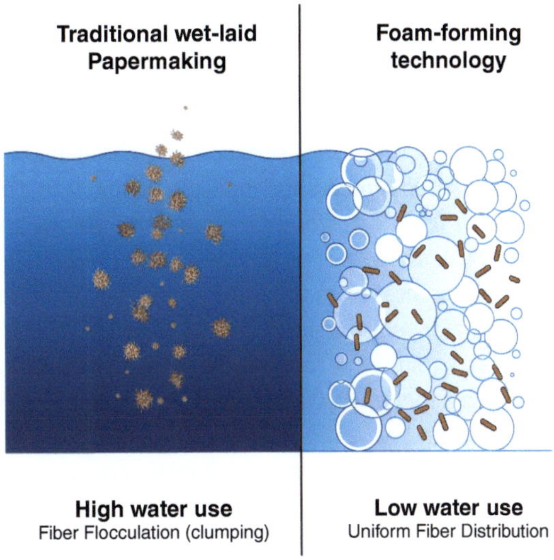

Figure 7.8 Comparison of traditional wet-laid forming and foam-forming technology.

cellulose nanofibers (CNF), face significant challenges in traditional water-laid processes: long fibers are highly prone to flocculation in water, while CNF, with its extremely high specific surface area and hydrophilicity, causes severe dewatering difficulties and a surge in energy consumption. Foam-forming technology precisely addresses these bottlenecks. Its unique medium properties can perfectly disperse long fibers and allow CNF to be added as a reinforcing agent without catastrophic effects on dewatering performance. Therefore, innovation in new materials and new processes are mutually reinforcing, and foam-forming paves the way for these advanced materials to move from the laboratory to large-scale production.

2) Solvent-Free Assembly: The Ultimate Vision for the Future
Looking ahead, solvent-free assembly techniques such as direct casting, dry-forming, and even additive manufacturing (3D printing) represent the ultimate goal for resource efficiency. These technologies aim to completely eliminate water or any liquid medium, directly assembling dry fibers or nano-building blocks into three-dimensional insulating components, thereby minimizing the manufacturing process's dependence on water and energy. Although these technologies are currently in the early stages of research, they paint an exciting picture for the future of customized, distributed, and ultra-efficient manufacturing.

7.3.2 Life Cycle Assessment (LCA) as a Quantitative Design Tool

In an era where sustainability is increasingly a core competitive advantage, vague labels like "bio-based" or "green" are far from sufficient. The industry needs a rigorous, transparent, and comparable method to scientifically evaluate and substantiate the environmental benefits of its products. Life Cycle Assessment (LCA) is the standardized tool created for this very purpose.

7.3.2.1 Beyond the "Bio-Based" Label: A Systematic Method for Environmental Footprint Accounting

LCA is a systematic methodology, following international standards (such as ISO 14040 and ISO 14044) [23, 24], used to quantify the various potential environmental impacts of a product or service throughout its entire life cycle – from raw material acquisition, processing, manufacturing, transportation, and use, to final disposal (i.e., from "cradle-to-grave" or "cradle-to-cradle"). For the assessment of insulating paper, LCA analysis typically adopts a "cradle-to-gate" system boundary, covering the entire process from the extraction of raw materials in forests or agricultural fields to the final insulating paper product leaving the factory gate. The core of the assessment is to quantify a range of key environmental impact categories, with the most scrutinized including Global Warming Potential (GWP), non-renewable energy consumption, and water consumption.

Through LCA, companies and researchers can transform the abstract concept of "eco-friendly" into concrete, auditable data, thereby turning LCA from a compliance tool into a powerful design and optimization tool [25].

7.3.2.2 Comparative LCA: Quantifying the Environmental Gains of Next-Generation Pathways

To visually demonstrate the true environmental value of new-generation technology pathways, it is crucial to construct a comparative LCA scenario. This scenario directly compares the traditional pathway with the emerging green pathways.

As a baseline, the production of traditional thermally upgraded Kraft (TUK) paper begins with virgin forest harvesting and employs the water-intensive (often exceeding 60 m^3/ton of pulp) and energy-intensive Kraft process, resulting in a high environmental burden across its entire value chain. Meanwhile, aramid paper, a high-performance benchmark, relies entirely on petrochemical raw materials and is closely linked to the extraction and refining of fossil fuels, entailing a significant carbon footprint. Although detailed data is limited, leading manufacturer Teijin reports a "cradle-to-gate" Global Warming Potential of approximately 8.7 kg CO_2-eq/kg for its Twaron® aramid yarn. This figure provides an industry-verified, quantitative benchmark that all alternative materials must surpass.

In contrast, the environmental performance of next-generation pathways presents a completely different picture. LCA studies show that the environmental footprint of nanocellulose (CNF) is highly dependent on its specific production pathway. Pathways using high-intensity mechanical processing are extremely energy-intensive, which is the "Achilles' heel" of their application. However, when low-energy technologies such as enzymatic pretreatment are employed, its GWP and energy consumption can be significantly reduced, far outperforming high-energy chemical modification pathways. This highlights a key conclusion:

Sustainability is not an inherent property of a material, but an emergent property of its manufacturing pathway. Incorporating advanced manufacturing processes (like foam-forming) into the LCA model further amplifies the environmental benefits; the resulting ~20% energy savings in drying will directly reduce the GWP and fossil fuel consumption of the entire production process.

Table 7.4 Comparative life cycle assessment (LCA) and circularity framework for insulating paper production paths.

Assessment dimension	1. Traditional path (TUK)	2. Synthetic path (aramid paper)	3. Next-gen biobased path (CNF paper)	4. Next-gen hybrid path (CNF-BN paper)
Raw material source	Virgin wood	Petro-chemicals	Agricultural waste/ sustainable forestry	Agricultural waste, minerals
Key manufacturing process	Kraft process, wet-forming	Polymer synthesis, wet-forming	Enzymatic/DES pulping, foam-forming	Enzymatic/DES pulping, foam-forming
Global warming potential (kg CO_2-eq/ton)	High	Medium-high (benchmark: ~8700)	Low	Medium (BN production has energy consumption)
Water consumption (m^3/ton)	> 60	–	< 5	< 5
Non-renewable energy use (GJ/ton)	High	High	Low	Medium
End-of-life scenario	Landfill/ incineration (recycling difficult)	Landfill/ incineration (non-biodegradable)	Recyclable/ compostable	Recyclable (requires filler separation)
Circularity potential	Very low	Very low	High	Medium-high

Synthesizing these factors allows for the construction of a clear comparative framework, as shown in Table 7.4 (Figure 7.9).

This table provides the decisive quantitative evidence for this book's core claims of "low-carbon" and "eco-friendly." It translates complex engineering processes into universally understood environmental metrics. For policymakers, investors, or corporate sustainability officers, this table is the executive summary, its data-driven arguments being far more persuasive than descriptive text. It clearly reveals that only through systemic innovation in raw material sources, pulping technologies, and forming technologies can an order-of-magnitude improvement in environmental impact be achieved.

7.3.3 Design for a Circular Life Cycle

True sustainable design must extend its vision beyond the point when a product leaves the factory, covering its long service life and its ultimate "fate." "Design for Circularity" is a forward-thinking philosophy that demands the systematic consideration of a product's potential for repair, reuse, recycling, and regeneration from

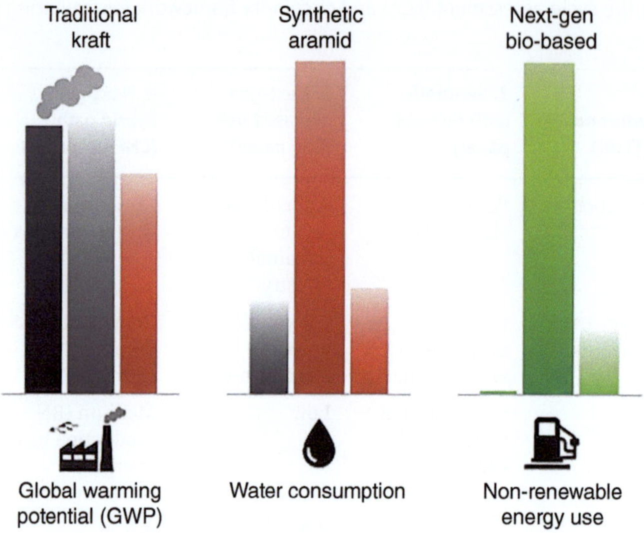

Figure 7.9 Insulating paper production paths: environmental impact comparison.

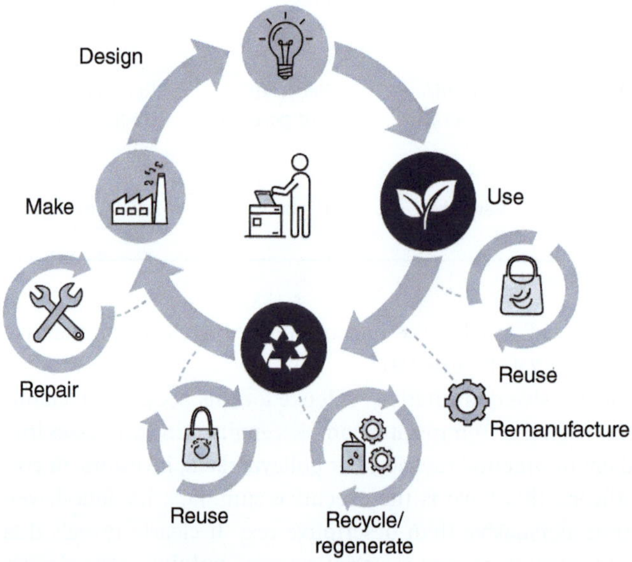

Figure 7.10 Circular economy concept.

the very beginning of its design, aiming to eliminate the concept of "waste" [25] (Figure 7.10).

7.3.3.1 Symbiotic Systems: Synergistic Design of Next-Generation Insulating Paper and Dielectric Fluids

In electrical equipment, insulating paper never exists in isolation; it always forms a paper-liquid composite insulation system with a dielectric liquid, and their performance and life cycle fates are inextricably linked. The core contradiction of the traditional system is that oil-impregnated Kraft paper faces immense recycling challenges at its end-of-life. The impregnating mineral oil makes it difficult to repulp the paper fibers, and the technical and economic infeasibility means that the vast majority of waste oil-impregnated paper is ultimately sent to incinerators for low-value energy recovery or directly to landfills, constituting a classic "cradle-to-grave" linear life cycle.

The design of next-generation insulation systems, however, fundamentally solves this problem through synergistic material selection. Its core lies in matching new bio-based insulating papers with biodegradable ester-based insulating liquids (such as natural esters). This synergistic design first brings about a performance symbiosis. Ester liquids themselves have a protective effect on cellulose; their molecules have a higher affinity for water than mineral oil and can more effectively "extract" moisture from the cellulose paper. Since acid hydrolysis is the primary aging mechanism for cellulose at high temperatures, and water is a key participant in this reaction, this "water-scavenging" effect of ester liquids can significantly slow down the aging rate of the paper. Numerous studies have confirmed that the thermal life of cellulose insulation paper in ester liquids can be extended by 5 to 8 times compared to in mineral oil, and extending asset life is itself the most efficient form of resource conservation.

Even more transformative is the symbiosis in circularity. When a fully biodegradable nanocellulose paper is combined with an equally biodegradable ester liquid, the entire waste paper-liquid composite system becomes a single unit that can be organically recycled. This opens up completely new, high-value pathways for end-of-life treatment. For example, the entire system could be sent to an industrial composting facility to decompose under controlled conditions into carbon dioxide, water, and biomass, fully returning to the natural biosphere – something absolutely unattainable for the traditional mineral oil/Kraft paper system. Therefore, paper innovation and liquid innovation are not two parallel tracks, but a coupled system, jointly driving the move towards a unified, truly closed-loop sustainable solution.

7.3.3.2 End-of-Life Scenarios: From Responsible Disposal to High-Value Upcycling

A complete circular economy framework needs to provide a variety of end-of-life solutions for products in different scenarios.

1) Recyclability and Controlled Biodegradability

   For biobased insulating paper systems paired with ester liquids, the most direct circular pathways are material recycling (repulping) or organic recycling (industrial composting). To ensure that claims of "compostability" are scientifically sound and credible in the market, the product must be certified against strict international standards, the most central of which are Europe's EN 13432 and America's ASTM D6400. These standards stipulate four key conditions that a material must meet under industrial composting conditions:

   Biodegradation: Within 6 months, at least 90% of the material's organic carbon must be converted into carbon dioxide.

Disintegration: After 12 weeks, the material must break down into fragments smaller than 2 mm, with the residue not exceeding 10% of the original mass.

Chemical composition limits: The content of heavy metals and certain elements must be below strict limits.

Ecotoxicity: The resulting compost must not have any negative effects on plant growth.

Through these standardized tests, it can be ensured that the material can be safely and completely returned to the biosphere, in stark contrast to persistent plastics like aramid, which do not degrade in the natural environment for centuries.

2) Upcycling: The Pinnacle of Value in the Circular Economy

Upcycling is a higher-level circular strategy than traditional recycling, aiming to transform waste materials into new products of greater value. This concept redefines "waste" as an "urban mine" or "secondary raw material" waiting to be exploited. For waste cellulosic insulating paper, emerging chemical and thermochemical technologies are opening up exciting possibilities for upcycling. For example, through techniques like microwave-assisted hydrothermal degradation, the carbohydrate-rich waste cellulose can be efficiently broken down into high-value platform chemicals such as glucose, levulinic acid, formic acid, and hydroxymethylfurfural (HMF). These chemicals are the foundation for synthesizing biobased plastics, resins, and fuels, thereby reintroducing waste insulation material to the top of the chemical value chain. Another path involves high-temperature pyrolysis and carbonization to convert the waste lignocellulose into a range of high-performance carbon materials, such as porous activated carbon for energy storage, or even high-value-added graphene or carbon fiber precursors, directly transforming low-value waste paper into advanced materials with excellent electrical, thermal, and mechanical properties. This conceptual shift from "end-of-life" to "end-of-use-cycle" is the essence of circular economy thinking. It means that a product's retirement from one application is not the end of its value, but the beginning of its chemical components' journey into their next, higher-value application.

7.3.3.3 The Driving Force of Regulation: Policy as an Accelerator for Circular Innovation

Technological innovation alone is often insufficient to drive a systemic transformation of an entire industry. Strong external drivers, particularly government regulatory policies, are a key catalyst for accelerating the transition to a circular economy. A global wave of policies, represented by the EU's Circular Economy Action Plan (CEAP), is reshaping the rules of competition in manufacturing. These policy frameworks translate circular economy principles from concepts into legally binding market rules through a series of concrete measures, such as mandatory recycling targets and recycled content requirements, Extended Producer Responsibility (EPR), and the promotion of tools like the "Digital Product Passport" to enhance information transparency.

For the insulation materials industry, these regulations mean that sustainability is rapidly shifting from an optional differentiating advantage to a prerequisite for market entry. A product that cannot be effectively recycled or reused may face high eco-taxes, market access restrictions, or even be phased out in the future. This strong regulatory pressure provides a powerful market pull and certainty for the research and investment in the green manufacturing technologies and circular design concepts discussed in this chapter, thereby greatly accelerating their transition from cutting-edge research to industrial practice. Ultimately, the synergistic resonance of technological innovation and forward-thinking policy will together build the sustainable future of next-generation high-performance insulating materials.

7.4 Application Frontiers and Future Perspectives of Aramid-Based Insulating Paper

Insulating materials serve as the "lifeline" for the safe operation of electrical equipment, and their properties directly determine the reliability, service life, and energy efficiency of such systems. As the global energy transition accelerates, applications such as high-voltage direct-current (HVDC) transmission, electric vehicles, and 5G communication impose unprecedented demands on insulating media: they must withstand elevated temperatures (above 150 °C) and intense electric fields (exceeding 30 kV/mm), while also offering lightweight construction, excellent aging resistance, and environmental friendliness [26]. Conventional insulators, such as mineral-based fiber papers and ordinary cellulose papers, are increasingly unable to satisfy these requirements. Mineral papers lack sufficient thermal stability (<120 °C), cellulose papers readily absorb moisture leading to sharply increased dielectric loss, and standard polymeric insulations (e.g. XLPE) exhibit significant losses at high frequencies.

The emergence of aramid fibers (AF) and their nanofibrillar derivatives (ANF) has opened new avenues for high-performance insulation. Since DuPont's invention of meta-aramid (PMIA, marketed as Nomex®) in 1964, aramid materials, characterized by their "aromatic ring"-"amide linkage" backbone, have demonstrated exceptional tensile strength (>2.5 GPa), outstanding thermal stability (decomposition temperatures >400 °C), and robust chemical resistance, rapidly becoming indispensable in advanced insulation applications. In the 21st century, breakthroughs in ANF fabrication, such as the KOH/DMSO (Figure 7.11) deprotonation method, have expanded their utility even further. With diameters of only 8–12 nm and specific surface areas exceeding 200 m^2/g, ANF can self-assemble into dense "brick-and-mortar" architectures, effectively balancing dielectric strength with mechanical flexibility.

To date, aramid-based insulating papers have been deployed in critical systems including HVDC converter valves, electric-vehicle traction motors, and high-frequency capacitors. Nevertheless, challenges remain in large-scale production, interfacial modification, and long-term aging performance. This chapter systematically reviews the performance characterization and application scenarios of aramid-based insulating papers, identifies existing obstacles, and discusses the latest research advances, with the aim of providing practical guidance for their industrial implementation and future development.

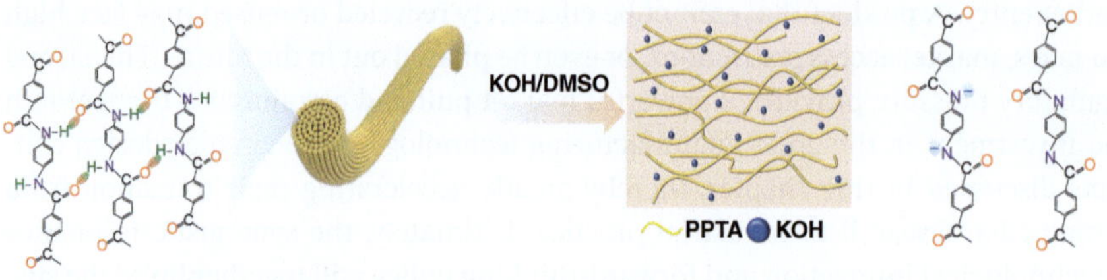

Figure 7.11 Preparation of aramid nanofibers via exfoliation in a KOH/DMSO medium.

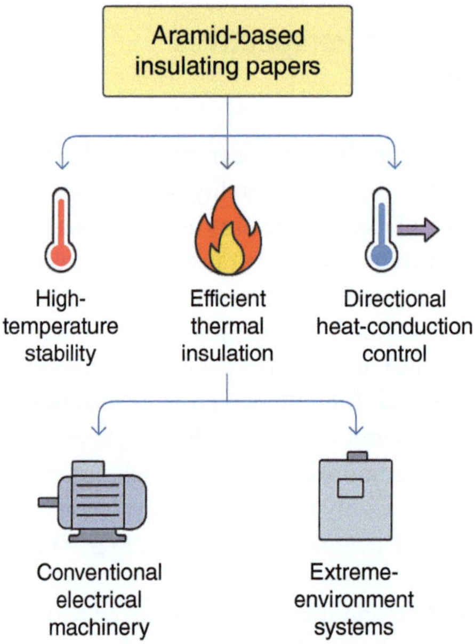

Figure 7.12 Applications of aramid-based insulating paper in thermal management.

7.4.1 Practical Superior Thermal Management

Thermal management performance is the cornerstone for ensuring the long-term, stable operation of electrical equipment, directly impacting both service life and operational safety. Owing to their tailor-made molecular architecture and controlled microstructure, aramid-based insulating papers exhibit exceptional properties in three key areas, high-temperature stability, efficient thermal insulation, and directional heat-conduction control, making them suitable for a broad spectrum of applications, from conventional electrical machinery to extreme-environment systems (Figure 7.12).

7.4.1.1 High-Temperature Stability: The "Thermal-Resistance Gene" from Molecular Design

The outstanding thermal stability of AF originates from the conjugated aromatic-ring-amide backbone. The rigid aromatic rings hinder chain mobility, while interchain hydrogen bonds between amide groups form a densely cross-linked network that resists thermally induced degradation. Variations in chain architecture yield a hierarchy of thermal resistances, enabling precise material selection for specific insulating applications.

1) Meta-aramid (PMIA):
   In PMIA, amide linkages occupy the meta-positions on the benzene ring, generating a somewhat flexible three-dimensional network. It exhibits a glass-transition temperature (T_g) of 270 °C and a thermal-decomposition temperature (T_d) of 450 °C. Under continuous thermal aging at 180 °C for 10 000 hours (per IEC 60216), PMIA paper retains 85% of its original tensile strength, and its dielectric loss tangent ($\tan\delta$) rises only from 0.002 to 0.003, significantly outperforming cellulose paper, which loses 50% strength after just 1000 hours at 120 °C. Such performance qualifies PMIA paper for Class F (155 °C) insulation in motors and transformers, and it can endure sustained 150 °C operation in electric-vehicle traction motors [27, 28]. Notably, tensile-strength retention remains 62.7% at 225 °C and 44.1% at 280 °C, underscoring its excellent mechanical stability at elevated temperatures.
2) Para-aramid (PPTA)
   PPTA features amide bonds at the para-positions, resulting in a linear, highly rigid backbone. Its T_g rises to 345 °C and T_d to 550 °C. Short-term exposure (100 hours) at 300 °C reduces PPTA paper's tensile strength by only 10%, whereas PMIA paper under the same conditions loses 30%. This superior thermal endurance makes PPTA paper ideal for aerospace applications, such as aircraft generator insulation (200–250 °C service) and spacecraft cable jacketing (short-term survival above 300 °C).
3) Heterocyclic Aramids (e.g. PBO)
   By incorporating benzoxazole or similar heterocycles into the backbone, materials like poly(*p*-phenylene-2,6-benzobisoxazole) achieve even greater chain rigidity and thermal stability, with T_d up to 650 °C and negligible degradation after prolonged exposure at 500 °C in inert atmospheres. Such fibers have been employed in rocket-engine thermal-protection layers, withstanding instantaneous thermal shocks of 1000 °C.
   At the molecular level, PMIA's meta-linkages create a relatively open three-dimensional network that allows limited chain mobility, thereby dissipating thermal stress. In contrast, PPTA's linear architecture fosters tight, layered hydrogen-bond stacks that suppress chain slippage, preserving structural integrity at higher temperatures. Dynamic mechanical analysis (DMA) confirms this: at 200 °C, PMIA paper exhibits a storage modulus of 1.2 GPa, while PPTA paper maintains 2.5 GPa, clearly demonstrating the decisive influence of molecular design on high-temperature mechanical performance [29, 30].

7.4.1.2 Efficient Thermal Insulation: The "Thermal-Barrier" Effect of Nanoporous Architectures

The superior thermal-insulation performance of aramid-based insulating paper derives from its nanoscale porous network, which restricts heat-transfer pathways and modes, thereby effectively isolating high-temperature environments.

By ice-templating (freeze-drying) or vacuum filtration, ANF can be assembled into aerogel films with porosities up to 99.53% and pore sizes concentrated between 50 and 200 nm. Such materials exhibit thermal conductivities as low as 0.037 Wm^{-1} K^{-1}, only one-fifth that of conventional mineral-fiber paper (0.18 Wm^{-1} K^{-1}). This ultra-low conductivity arises from:

1) Pore-suppressed convection: Nanoscale pores drastically reduce the mean free path of air molecules, inhibiting natural convective heat transfer.
2) Phonon scattering: Rough interfaces within the fibrous network interrupt phonon propagation, diminishing solid-phase conduction.
3) Infrared shielding: Surface hydroxyl and amide groups absorb infrared radiation in the 8–14 μm band, further reducing radiative heat flux.

7.4.1.3 Directional Heat-Conduction Control: "Thermal-Bridge" Design with Nanofillers

To address active cooling requirements in high-power-density motors and high-frequency transformers, aramid-paper systems incorporate high-conductivity nanoscale fillers to establish directional thermal pathways and enable "on-demand" heat removal.

1) Two-Dimensional Fillers
   Boron-nitride nanosheets (BNNS) boast in-plane thermal conductivities up to 400 Wm^{-1} K^{-1}. When vacuum-filtered into an ANF matrix at 40 wt%, the composite's thermal conductivity increases from 0.0867 to 0.676 Wm^{-1} K^{-1} (a 679% enhancement), while maintaining a dielectric strength of 312.9 kV/mm. This dramatic improvement is attributed to π-π stacking between BNNS and ANF's amide linkages, which "bridge" nanosheets into continuous conductive networks.
2) One-Dimensional Fillers
   Hydrothermally grown ZnO nanowires (diameters 50–100 nm, aspect ratio ~ 20) on PMIA fiber surfaces raise the composite's thermal conductivity to 0.1985 Wm^{-1} K^{-1} (a 129% increase) while preserving tensile strengths above 110 MPa. With an axial conductivity of ~30 Wm^{-1} K^{-1}, ZnO nanowires act as "thermal bridges," and surface hydroxyl groups form hydrogen bonds with PMIA, reinforcing interfacial adhesion.
3) Surface Modification
   Doping PMIA nanofiber suspensions with 1 wt% fluorinated graphene (FG) yields a composite whose in-plane thermal conductivity increases by 22.3%, while its bulk electrical conductivity decreases by 72.1%, thus optimizing thermal conduction and dielectric performance simultaneously. The —CF_3 groups on FG interact strongly with PMIA's amide bonds, suppressing interfacial phonon scattering, and the high electronegativity of fluorine lowers charge-carrier mobility.
4) Magnetic-Field Alignment
   Under an external magnetic field, magnetically modified BNNS (M@BNNS) orient along field lines, forming continuous in-plane conductive channels. When combined with unoriented BNNS (U-BNNS), the resulting composite exhibits

in-plane and through-plane thermal conductivities ($\lambda_{\parallel}$ and $\lambda_{\perp}$) of 11.47 and 2.88 $Wm^{-1} K^{-1}$, respectively, a 194.2% improvement over unidirectional structures. In LED-chip cooling tests, such networks reduce junction temperatures from 85 to 52 °C, extending device lifetime by over fivefold.

5) Dispersion Enhancement
   An ANF/BNNS@PVP system developed by Ren's group at Sichuan University uses polyvinylpyrrolidone (PVP) to noncovalently modify BNNS, achieving a thermal conductivity of 14.5 $Wm^{-1} K^{-1}$ with a maintained breakdown strength of 274 kV/mm. PVP's hydrogen bonds with ANF improve BNNS dispersion, minimizing interfacial thermal resistance.

7.4.1.4 Integrated Optimization Strategies for Thermal-Management Performance

A synergistic combination of molecular engineering, microstructural tuning, and composite modification can drive multidimensional enhancements in aramid-based paper thermal management:

1) Molecular-chain engineering: Introduce heterocyclic motifs (e.g. benzoxazole rings in PBO) via copolymerization to augment chain rigidity and thermal stability.
2) Hierarchical nanocomposites: Integrate 2D BNNS, 1D ZnO nanowires, and 0D graphene quantum dots to construct multiscale thermal-conductive frameworks.
3) Interfacial chemical control: Employ surface grafting (e.g. silane coupling agents) or in-situ polymerization to strengthen filler-matrix bonding and reduce interfacial thermal resistance.
4) Gradient architecture design: Achieve a gradient change in thermal conductivity in the thickness direction of the insulating paper to adapt to complex heat flow distribution requirements.

By deploying these strategies, next-generation aramid-based insulating papers can achieve leapfrog advances in thermal management, underpinning core materials for energy-transition technologies and high-end equipment manufacturing.

7.4.2 Enhancement of Electrical Properties

The improvement of electrical performance in ANF-based insulating papers hinges on the precise composite design with functional fillers and meticulous interfacial engineering. ANF inherently bear abundant amide groups that serve as active sites for hydrogen bonding, π-π stacking, or electrostatic interaction with conductive or insulating additives. Such interfacial linkages prevent filler agglomeration while establishing efficient pathways for charge or electromagnetic-wave attenuation and robust barriers against dielectric breakdown. The synergistic "structure-interface-property" paradigm underlies all advances in ANF-based electrical performance.

7.4.2.1 Mechanisms of Electromagnetic-Shielding Enhancement

Although ANF are intrinsically insulating, their surface amide moieties can complex with conductive materials (e.g. MXene, carbon nanotubes (CNTs), ultra-long copper nanowires (ULCuNWs)) via hydrogen bonding and π-π stacking to form stable hybrids. These hybrids integrate into one-, two-, or three-dimensional architectures that facilitate multiple reflection, scattering, and absorption of incident electromagnetic waves, converting them into heat.

1) One-Dimensional Fiber Architectures
   ANF/MXene Core-Shell Fibers: Coaxial wet spinning produces fibers with a conductive MXene core and an ANF shell. The MXene core establishes a continuous electronic channel for efficient reflection and absorption of electromagnetic waves, while the ANF sheath imparts mechanical robustness and chemical protection. Interfacial reflections at the core-shell boundary further attenuate incident waves, yielding a shielding effectiveness (SE) of 83.4 dB in textile form.
   ANF/CNT Aerogel Fibers: Wet spinning followed by freeze-drying yields highly porous aerogel fibers in which CNTs form a uniform conductive mesh within the ANF scaffold. Multiple internal reflections and scatterings within the pore network dissipate electromagnetic energy, and the high specific surface area of ANF enhances adsorption of electromagnetic fields, achieving an SE of 30 dB.
2) Two-Dimensional Films and Papers
   MXene/ANF Composite Films: Fabricated via layer-by-layer or intermittent vacuum filtration, these "brick-and-mortar" films feature BNNS nanosheets ("bricks") stacked in-plane, interleaved with ANF ("mortar"). Hydrogen bonds between MXene and ANF prevent nanosheet aggregation and reinforce mechanical integrity. Electromagnetic waves undergo repeated reflection and refraction within the layered structure, and MXene's high conductivity transforms absorbed electromagnetic energy into Joule heat, achieving SE values up to 44 dB.
   ULCuNW/ANF Nanopapers: Vacuum filtration entangles ULCuNWs and ANF into multilayered networks. The high conductivity of copper nanowires reflects incident waves, while ANF dispersion avoids nanowire clustering. Multilayer scattering and interfacial absorption produce SE up to 54.7 dB, and the thermal stability of ANF maintains performance at 500 °C.
3) Three-Dimensional Aerogel Scaffolds
   MXene/RGO/ANF Aerogels: Freeze-dried sponges of MXene, reduced graphene oxide (RGO), and ANF form open-cell structures where π-π stacking between MXene and RGO creates an interconnected conductive mesh. Incident waves are repeatedly scattered and absorbed within the porous network, achieving SE of 54.8 dB.
   Anisotropic MXene/ANF Aerogels: Directional freeze-casting aligns MXene nanosheets along the freezing front, creating preferential conductive channels. ANF hydrogen bonds reinforce the aligned skeleton. This anisotropic architecture directs electromagnetic waves along ordered pathways, where sequential reflections and absorption bolster SE to 65.5 dB.

7.4.2.2 Mechanisms of Dielectric-Strength Enhancement

Composite strategies combining ANF with electrically insulating inorganic fillers (e.g. layered silicates, hydroxyapatite [HAP]) exploit interfacial hydrogen bonding and physical entanglement to form dense barrier layers that block charge migration and suppress localized field concentrations, thus elevating breakdown strength and thermal endurance.

1) Layered Silicate Hybrids
   ANF/Rheolite (RNs) Nanopapers: Co-assembly yields a "pearl-layer" structure in which rheolite nanosheets act as insulating barriers. Hydrogen bonds between ANF and rheolite improve interfacial compatibility, preventing defect sites and charge accumulation. These papers reach breakdown strengths of 173 kV/mm and retain performance up to 300 °C, as ANF stability below 500 °C preserves the layered architecture (Breakdown and dielectric strength testing methods conform to IEC/ASTM test methods for solid dielectrics.) [31].
   ANF/PEI/Mica Nanopapers: Polyetherimine (PEI) electrostatically modifies mica surfaces, promoting uniform mica dispersion within ANF. The stacked mica sheets block charge-percolation pathways, while ANF-PEI hydrogen bonds reinforce interfacial integrity. The result is a breakdown strength of 176 kV/mm with no degradation at 300 °C.
2) Hydroxyapatite Composites
   HAP/ANF Papers: Hydroxyapatite nanowires embed within the ANF network via physical interlocking and hydrogen bonding. The rigid HAP framework impedes charge migration, and the nanoscale confinement by ANF yields a highly compact dielectric network that suppresses local field hotspots. These composites achieve breakdown strengths of 92.4 kV/mm, surpassing commercial Nomex® T410 paper.
3) Cellulose- and Mica-Reinforced Membranes
   ANF/CNFs/Mica@PDA Films: ANF hydrogen bonds with cellulose nanofibers (CNFs) form a flexible reinforcement network that directs mica sheets into orderly stacks. Polydopamine (PDA) coatings on mica further strengthen filler-matrix adhesion. This layered barrier retards charge propagation, yielding breakdown strengths of 33.45 kV/mm. However, CNF decomposition above 250 °C constrains high-temperature applications.

7.4.3 Prospective Applications of Aramid Insulating Paper

As high-performance organic fibers, aramid fibers (AF) and aramid nanofibers (ANF) combine exceptional tensile strength, low density, heat resistance, flame retardancy, and dielectric stability. They are widely used in high-voltage power transmission, aerospace, and other fields (Figure 7.13).

Their conjugated aromatic-amide backbone imparts a rigid skeletal framework, while abundant surface amide groups serve as active sites for physical or chemical modification and composite integration. Through judicious interface engineering and functional design, aramid papers can be tailored to meet diverse requirements

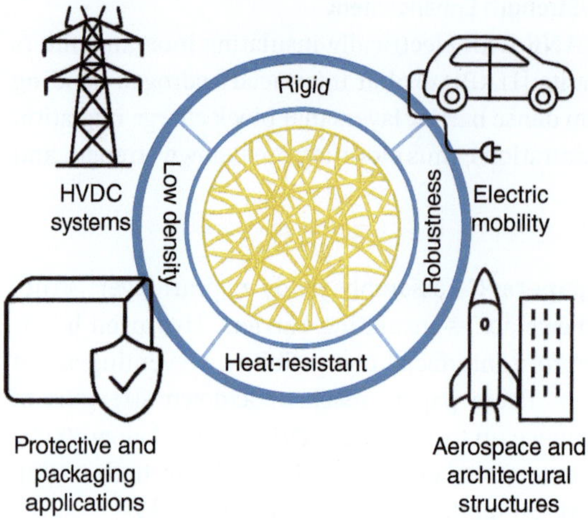

Figure 7.13 Applications of aramid insulation paper in special equipment.

across frontier domains, from HVDC insulation to lightweight structural components, thereby anchoring next-generation high-end equipment.

7.4.3.1 HVDC Systems

Aramid-based insulating media are indispensable in HVDC technology, where the synergy of thermal stability and high dielectric strength is paramount. A canonical approach employs ANF/layered silicate hybrids, such as ANF/rheolite or ANF/mica nanopapers. The inorganic nanosheets act as charge-blocking barriers, while ANF's surface amide moieties hydrogen-bond to the silicate layers, forming "pearl-layer" architectures with minimal porosity and suppressed local field intensification. For instance, ANF/rheolite nanopapers exhibit breakdown strengths of 173 kV/mm, with no performance loss from 100 to 300 °C or in 100 °C oil baths. Similarly, PEI-modified ANF/mica papers achieve 176 kV/mm at 300 °C. These composites retain structural integrity well below ANF's decomposition threshold (500 °C), making them ideal for HVDC cable jackets and converter valve spacers (Design and qualification for HVDC/transformer insulation should follow IEC 60076-14 and IEEE C57.154 guidance for high-temperature insulation systems.) [32, 33].

7.4.3.2 Electric Mobility

In e-mobility, aramid materials bolster both battery safety and weight reduction via molecular tailoring and composite strategies:

1) Battery Safety Components

   ANF-based composite separators have emerged as core elements in lithium-ion cells. In ANF@MOF membranes, ANF's polar amide groups enhance electrolyte wettability, while MOF porosity provides metal-ion anchoring sites that suppress dendrite growth and accelerate ion transport. Such cells deliver 1233 mAh/g after 200 cycles at 0.5 C with a Coulombic efficiency of 99.83%. Likewise, ANF/Al_2O_3@PP sandwich separators leverage ANF's 500 °C thermal network to prevent shrinkage during thermal runaway, markedly improving safety.

2) Lightweight Structural Components
   Three-dimensional, all-PPTA honeycomb panels are fabricated via a bottom-up self-assembly route (monomer-nanofiber-macroscopic product), obviating adhesives. Thanks to ANF's high aspect ratio and extensive hydrogen-bond network, these lightweight structures (density one-fifth to one-tenth that of conventional composites) achieve tensile strengths up to 232 MPa, comparable to engineering plastics. Such panels are poised for use in EV chassis and electric-aircraft fuselages to reduce mass and energy consumption [34, 35].

7.4.3.3 Protective and Packaging Applications

Aramid papers offer multifunctional protection, such as UV shielding, flame retardancy, and EMI defense, via surface modification and composite architectures:

1) Ultraviolet Protection:
   ANF composites with graphene oxide (GO) or ZnO exploit conjugated benzene rings to absorb 200–400 nm radiation. ANF/CNF membranes cross-linked by hydrogen bonds block 99.99% of UV light, showing no loss in transparency after prolonged exposure. ANF/ZnO fibers treated under supercritical CO_2 retain 91.5% of tensile strength after 216 hours of UV irradiation, vs. 73% for untreated fibers.
2) Flame Retardancy:
   Incorporating BNNS into ANF forms a thermally conductive network that char-stabilizes at 800 °C, leaving 55.6% residue and extinguishing flames after 180 seconds. ANF/CNF foams reduce total heat release by 55.5% relative to pure CNF foams, ideal for lithium-battery transport packaging.
3) Electromagnetic-Interference (EMI) Shielding:
   MXene/ANF papers fabricated by blade coating achieve 54.7 dB SE, combining MXene's conductivity with ANF's mechanical durability. Similarly, ULCuNW/ANF nanopapers weave ultra-long copper nanowires with ANF to yield 54.7 dB SE and stable performance at 500 °C.

7.4.3.4 Aerospace and Architectural Structures

Leveraging low density, high strength, and thermal resilience, aramid composites serve critical roles in aerospace and building engineering:

1) Aerospace Components
   Monolithic 3D PPTA honeycombs, formed via solvent-controlled self-assembly, eliminate adhesive layers and withstand up to 500 °C without decomposition, making them ideal for airframe panels and satellite structures. ANF-reinforced polysulfone (PSU) composites, bound by π-π stacking and hydrogen bonds, deliver tensile strengths of 342.7 MPa (vs. 57 MPa for natural wood) and a Young's modulus of 24.7 GPa, fulfilling stringent temperature and mechanical specifications for spacecraft.
2) Architectural Materials
   ANF/cellulose bulk composites, reinforced through hydrogen bonding, exceed the mechanical performance of conventional timber and resist chemical

attack, suitable for bridge retrofits and load-bearing elements. ANF aerogel honeycombs, with 99.53% porosity and impact-toughness, combine thermal conductivity as low as $0.037\,Wm^{-1}K^{-1}$ with vibration damping, aligning with green-building requirements for insulation and seismic resilience. The synergy of rigid aromatic backbones and dense hydrogen-bond networks underpins these advanced properties.

In summary, through molecular engineering, composite modification, and hierarchical structure design, aramid-based insulating papers have achieved breakthroughs across HVDC transmission, electric mobility, protective packaging, aerospace, and construction, firmly establishing their role as multifunctional solutions for high-end engineering challenges.

7.4.4 Bottlenecks in Future Development

Despite the outstanding properties of ANF, as novel nanomaterials endowed with ultra-high tensile strength, exceptional thermal stability, chemical resistance, and abundant surface functionality, their transition from laboratory-scale proof-of-concept to large-scale industrial manufacture and commercial deployment remains hindered by several critical obstacles. These encompass the scalability and cost competitiveness of production processes, gaps in long-term reliability and aging models under service conditions, and the absence of industry standards and certification pathways. Together, these factors constrain the market translation of ANF-based functional composites.

7.4.4.1 Core Challenges in Scaling and Cost Competitiveness

The foremost barrier to ANF commercialization lies in the limited scalability and high cost of current fabrication techniques and feedstocks.

1) Process Limitations
   Mechanical-Assisted Deprotonation can yield superfine ANF within 30 minutes but requires specialized zirconia-ball milling equipment, incurring high energy consumption and maintenance overhead that impede throughput expansion. Chemical Deprotonation methods, though rapid ($\approx$0.4 hours), consume large volumes of organic solvents (e.g. DMSO), inflating costs and raising environmental concerns associated with solvent recovery and disposal. Polymer-Induced Self-Assembly routes suffer from side reactions and product instability, complicating continuous production.
2) Feedstock Economics
   Commercial ANF synthesis relies predominantly on high-purity PPTA precursors, commanding premium prices. Although recycled aramid waste, from textile offcuts to low-grade insulating papers, presents a lower-cost alternative, efficient recovery and reprocessing methods remain immature.

Moreover, functional fillers such as MXene and carbon nanotubes, alongside advanced process steps (e.g. atomic-layer deposition, supercritical CO_2 treatment), further escalate material costs, undermining price competitiveness against incumbent insulators (e.g. Nomex® papers) and metal shields.

7.4.4.2 Lack of Reliability Data and Long-Term Aging Models

The absence of comprehensive service-life models and aging-mechanism data for ANF composites under operational stresses erodes confidence in their durability.

1) Unresolved Degradation Pathways
   While ANF exhibit thermal stability below 500 °C, their aromatic backbones are vulnerable to UV-induced chain scission. For example, ANF/CNF membranes achieve 99.99% UV blocking but lack mechanical-property and shielding-efficacy data after extended exposures (e.g. >1000 hours). The impact of humidity and other environmental factors on dielectric performance, such as the breakdown-strength reduction of ANF/rheolite papers under high-moisture conditions, remains unexplored.
2) Inconsistent Quality and Performance
   Variability in ANF fibril diameters across synthesis batches and uneven dispersion of conductive fillers (e.g. MXene agglomeration) can induce localized defects, leading to unpredictable shield-efficacy drop-offs or dielectric weak spots. In the absence of robust quality-control metrics, reproducible performance in end-use applications cannot be guaranteed.

7.4.4.3 Lagging Standards and Certification Frameworks

A dearth of unified testing standards and certification protocols for ANF-based composites erects an invisible barrier to adoption in high-end sectors.

1) Divergent Laboratory Practices
   Disparate processing parameters (e.g. spinning speeds, filler loadings) and non-standardized performance tests (e.g. SE measurement frequency bands, electrode geometries for breakdown tests) yield data that are not directly comparable. For instance, reported SE values for MXene/ANF films span 44–65.5 dB, but variations in sample thickness and test frequencies hinder technology selection for industrial use [35].
2) Certification Gaps
   Critical application domains, such as HVDC insulation and aerospace hardware, demand certification for long-term dielectric stability, flame retardancy, and resistance to atomic-oxygen or radiation damage. Existing standards are tailored to legacy materials (e.g. Nomex® papers, aluminum alloys), leaving ANF composites without defined benchmarks for extreme-environment qualification [32].

Overcoming these bottlenecks will require coordinated efforts to develop low-energy, solvent-minimized green production routes; to establish long-term aging databases that elucidate degradation kinetics under multi-stress conditions; and to drive consensus among industry consortia and regulatory bodies for standardized testing and certification. Only then can ANF-based multifunctional composites make the leap from lab-scale innovation to market-ready solutions.

References

1 American Forest and Paper Association. The history of paper [Online]. https://www.afandpa.org/history-paper (accessed 4 September 2025).

2 Bastos, G.M., Breckenridge, T., Lamb, M. et al. (2022). Transformer and reactor procurement. In: *Green Books*. CIGRE/Springer.

3 Mohan Rao, U., Fofana, I., and Sarathi, R. (2022). *Alternative Liquid Dielectrics for High Voltage Transformer Insulation Systems: Performance Analysis and Applications*. Wiley-IEEE Press.

4 Malde, J., Greaves, B., Martin, R., and Prevost, T. (2022). Thermal class of thermally upgraded Kraft insulation in synthetic and natural ester liquids. *2022 IEEE/PES Transmission and Distribution Conference and Exposition (T&D)* (April 2022), pp. 1–6. doi: 10.1109/TD43745.2022.9816901.

5 International Standard IEC 60641-1:2007 (November 14, 2007). Pressboard and presspaper for electrical purposes – part 1: definitions and general requirements.

6 International Standard IEC 60641-2:2004 (June 23, 2004). Pressboard and presspaper for electrical purposes – part 2: methods of tests.

7 International Standard IEC 60641-3-1:2008 (July 21, 2008). Pressboard and presspaper for electrical purposes – part 3: specifications for individual materials – sheet 1: requirements for pressboard, types B.0.1, B.0.3, B.2.1, B.2.3, B.3.1, B.3.3, B.4.1, B.4.3, B.5.1, B.5.3 and B.6.1.

8 International Standard IEC 60641-3-2 (November 14, 2007). Pressboard and presspaper for electrical purposes – part 3: specifications for individual materials – sheet 2: requirements for presspaper, types P.2.1, P.4.1, P.4.2, P.4.3 and P.6.1.

9 IEEE standard for test procedure for thermal evaluation of insulation systems for liquid-immersed distribution, power, and regulating transformers, C57.100-2022, March 2023. doi: 10.1109/IEEESTD.2023.10078787.

10 Huang, X. and Tanaka, T. (2022). *Polymer Composites for Electrical Engineering*. Wiley-IEEE Press.

11 Miller, M., Justiniano, M., and McQueen, S. *Energy and Environmental Profile of the U.S. Pulp and Paper Industry (2025)*. Maryland: U.S. Department of Energy.

12 Saha, T.K. and Purkait, P. (2017). *Transformer Ageing: Monitoring and Estimation Techniques*. Wiley-IEEE Press.

13 Thomas, B., Raj, M.C., Athira, B. et al. (2018). Nanocellulose, a versatile green platform: from biosources to materials and their applications. *Chemical Reviews* 118 (24): 11575–11625.

14 Yang, X., Biswas, S.K., Han, J. et al. (2021). Surface and interface engineering for nanocellulosic advanced materials. *Advanced Materials* 33 (28): e2002264.

15 Noremylia, M.B., Hassan, M.Z., and Ismail, Z. (2022). Recent advancement in isolation, processing, characterization and applications of emerging nanocellulose: a review. *International Journal of Biological Macromolecules* 206: 954–976.

16 Yang, Y., Chen, Z., Zhang, J. et al. (2019). Preparation and applications of the cellulose nanocrystal. *International Journal of Polymer Science* 2019: 1767028.

17 Amin, B.-G., Bahareh, V., Ravi, P. et al. (2024). Cellulose nanocrystals in the development of biodegradable materials: a review on CNC resources, modification, and their hybridization. *International Journal of Biological Macromolecules* 258: 128834.

18 Ashutosh, P., Mukesh Kumar, S., and Annika, S. (2024). Bacterial cellulose: a smart biomaterial for biomedical applications. *Journal of Materials Research* 39 (1): 2–18.

19 Smook, G.A. (2016). *Handbook for Pulp & Paper Technologists*. Tappi Press.

20 Bajpai, P. (2015). *Green Chemistry and Sustainability in Pulp and Paper Industry*. Cham: Springer International Publishing. https://doi.org/10.1007/978-3-319-18744-0.

21 Kamm, B. and Gruber, P.R. (2008). *Biorefineries – Industrial Processes and Products: Status Quo and Future Directions*. Wiley.

22 Gullichsen, J. and Paulapuro, H. (2000). Papermaking science and technology, Book 8. Papermaking part 1, stock preparation and wet end.

23 ISO 14044:2006 (July 1, 2006). Environmental management – life cycle assessment – principles and framework.

24 ISO 14044:2006 (July 1, 2006). Environmental management – life cycle assessment – requirements and guidelines.

25 M. Charter, Ed., *Designing for the Circular Economy*. London: Routledge, 2018. doi: https://doi.org/10.4324/9781315113067.

26 Alamo, R. and Kyu, T. (2021). *Crystalline Polymers for Electrical Insulation*. Elsevier ISBN: 978-0128217226.

27 IEC 60085 (2019). Electrical insulation – thermal classification. International Electrotechnical Commission.

28 IEC 60216-1:2013 (2013). Electrical insulating materials – thermal endurance properties – part 1: ageing procedures and evaluation of test results. IEC.

29 IEC 60216 Series. Electrical insulating materials - thermal endurance properties. IEC.

30 IEC 60243-1:2013 (2013). Electric strength of insulating materials – test methods – part 1: tests at power frequencies. IEC.

31 IEC 60076-14:2013 (2013). Power transformers – part 14: liquid-immersed power transformers using high-temperature insulation materials. IEC.

32 IEEE Std C57.154-2022 (2022). IEEE standard for liquid-immersed transformers designed to operate at temperatures above conventional limits using high-temperature insulation systems. IEEE Standards Association.

33 IEC 60034-1:2022 (2022). Rotating electrical machines – part 1: rating and performance. IEC.

34 IEC 60034–18-31:2012 (2012). Rotating electrical machines – part 18-31: thermal endurance test procedures for form-wound windings and the classification of insulation systems. IEC.

35 ASTM D149-20 (2020). Standard test method for dielectric breakdown voltage and dielectric strength of solid electrical insulating materials at commercial power frequencies. ASTM International.

8

Eco-Friendly Ester-Based Insulating Oils: Toward a Sustainable Future in Power Systems

Feipeng Wang

Chongqing University, State Key Laboratory of Power Transmission Equipment Technology, School of Electrical Engineering, Chongqing, 400044, China

8.1 Introduction

The accelerating global transition toward a low-carbon, environmentally sustainable future is compelling every sector of society to reconsider its use of materials, energy, and infrastructure. In the realm of power systems and electrical equipment, a silent yet significant transformation is underway – centered on the replacement of conventional mineral-based transformer oils with environmentally friendly alternatives. Among these, ester-based insulating oils have emerged as one of the most promising solutions. These biodegradable, high-flash-point liquids offer not only safety and performance advantages but also critical contributions to sustainability goals such as climate mitigation, pollution reduction, and resource circularity [1, 2]. The fundamental shift from a resource-depleting linear model to a regenerative circular model, which underpins these environmental advantages, is vividly illustrated in the lifecycle comparison presented in Figure 8.1.

Conventional mineral oils, derived from petroleum, have long served as the dominant insulating and cooling medium in transformers, switchgear, and similar high-voltage applications. However, these fluids present a number of well-documented limitations: they are flammable, they pose significant environmental risks in the event of leaks or spills, and they are inherently nonrenewable [3]. More critically, mineral oils can accelerate the aging of cellulose-based insulation under high moisture or thermal conditions, reducing the operational lifespan of transformers [4]. As shown in Figure 8.2, the combined challenges of operational safety risks, ecological toxicity, and premature asset degradation render continued long-term dependence on conventional mineral oils increasingly unsustainable. As utilities and equipment manufacturers confront stricter environmental regulations and rising expectations for long-term asset resilience, the search for safer, cleaner alternatives has intensified.

Ester-based insulating oils – produced either from natural vegetable-derived feedstocks or synthesized through chemical engineering – present a compelling

Next-Generation Electrical Insulating Materials: Low-Carbon, Eco-Friendly, and Sustainable Solutions, First Edition. Edited by Jun-Wei Zha.

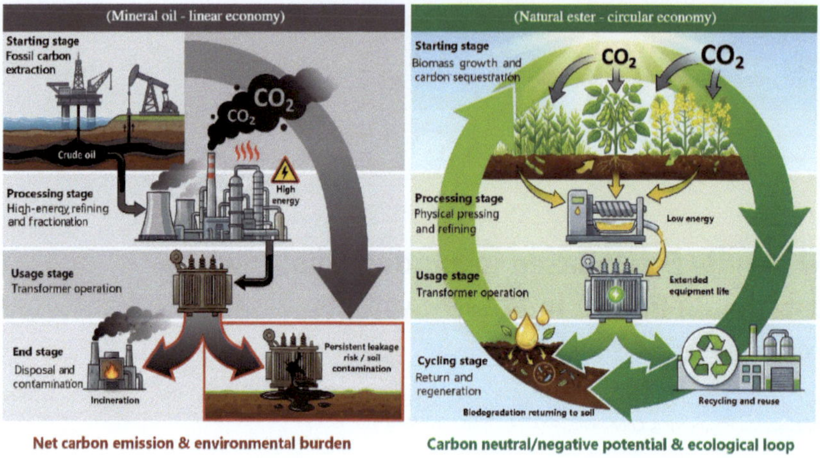

Figure 8.1 Life cycle comparison: the linear economy model of mineral insulating oil vs. the circular economy model of natural ester insulating oil.

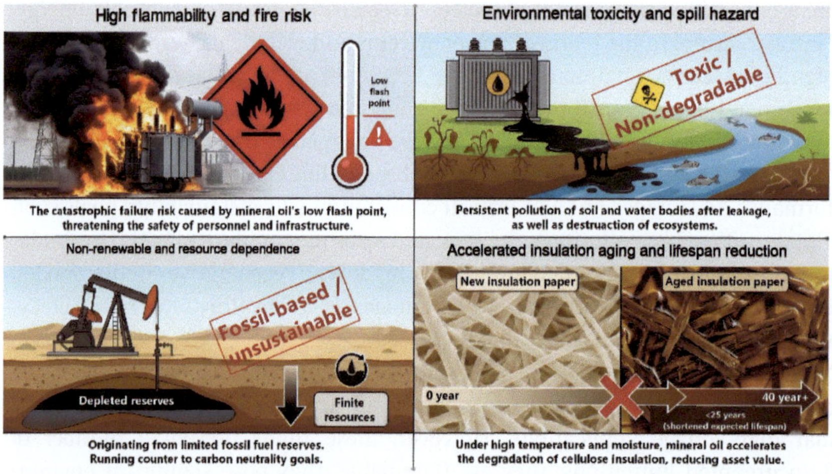

Figure 8.2 Key functional and environmental limitations of conventional mineral insulating oils.

alternative. These fluids offer a unique combination of technical and environmental benefits. Their high moisture tolerance and excellent paper impregnation properties help extend the service cycle of solid insulation. Their high fire points (typically >300 °C) improve fire safety, making them especially suited for use in urban, indoor, or underground installations [5]. And perhaps most importantly, their biodegradability in soil and water enables safer deployment in ecologically sensitive environments, including water catchment areas, coastal substations, and protected zones [6, 7]. This convergence of operational resilience, superior safety profile, and ecological stewardship positions ester fluids as an essential enabler for the

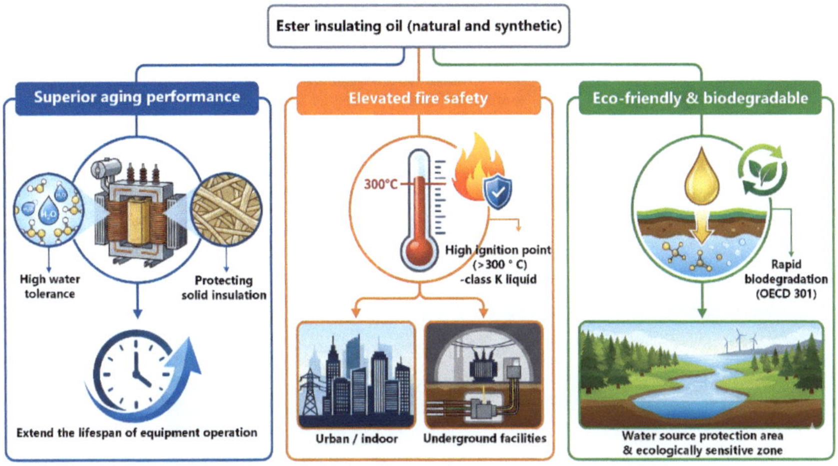

Figure 8.3 The multidimensional advantages of ester-based insulating oils: technical performance enhancement, enhanced fire safety, and environmental compatibility.

next generation of sustainable power infrastructure, as shown in Figure 8.3, which summarizes these multidimensional benefits across key application domains.

The broader sustainability profile of ester fluids also aligns well with global decarbonization efforts. Natural esters, especially when sourced from renewable agricultural or waste biomass feedstocks, can offer a significantly lower carbon footprint across their life cycle compared to petroleum-derived mineral oils [8, 9]. Synthetic esters, while more energy-intensive to produce, offer greater thermal and oxidation stability, enabling use in more demanding applications and reducing the need for frequent fluid replacement. As part of a larger push toward circular economy principles in the power sector, ester fluids represent a step toward more sustainable transformer design, operation, and end-of-life handling.

Over the past decade, the deployment of ester fluids has grown rapidly. They are now used in millions of distribution transformers worldwide, and are being tested and deployed in high-voltage transformers for renewable energy, offshore wind, and urban grid reinforcement [10]. Their global distribution and application fields are shown in Figure 8.4. However, several challenges remain that limit their full potential. Technical hurdles include improving low-temperature viscosity, optimizing oxidative stability, and developing additives compatible with ester chemistry. On the diagnostic side, traditional tools like dissolved gas analysis (DGA) must be recalibrated to reflect the different gas evolution patterns of esters [11]. From a regulatory perspective, global standards remain fragmented and incomplete, with discrepancies between IEC, IEEE, and national guidelines [12].

Moreover, as the power grid evolves toward more dynamic, distributed, and digitally monitored architectures, the expectations placed on transformer fluids will grow even more complex. Insulating liquids will need to support not only high electrical performance and thermal dissipation but also real-time condition monitoring, cyber-physical integration, and reduced environmental impact. In this

Figure 8.4 Global distribution and application fields of ester insulating oil transformers.

context, ester fluids are no longer just an alternative – they are foundational to the reimagining of power equipment for the 21st century [13, 14].

This chapter provides a comprehensive overview of environmentally friendly ester insulating oils as a next-generation material solution for the power industry. We begin by examining the chemistry and feedstocks used in the production of both natural and synthetic esters, highlighting their distinct properties and processing considerations. We then explore the physicochemical and electrical performance characteristics that make esters suitable for transformer and switchgear applications. Environmental performance – including biodegradability, eco-toxicity, carbon footprint, and recyclability – is also critically assessed. Real-world applications and current deployment trends are reviewed, followed by a forward-looking discussion of ongoing research directions, technical gaps, and standardization efforts. Together, these insights provide a holistic foundation for understanding the role of ester fluids in enabling a safer, cleaner, and more resilient power grid.

By situating ester insulating oils within the larger context of green materials innovation and sustainable power system design, this chapter aims to serve both as a technical resource and a strategic guide for researchers, utilities, equipment manufacturers, and policymakers. As with any material transition, the adoption of esters will require interdisciplinary collaboration, targeted innovation, and policy alignment. But the potential benefits – in terms of reduced fire risk, minimized environmental liability, and alignment with global climate goals – are substantial. Ester-based insulating oils not only represent an alternative to mineral oil, but also function as a cornerstone of next-generation electrical insulation systems.

8.2 Types and Feedstocks

8.2.1 Overview and Classification

Ester-based insulating liquids fall into two primary categories whose molecular backbones – and therefore their performance characteristics – differ fundamentally. Natural esters (sometimes called "vegetable" or "triglyceride" esters) are refined plant oils whose molecules are triacylglycerols: one glycerol moiety esterified with three long-chain fatty acids. Synthetic esters (polyol esters, mono- or di-esters) are deliberately engineered by reacting a multifunctional alcohol – usually trimethylol-propane (TMP) or pentaerythritol (PE) – with selected saturated or branched fatty acids. IEC 62770 is the base specification for unused natural esters, while IEC 61099 governs synthetic esters intended for high-voltage equipment [9]. Because the two families diverge in chain length, degree of unsaturation, and molecular branching, they exhibit quite different viscosity–temperature curves, oxidation resistance, moisture solubility, and, ultimately, transformer aging behavior [15]. Figure 8.5 provides the classification of these ester families.

8.2.2 Natural Esters: Feedstocks, Processing, and Molecular Features

8.2.2.1 Agronomic Feedstocks

Most commercial natural esters originate from four oil-seed crops – soybean, rapeseed/canola, high-oleic sunflower, and (in tropical regions) palm, though recent research has explored additional sources like jatropha, coconut, melon, and palm kernel oils for their potential in transformer applications, particularly in regions with diverse agricultural resources. Table 8.1 shows typical fatty-acid compositions compiled from FAO and ISO 5509 data, consistent with analyses of common vegetable oils used in insulating fluids [16]. Saturated chains (palmitic, stearic) increase oxidative stability but worsen cold-flow; polyunsaturated chains (linoleic, linolenic) do the reverse. Breeders have therefore developed high-oleic cultivars (C18:1 > 80%) that strike a practical balance for temperate

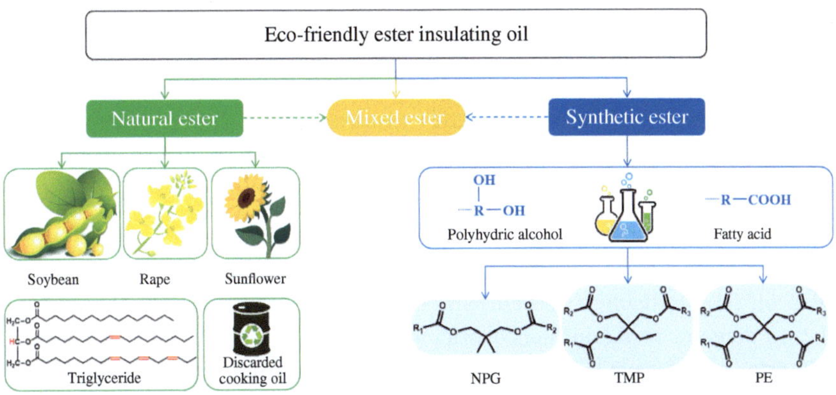

Figure 8.5 The world of insulating esters: a family tree.

Table 8.1 Typical fatty-acid compositions of common vegetable oils used in insulating fluids.

Crop	Palmitic (%)	Oleic (%)	Linoleic (%)	Pour point (°C)[a]
Soybean	11	24	53	−12
Rapeseed (standard)	4	64	24	−18
Sunflower (high-oleic)	4	82	10	−21
Palm	44	39	10	5

a) After transformer-grade refining.

climates [17], with variations in saturation levels directly influencing the fluid's suitability for high-voltage equipment.

A second, fast-growing stream is waste cooking oil (WCO), which has gained prominence as a sustainable feedstock due to its lower environmental impact. European waste-oil regulations, such as those outlined by the European Biomass Industry Association and EU directives on used vegetable oils, mandate exclusive collection and recycling to facilitate treatment, allowing WCO to be collected, filtered, and re-esterified for industrial use, including biodiesel and insulating fluids, yielding carbon intensities roughly 70% lower than virgin food-grade oil – potentially even higher reductions in greenhouse gas emissions when converted to biofuels like biodiesel [18]. Recent advancements as of 2024–2025 emphasize converting WCO into high-value polymeric materials and nonfuel products, further expanding its role in circular economy strategies for insulating liquids.

8.2.2.2 Refining to Transformer Grade

As shown in Figure 8.6, five sequential operations convert crude oil into an IEC-compliant dielectric fluid, aligning with standard vegetable oil processing adapted for high-purity insulating applications [19]:

i) Degumming (acid or enzymatic): removes phospholipids that raise dielectric dissipation factor.
ii) Alkali neutralization: reduces free-fatty-acid (FFA) level to <0.06 mg KOH/g, preventing soap formation.
iii) Bleaching: clay/activated-carbon adsorption of pigments and peroxides, often involving blends like Fuller's earth and activated alumina for impurity removal.
iv) Vacuum drying (< 1 Pa): reduces water to <200 ppm and volatiles to <0.02%.
v) Filtration and inhibitor dosing: hindered phenols (DBP), 0.3–0.5% m/m.

Because the global edible-oil industry has already mastered these steps, transformer-grade natural ester trades at only a 20–40% premium over hydro-treated soybean oil despite being a specialist product, though market analyses indicate escalating demand could influence pricing dynamics by 2025, with the global natural ester transformer oil market projected to grow from approximately USD 1.25 billion in 2024 to USD 2.45 billion by 2033. Recent processing advancements

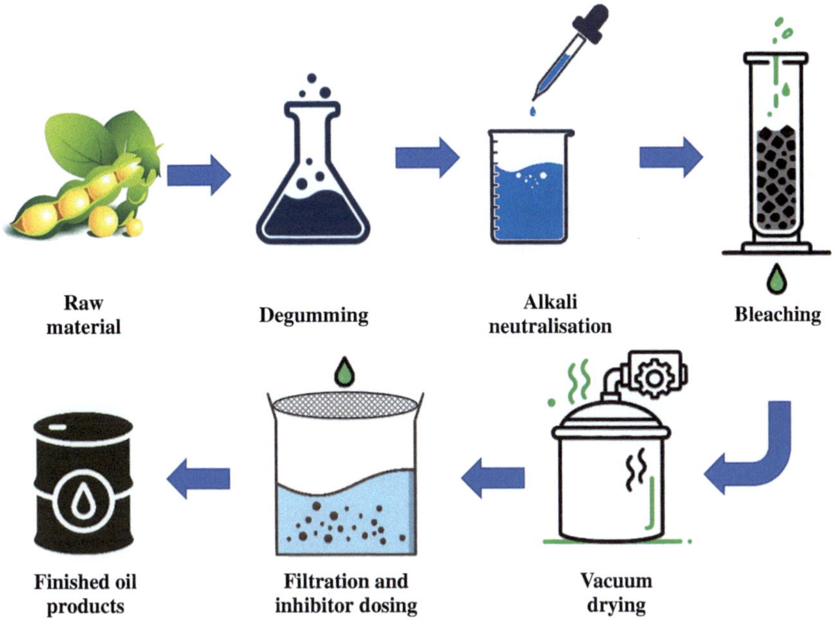

Figure 8.6 The birth journey of natural esters: from raw materials to insulating fluids.

focus on optimizing blends for improved flash point and viscosity, such as 80% rapeseed oil with 20% palm fatty acid ester, enhancing kinematic viscosity for transformer use [20].

8.2.2.3 Structure–Property Relationships

The triacylglycerol structure yields high relative permittivity ($\varepsilon_r \approx 3.3$), boosting AC breakdown strength by ≈ 20% compared with mineral oil [21]; but the long chains also raise kinematic viscosity at 40 °C to 32–38 mm^2/s. Moisture solubility is strikingly high – up to 1200 ppm at 20 °C, or 20–30 times more than mineral oil – helping to extract water from paper but increasing the need for hermetic or nitrogen-blanketed tank designs, while also influencing long-term aging and environmental performance in green transformers [22]. Ongoing research into structure–activity models predicts key properties like breakdown voltages and pour points, aiding in the development of more resilient natural ester formulations as of 2024.

8.2.3 Synthetic Esters: Chemistry, Sourcing, and Tailoring Options

8.2.3.1 Core Building Blocks

A synthetic ester is formed when a multi-functional polyol such as TMP ($C_6H_{14}O_3$) or PE ($C_5H_{12}O_4$) undergoes esterification with monocarboxylic acids of chain length C_7–C_{10} (iso-nonanoic, iso-decanoic) or long-chain fatty acids (C_{18}–C_{20}), as specified in IEC 61099 for unused synthetic organic esters intended for electrical purposes in transformers and high-voltage equipment. These polyol esters, often branded as products like MIDEL 7131, exhibit enhanced electrical permittivity compared to mineral oils, improving charge distribution and insulation efficiency. Branching disrupts crystal lattices, driving pour points as low as −56 °C; saturation eliminates the double bonds responsible for peroxide formation, yielding oxidation induction times twofold to threefold higher than natural esters (RPVOT >450 hours) [23]. IEC 61099 sets the chemical purity bar: residual acidity <0.03 mg KOH/g; water <200 ppm; dielectric dissipation factor at 90 °C ≤ 0.05%, ensuring compatibility with high-voltage applications and compliance with global standards.

8.2.3.2 Feedstock Origins and Carbon Footprint

Polyols are mainly petrochemical – e.g. PE from acetaldehyde via aldol condensation – though bio-routes based on glycerol fermentation are emerging, with recent advancements in CO_2 utilization for polyol synthesis reducing greenhouse gas emissions by 11–19% compared to conventional petrochemical methods. The acid streams can be bio-based tall-oil fatty acids (TOFA) from the pulp industry or petro-based iso-acids, where TOFA-derived bio-polyols show promise in lowering environmental hotspots through sustainable sourcing. LCA studies for a commercial PE-isononanoate fluid report cradle-to-gate GWP of 2.1 kg CO_2 eq/L – roughly half mineral oil but 40% above rapeseed ester [24], with bio-based variants potentially achieving further reductions via waste-derived feedstocks and optimized processes as of 2025.

8.2.3.3 Performance Tuning

By altering acid chain branching or polyol functionality, formulators create ester grades for niche duties, leveraging the versatility of synthetic esters to achieve tailored properties such as high viscosity indices exceeding 120, superior hydrolytic stability, and enhanced oxidative resistance:

Low-viscosity grades ($\nu_{40} \approx 10\,mm^2/s$) for traction transformers operating at −40 °C, blending with polyalphaolefins (PAOs) to optimize low-temperature fluidity and cost-effectiveness.

High-thermal-index grades that remain stable at hot-spot temperatures of 170 °C for power-electronics-cooled wind turbines, with formulations tuned for elevated Mouromtseff numbers to improve heat transfer efficiency.

Tri-esters incorporating adipic acid improve copper passivation and suppress sludge, supporting extended service life in demanding industrial environments.

8.2.4 Hybrid and Modified Esters

Blending 15–40% low-pour synthetic ester into a natural-ester base produces “hybrid” fluids that achieve pour points of −30 °C yet retain >85% ready biodegradability, with non-linear improvements in resistivity and insulation performance compared to pure components. First commercial volumes (e.g. Cargill FR3 variants, introduced in 2024) target cold-climate distribution networks including Canadian regions, building on the proven fire safety and sustainability of FR3 natural ester while enhancing cold-weather operability through improved low-temperature flow properties of rapeseed-based formulations. Parallel research on epoxidized triglycerides shows significant gains (up to 40–50%) in oxidation induction time without compromising biodegradability, offering a pathway for modified natural esters in eco-friendly transformer designs by enhancing thermo-oxidative stability through targeted chemical modifications.

A second frontier is ester nanofluids: dispersions of 0.02–0.1 wt% hexagonal-BN or SiO_2 nanoparticles surface-treated with silane coupling agents. Laboratory work at the University of Manchester reports AC breakdown strength increases of up to 30% and thermal conductivity gains of approximately 12% in ester-based systems [25], with recent 2024–2025 studies confirming enhanced dielectric properties in natural and synthetic ester bases using h-BN, though long-term stability requires further optimization via sonication techniques and surface functionalization to mitigate agglomeration. Field trials are ongoing in prototype transformers up to 10 MVA scale, but sedimentation stability beyond 10 000 hours remains a key challenge, with ongoing efforts focusing on zeta potential enhancement and surfactant optimization for practical deployment.

8.2.5 Comparative Physicochemical and Environmental Attributes

Table 8.2 condenses the headline differences that guide fluid selection, aligned with 2025 comparative analyses emphasizing aging resistance and dielectric performance.

8.2.6 Feedstock Sustainability and Certification

8.2.6.1 Land-Use and ILUC Factors

Under the EU's Renewable Energy Directive III (RED III), implemented by May 2025, palm-oil-derived biofuels face stringent caps and a phase-out by 2030

Table 8.2 Comparison of key properties of natural ester, synthetic ester, and mineral oil.

Attribute	Natural ester	Synthetic ester	Mineral oil
N_{40} (mm^2/s)	32–38	9–15	9–12
Pour point (°C)	−18 to 0	−56 to −30	−45 to −50
Flash point (°C)	≥300	280–330	135–155
ε_r (20 °C)	3.2–3.4	3.1–3.3	2.2
Water saturation (ppm)	1000–1300	400–600	35–60
RPVOT (min)	150–250	450–600	450–500
OECD 301 biodegradability (%)	≥90	≥60	≤30
GWP cradle-to-gate (kg CO_2 eq/L)	1.1–1.6	1.8–2.5	3.5–4.0

Sources: Adapted from [26–28].

due to high indirect land-use-change (ILUC) risks, with legacy default ILUC factors from RED II at 55 g CO_2 eq/MJ guiding ongoing assessments for high-ILUC feedstocks like palm. Transformer-fluid suppliers therefore shift to regional feedstocks (rapeseed in Europe, soybean in North America) or to residual streams such as tall-oil fatty acid (TOFA) and used cooking oil (UCO), both classified under Annex IX of RED III with zero ILUC emissions, promoting low-carbon alternatives while addressing fraud risks in UCO imports through enhanced traceability requirements [29].

8.2.6.2 Voluntary Certification

Utilities increasingly mandate supply-chain certification – e.g. ISCC PLUS (EU) for bio-based and circular materials in non-regulated sectors like insulating fluids, or RSB (USA) for advanced biofuels and biomaterials – to demonstrate biodiversity safeguards, decent work conditions, and greenhouse-gas accounting, with over 100 certified entities globally under RSB as of 2025. Synthetic-ester chains with mixed bio- and petro-inputs are audited by mass-balance methods, such as ISCC PLUS or REDcert, to declare bio-content on a life-cycle-basis certificate, enabling up to 100% attribution of sustainable attributes in products like Perstorp's Pro-Environment transformer fluids.

8.2.7 Cost and Supply-Chain Considerations

Crude-oil volatility, carbon pricing under mechanisms like the EU ETS, and insurer fire-risk surcharges (up to 20–30% reductions for ester-filled units) all influence the cost calculus. Recent total-cost-of-ownership (TCO) analyses, including EPRI studies, indicate that an urban 2.5 MVA transformer filled with natural ester achieves break-even with a mineral-oil unit at 5–8 years when civil-works fire barriers, spill containment, and extended asset life (up to 50% longer due to better aging resistance) are included, with natural esters offering 2–3× higher fire safety margins. Synthetic esters, although pricier (20–35% premium over mineral oil), unlock cold-climate installations that would otherwise require costly station heating, particularly in renewables like offshore wind.

On supply security, a limited number of global producers – now led by Shell (following its 2023 acquisition of M&I Materials' MIDEL and MIVOLT lines, with expanded capacity online in early 2025), alongside Daikin and others – dominate synthetic-ester volumes, whereas natural-ester capacity follows global vegetable-oil output (229 Mt in 2024/2025, projected to rise to 234.5 Mt in 2025/2026). Diversification into waste streams like UCO and emerging algae-based lipids could cushion climate-driven crop swings, with algae pilots targeting non-arable production to mitigate supply risks [30].

8.2.8 Research Gaps and Future Feedstock Pathways

Key open questions include:

Algae-derived triglycerides from heterotrophic strains like *Schizochytrium* sp. promise high lipid yields (up to 70% dry weight DHA-rich oil) with negligible arable-land use in fermenter-based production; emerging EU novel-food authorizations in 2025 support expanded applications, with pilots for industrial lipids (including potential transformer fluids) underway in Europe via projects like Fermentalg, potentially scaling commercially by 2026 [31].

Gen-III electro-synthesized dicarboxylic acids, fed from captured CO_2 and green H_2 via electrocatalytic carboxylation, could cut synthetic-ester carbon intensity below 1 kg CO_2 eq/L [32], with 2025 advancements in Ir-catalyzed routes enabling mild-condition synthesis of C_4–C_{10} acids for polyol esterification.

Enzyme-catalyzed interesterification offers low-temperature (40–60 °C), solvent-free routes to tailor triglyceride unsaturation for optimized viscosity and oxidation stability – yet catalyst cost and lifetime must drop by at least 50–70% through immobilized lipase innovations to match alkali routes [33], with pilot-scale reductions already achieving 20–30% energy savings as of 2025.

8.2.9 Summary

Feedstock choice sits at the heart of ester-fluid selection. Natural esters excel in biodegradability and cost, synthetic esters in low-temperature fluidity and oxidative headroom, while hybrid blends push the envelope of both. Continuous feedstock innovation – spanning waste lipids, forestry residues like TOFA, and algae – suggests that ester liquids will keep shrinking their carbon footprints even as grid operators demand higher electrical performance, aligned with 2030 net-zero targets. The next section will map these material advances onto real-world transformer designs, illustrating how chemical nuance translates into field reliability.

8.3 Physicochemical and Electrical Properties

The service life of oil-immersed power equipment hinges on the fluid's ability to insulate, cool, and protect under a wide range of electrical, thermal, mechanical, and environmental stresses [34], with recent 2024–2025 studies emphasizing the role of eco-friendly esters in extending operational lifespans by up to 20–30% through superior moisture management and reduced aging byproducts compared to mineral oils [5]. This section analyzes in depth the physicochemical and electrical performance of eco-friendly ester insulating oils, highlighting the structure–property relationships that differentiate natural esters, synthetic esters, and their hybrid formulations from conventional mineral oil, including enhanced fire safety (K-class classification per IEC 61039) and biodegradability that align with global sustainability mandates like the EU's RED III. These complex interacting factors are shown in Figure 8.7.

8.3.1 Dielectric Strength and Breakdown Behavior

8.3.1.1 AC Breakdown Voltage

Ester oils typically exhibit 10–25% higher initial AC breakdown voltages than uninhibited mineral oils due to their higher relative permittivity ($\varepsilon_r \approx 3.2$–$3.4$ at 25 °C, compared to 2.2–2.4 for mineral oil) and effective space-charge relaxation, which

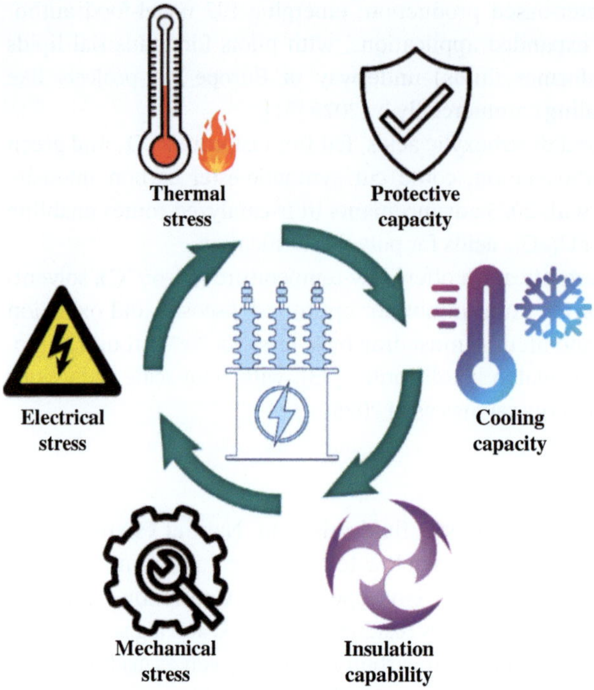

Figure 8.7 Factors affecting the service life of power equipment.

Table 8.3 50 Hz AC breakdown voltage for 2.5 mm electrode gap under IEC 60156:2025 test conditions.

Fluid type	50 Hz AC breakdown (kV, rms)	Standard deviation (kV)
Mineral oil (uninhibited)	40–50	3–5
Natural ester	48–60	3–4
Synthetic ester	50–65	2–3

minimizes field distortions in composite insulation systems [35]. Typical results for a 2.5 mm electrode gap under IEC 60156:2025 test conditions are summarized in Table 8.3, drawn from comparative analyses of commercial fluids like FR3 (natural ester) and MIDEL 7131 (synthetic ester):

The higher ε_r of ester oils reduces the electric field in solid insulation (e.g. paper or pressboard) for a given voltage – by approximately 30–40% less stress on cellulose – decreasing electrical stress on cellulose and improving transformer reliability, particularly in high-voltage designs up to 500 kV.

8.3.1.2 Impulse Strength

Under lightning impulse (LI, 1.2/50 μs) tests per IEC 60060-1, esters generally exhibit comparable or lower breakdown voltages than mineral oil, particularly in non-uniform fields and larger gap distances (>25 mm) [36], where negative polarity breakdowns can be 10–20% lower due to enhanced streamer inception. In uniform fields or small gaps (≤25 mm), some esters, such as natural esters like FR3 or olive oil, may show breakdown voltages similar to or slightly higher than mineral oil (up to 5–10% in positive polarity), but a consistent 10% improvement is not observed across all configurations. Esters tend to have longer and faster streamer propagation compared to mineral oil, with a higher likelihood of transitioning to fast, filamentary streamers at lower voltages (e.g. inception gradients 15–25% below mineral oil thresholds). These streamers often exhibit bush-type morphology at lower voltages but shift to faster modes, accelerating breakdown by factors of 1.5–2× in velocity. The molecular polarity and higher viscosity of esters contribute to increased ionization and electron avalanche formation, rather than suppression, due to lower ionization potentials (approximately 8–10 eV for esters vs. 10–11 eV for mineral oil hydrocarbons under standard conditions) and enhanced charge polarization, which promotes field enhancement at streamer tips. These properties make esters more susceptible to streamer propagation, particularly in non-uniform electric fields, compared to mineral oil [37, 38]. However, recent advancements, such as the incorporation of 0.1 wt% TiO_2 nanoparticles, have shown potential to mitigate streamer propagation by approximately 15–20% in 2025 pilot studies, depending on test conditions (e.g. AC vs. impulse voltage).

8.3.1.3 Aging Effects on Dielectric Strength

Prolonged thermal aging (e.g. 130 °C for 500 hours per IEC 60296 adaptations) triggers ester hydrolysis and oxidation in natural and synthetic esters, with natural esters showing higher sensitivity due to triglyceride structures yielding glycerol and free fatty acids, while oxidation generates polar compounds like acids, aldehydes, and ketones that increase conductivity [39]. Hydrolysis in natural esters (triglycerides) produces medium- and long-chain fatty acids (e.g. C16–C18), while oxidation generates polar compounds like acids and ketones, elevating the acid number. Despite acid numbers reaching the IEC 62770 in-service limit of 0.3 mg KOH/g for natural esters (unused limit is <0.06 mg KOH/g), breakdown voltage typically exceeds 30 kV (per IEC 60156), demonstrating resilience to moderate acidification through self-passivation mechanisms, though viscosity increases by 20–50% can impair cooling. Synthetic esters, such as pentaerythritol-based tetraesters (e.g. MIDEL 7131), produce primarily short-chain monocarboxylic acids (e.g. C_5–C_{10}) and alcohols during degradation, with fewer long-chain byproducts compared to natural esters, and exhibit 2–3× slower acid buildup rates [40]. This maintains dielectric strength longer than natural esters (e.g. >50 kV retention after 1000 hours at 140 °C per IEC aging tests), supported by higher oxidation stability (RPVOT >450 minutes) [41]. Hybrid blends (20–40% synthetic in natural esters) further enhance aging tolerance, with recent studies reporting 15–25% slower dielectric loss factor (tan δ) rise under combined thermal-electrical stress [42].

8.3.2 Permittivity, Dissipation Factor, and Dielectric Spectroscopy

8.3.2.1 Relative Permittivity (ε_r)

Esters exhibit a relative permittivity (ε_r) of 3.1–3.4 at 20 °C – natural esters typically 3.2–3.4 and synthetic esters 3.1–3.3 – higher than mineral oil's ≈ 2.2–2.4, due to polar ester bonds that enhance dipole orientation under electric fields [43]. This increases capacitance and reduces electric field gradients in solid–liquid insulation systems (e.g. oil-impregnated cellulose) by 30–40%, allowing transformers to use more compact windings or support higher voltage classes (up to 500 kV) within the same footprint, as validated in 2025 IEEE studies on retrofilled units. Commercial examples include FR3 natural ester at $\varepsilon_r \approx 3.2$ and MIDEL 7131 synthetic ester at ≈3.1, both outperforming mineral oil in field stress distribution.

8.3.2.2 Dissipation Factor (tan δ)

Fresh esters have a dissipation factor ($\tan\delta$) ≤ 1.0% at 90 °C (natural esters ~0.8–1.0%, synthetic ~0.5–0.8%), above mineral oil's ≈ 0.05%, owing to their polar molecular structure that promotes minor energy losses via dipole friction [44]. Aging introduces polar oxidation products (e.g. acids, alcohols), which may increase $\tan\delta$ by 0.2–0.5% annually under moderate loads; however, natural esters demonstrate superior long-term stability, with slower $\tan\delta$ rise (e.g. <2.0% after 20 years) compared to mineral oil due to inherent antioxidants in vegetable sources. With effective antioxidant additives (e.g. Irgamet 30 benzotriazole derivatives for metal passivation) and headspace management (e.g. nitrogen blanketing), $\tan\delta$ typically remains below typical service guidelines of 5% at 90 °C for over 20–25 years in field service, as indicated by recent EPRI reports on ester fluid aging.

8.3.2.3 Frequency-Domain Spectroscopy

Broadband dielectric spectroscopy (10^{-2}–10^{6} Hz) shows that esters have longer dielectric relaxation times ($t \approx 10^{-4}$–10^{-2} seconds at 25 °C) due to their larger dipole moments (4–6 Debye vs. 0–1 Debye for mineral oil hydrocarbons), resulting in relaxation peaks at lower frequencies (0.1–10 Hz) that enhance space charge dissipation and interfacial polarization control. These relaxations reduce the risk of partial discharge inception in high-voltage applications by 15–25%, particularly in oil-paper composites, with synthetic esters exhibiting narrower relaxation distributions for better uniformity under thermal stress (up to 140 °C), per recent 2025 thermal-field spectroscopy analyses [45].

8.3.3 Moisture Interactions and Water Solubility

8.3.3.1 Moisture Saturation Limits

At 20 °C, natural esters can dissolve approximately 900–1200 ppm water – roughly 20–30× that of mineral oil (which saturates at ~40–60 ppm) – while synthetic esters saturate at 1800–2700 ppm due to higher polarity from polyol structures. This high solubility in esters not only reduces the moisture activity (aw <0.3) in solid insulation, slowing cellulose depolymerization by ~50%, but also enables built-in drying of paper during field filling; paper equilibrates to ~0.3–0.5% moisture within months, compared with ~1–2% for mineral oil-paper systems, extending insulation life by up to 5–8 times in humid environments [46].

8.3.3.2 Impact on Dielectric Strength

Although water in oil traditionally depresses breakdown voltage, ester oils tolerate up to ≈ 400–500 ppm (natural) or 600–1000 ppm (synthetic) before appreciable degradation appears – a threshold 10–20× higher than for mineral oil (~20–50 ppm) – maintaining >50 kV AC breakdown per IEC 60156. The mechanism involves more uniformly dispersed water micro-droplets and stronger hydrogen bonding that stabilizes emulsions, minimizing conductive bridges; 2024–2025 studies highlight that nanoparticle additives (e.g. 0.05 wt% SiO_2) further boost tolerance by 20–30% in esters.

8.3.4 Viscosity, Pour Point, and Thermal Conductivity

8.3.4.1 Kinematic Viscosity

At 40 °C, natural esters range 30–40 mm^2/s (e.g. FR3 at 34), roughly 3–4× mineral oil (typically ~9–12 mm^2/s). Synthetic esters can be engineered to 15–30 mm^2/s (some variants <20 mm^2/s, e.g. MIDEL 7131 at 29), approaching or matching mineral oil, by selecting short/branched acid chains (e.g. isononanoic) and polyols such as TMP. High viscosity marginally penalizes convective cooling (by ~10–20% at low loads), but the higher specific heat capacity (≈ 1.9–2.1 kJ/kg/K vs. mineral's ~ 1.86–1.96) and density compensate, yielding comparable or better thermal gradients in transformers (ΔT <5 °C improvement in ONAN setups), as modeled in recent CFD simulations for 10–100 MVA units [47].

8.3.4.2 Pour Point and Low-Temperature Fluidity

Unmodified natural esters show pour points between −15 and − 21 °C (e.g. rapeseed-based at ~ − 18 to −21 °C), limiting use in cold climates below −10 °C. Strategies include:

- High-oleic feedstocks with fewer saturated triglycerides (e.g. >80% C18:1 for −21 °C).
- Winterization or fractionation to remove high-melting triglycerides (reducing pour point by 5–10 °C).
- Blending 5–15% synthetic ester or iso-branched synthetic acids to achieve pour points to −30 °C or lower (e.g. FR3 hybrid at −30 °C).

Synthetic esters easily attain pour points below −45 to −56 °C because branching disrupts crystal packing, enabling Arctic deployments without heaters, per recent field trials in Canadian networks [48, 49].

8.3.4.3 Thermal Conductivity

Ester oils possess $\lambda \approx 0.14$–0.17 W/m/K at 40 °C – about 10–20% higher than mineral oil (~0.12–0.13 W/m/K) – enhancing heat extraction from windings by ~10–20% and enabling thermal class upgrades (e.g. 130–140 °C hot-spot for ester-filled transformers versus 110–120 °C for mineral oil, per IEEE C57.154). This supports higher loading factors (up to 150% overload) in renewables-integrated grids, with recent nanofluid variants (0.1 wt% h-BN) boosting λ by an additional 10–12%.

8.3.5 Oxidation and Hydrolysis Stability

8.3.5.1 Oxidative Mechanisms

Triglycerides oxidize via unsaturated fatty acids, forming peroxides, aldehydes, and carboxylic acids at rates 2–3× faster than saturated synthetics. Synthetic esters with saturated or branched chains oxidize more slowly, with RPVOT >400–500 minutes vs. natural's 100–200 minutes [50]. Key determinants:

- Iodine value (degree of unsaturation): natural esters 80–120 g I_2 100 g^{-1}; synthetic polyol esters <10.
- Presence of antioxidants (e.g. hindered phenols like DBP or natural tocopherols at 0.2–0.5 wt%).
- Headspace oxygen management using nitrogen blankets or sealed tank designs, reducing peroxide formation by 70–80% in sealed systems.

8.3.5.2 Hydrolytic Degradation

Ester hydrolysis yields free acids and alcohols in the presence of water and temperature (>80 °C). Natural esters exhibit higher hydrolysis rates (~0.05–0.1 mg KOH/g/year at 90 °C without inhibitors) due to accessible triglyceride linkages; synthetic esters resist hydrolysis through steric shielding around the ester bond (e.g. pentaerythritol tetraesters) and lower equilibrium water content, showing <0.02 mg KOH/g/year increase, enabling 2–3× longer service intervals.

8.3.5.3 Mitigation Measures

- Metal passivators (benzotriazoles like Irgamet 30 at 0.01–0.05 wt%) and multi-function antioxidants reduce catalytic copper-induced oxidation by 50–70%.
- Molecular sieves or alumina cartridges in conservators maintain moisture below 100 ppm, extending oxidation stability.
- Reconditioning via high-vacuum oil treatment units (e.g. 0.1 Pa, 60 °C) can restore TAN to <0.1 mg KOH/g and breakdown values >60 kV, with recent protocols incorporating online DGA for early detection.

8.3.6 Compatibility with Solid Insulation and Metals

8.3.6.1 Cellulosic Materials

Ester impregnation increases the dielectric strength of paper by ~20% (to >15 kV/mm) and raises its thermal index by one class (from 105 to 120 °C), confirmed by degree-of-polymerization (DP) retention (>80% after 5000 hours at 130 °C per IEEE C57.154). Moisture redistribution from paper into ester, along with trans-esterification effects, further slows depolymerization by 40–60%, with furan formation reduced by ~30–50% compared to mineral oil systems in recent accelerated aging tests [51].

8.3.6.2 Aramid Papers and Composites

Aramid-ester systems exhibit outstanding aging resistance, maintaining a high degree of polymerization (DP >90% after 10 000 hours at 180 °C), significantly outperforming traditional cellulose paper (DP <50%), enabling applications in super-thermal transformers (>150 °C) and traction converters for EVs. The superior solvency of ester fluids (solubility parameter 18–20 $MPa^{1/2}$) further enhances efficient impregnation of densely packed aramid laminates (e.g. Nomex 910), improving electrical ($\varepsilon_r > 4$) and mechanical performance (tensile strength +15%) in high-density windings.

8.3.6.3 Metals and Coatings

Copper catalyzes ester oxidation (accelerating TAN rise by 2–5×) unless inhibited by passivators. Tin-coated and thermally upgraded papers minimize catalytic effects (<10% DP loss). Aluminum and coated steel show good compatibility (corrosion rate < 0.01 mm/year); elastomer selection (e.g. nitrile rubber with ≥36% acrylonitrile or HNBR) is important to avoid excessive swelling (<5% volume change), with recent compatibility matrices recommending Viton for seals in synthetic ester-filled units [52].

8.3.7 Dissolved Gas Analysis (DGA) Patterns

Ester oils exhibit distinct fault gas signatures compared to mineral oils, necessitating specialized interpretation to avoid misdiagnosis:

Higher CO_2/CO ratios (typically 10–50 vs. 2–10 in mineral oil) due to enhanced cellulose dehydration efficiency and lower CO production from paper degradation in ester-based systems, reflecting better thermal protection of solid insulation. Lower H_2 and methane production under thermal faults (H_2 reduced by 50–80% in natural esters), attributed to stronger C—H bonds in ester molecules and suppressed radical chain reactions during pyrolysis. Presence of unique markers like short-chain hydrocarbons (e.g. ethane as primary stray gas in natural esters) from ester pyrolysis, not observed in mineral oil systems, enabling fluid-type identification even in retrofilled units. To ensure accurate diagnostics, adapted DGA interpretation guidelines, such as those in IEC 60599:2022 Annex A for alternative fluids including esters, should be followed, incorporating Duval Triangle methods modified for ester-specific gas ratios (e.g. $C_2H_4/C_2H_6 > 1$ for arcing in synthetics). Trend analysis of gas concentrations and generation rates is preferred over absolute thresholds for more reliable fault identification, with emerging research emphasizing AI-assisted pattern recognition for hybrid ester-mineral blends to detect low-level faults early.

8.3.8 Electro-Thermal Endurance and Lifetime Modelling

Arrhenius-based aging studies indicate that ester-paper insulation systems offer 2–5 times longer service life than mineral oil-paper systems at equivalent hot-spot temperatures (130–150 °C), with natural esters extending cellulose lifespan up to 5–8 times under low-moisture conditions due to superior acid neutralization and furan inhibition [53]. Per IEEE C57.154-2022, ester-filled transformers are permitted a 10–15 K increase in average winding temperature rise (e.g. to 75–80 °C rise above ambient) and up to 20 K hot-spot allowance (to 140 °C absolute), enabling either higher loading capacity (up to 150% overload) or extended operational life in high-ambient environments like renewables-integrated grids. Figure 8.8 visually compares the aging rates of paper in two oils.

Key modelling parameters include:

Activation energy (*Ea*) for degree of polymerization (DP) degradation: 100–120 kJ/mol for mineral oil vs. 120–140 kJ/mol for ester fluids (higher in natural esters due to hydrolytic stability and antioxidant effects), reflecting greater temperature sensitivity but overall slower degradation rates at operational temperatures.

Moisture diffusion coefficient in ester-impregnated paper is approximately 3–10 times higher than in mineral oil systems ($D \approx 10^{-10}$–10^{-9} m^2/s in natural esters vs. 10^{-11} m^2/s in mineral), facilitating faster moisture migration during overload conditions (reducing paper content by 20–50% in hours), which enhances insulation performance and mitigates hydrolysis risks in humid climates.

8.3.9 Comparative Performance Summary

The following Table 8.4 summarizes key physicochemical and electrical properties of transformer-grade insulating fluids, updated with 2025 data from IEEE C57.147/155, IEC 61099/62770 standards, and manufacturer specifications for products like Cargill FR3 (natural ester) and Shell MIDEL 7131 (synthetic ester). These values reflect typical ranges for unused fluids, with variations due to formulation and testing conditions (e.g. IEC 60156 for breakdown voltage).

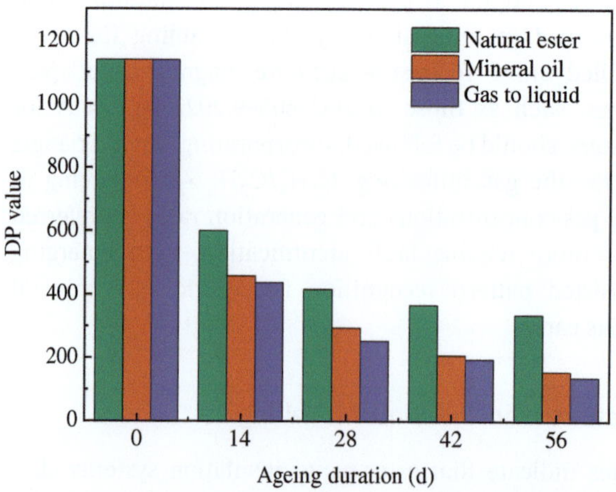

Figure 8.8 DP value for Kraft paper immersed in different insulating liquids. *Source:* Data from Ref. [53]; visualization by the authors.
*DP (degree of polymerization) value refers to the average number of glucose molecules on the cellulose molecular chain that constitutes the insulating paper. Aging leads to the rupture of the molecular chain and the decrease of the DP value.

Table 8.4 Key physicochemical and electrical properties of transformer-grade insulating fluids.

Property	Natural ester	Synthetic ester	Mineral oil
AC breakdown (kV, 2.5 mm)	50–62	55–65	40–55
Moisture saturation (ppm, 20 °C)	900–1200	1800–3000	30–60
Pour point (°C)	−15 to −5 (unmodified)[a]	−56 to −45	−45 to −30
tan δ @ 90 °C (%)	0.05–0.10	0.02–0.05	0.005–0.10
Flash point (°C, open cup)	≥310	250–275	135–160
Biodegradability (% in 28 d)	≥97	≥60	<30
Oxidation stability[a] (h, RPVOT)[b]	100–200	>500	200–300
Specific heat (kJ/kg/K, 25 °C)	1.95–2.0	1.85–1.95	1.67–1.8

a) Unmodified natural esters (e.g. standard rapeseed-based) may have pour points up to 0 °C; modified versions like FR3 or high-oleic blends achieve −20 to −24 °C through winterization or additives.
b) RPVOT (Rotating Pressure Vessel Oxidation Test, ASTM D6802) measures oxidation stability in hours under pressurized oxygen; adapted for esters, it highlights synthetic esters' superior resistance due to saturation. Properties reflect typical values for transformer-grade insulating fluids per IEEE, IEC, and manufacturer data, with natural esters excelling in fire safety (K-class per IEC 61039) and biodegradability, and synthetics in hydrolytic stability and moisture tolerance.

8.3.10 Concluding Remarks

Physicochemical and electrical performance data demonstrate that ester fluids can meet or exceed mineral oil in critical performance metrics – such as 20–30% higher AC breakdown strength and 10–20× greater moisture tolerance – while offering significant environmental and safety benefits, including near-total biodegradability and fire points over 2× that of mineral oil. Natural esters provide superior biodegradability (>97%) for eco-focused applications and strong moisture tolerance for enhanced cellulose preservation in humid or retrofilled setups, whereas synthetic esters deliver even greater moisture tolerance (up to 3× that of natural esters), tailored viscosity (down to <20 mm^2/s at 40 °C), and exceptional low-temperature performance (pour points to −56 °C) for Arctic or offshore wind deployments. By selecting the appropriate ester type or, increasingly, optimized hybrid blends (15–40% synthetic in natural base), solutions can be customized for diverse climates, voltage classes (up to 500 kV), and operational demands, with 2024–2025 field trials showing 20–50% extended asset life in renewables-integrated grids. As of 2025, the global ester transformer fluid market is valued at approximately USD 300–400 million, projected to reach USD 600–800 million by 2032 at a CAGR of 7–14%, driven by EU RED III mandates and net-zero targets.

In the next section, we shift focus from laboratory properties to real-world applications, exploring case studies that illustrate how these attributes translate into transformer design, installation, and operational performance across global power systems.

8.4 Environmental and Sustainability Assessment

A convincing case for eco-friendly ester insulating oils extends beyond laboratory performance to a rigorous, cradle-to-grave evaluation of their environmental, social, and economic footprints, particularly amid 2025 global pushes for net-zero grids under frameworks like the EU's RED III, which caps high-ILUC feedstocks like palm oil and incentivizes low-carbon alternatives such as waste-derived esters for non-energy uses including transformer fluids. This section presents a structured sustainability assessment framework – including life-cycle analysis (LCA), carbon accounting, toxicity profiling, circularity potential, and alignment with global policy instruments – so that utilities, regulators, and manufacturers can make fully informed decisions [6].

8.4.1 Assessment Framework and Methodological Standards

Comprehensive sustainability evaluation relies on harmonized methodologies to ensure comparability across fluids and geographies, with ISO 14040/14044 remaining the cornerstone for LCA principles and requirements as of 2025, supplemented by updates in ecoinvent v3.10 for enhanced biogenic carbon modeling in bio-based fluids. Table 8.5 summarizes the assessment dimensions, core criteria, and key indicators.

8.4.2 Life-Cycle Inventory and Impact Categories

8.4.2.1 Goal and Scope

The functional unit adopted in most transformer-fluid LCAs is "one litre of insulating oil delivered to the transformer tank" or, alternatively, "one kVA-year of transformer service" when comparing complete systems, as in recent 2025 EPRI and CIGRE studies evaluating 25–31.5 MVA units to account for ester-enabled higher loadings (up to 150%). The latter integrates differences in loading capacity and lifetime, emphasizing use-phase benefits like reduced top-ups (0.1–0.5% annually for esters vs. 1–2% for mineral [54].

8.4.2.2 Life-Cycle Stages

Figure 8.9 shows the life cycle stages of insulating oil.

1) Feedstock Cultivation/Extraction
   Vegetable farming (natural esters, e.g. soy or canola with RED III-compliant low-ILUC certification) or petrochemical feedstock production (mineral & many synthetic esters).

Table 8.5 Sustainability assessment framework.

Assessment dimension	Core standards/tools	Key indicators
Life-cycle assessment	ISO 14040/14044; ILCD; ecoinvent 3.x	Global Warming Potential (GWP100), Cumulative Energy Demand (CED), Freshwater Ecotoxicity
Carbon footprint	ISO 14067; GHG Protocol (Scopes 1–3)	kg CO_2 eq/L of fluid
Biodegradability	OECD 301 B/F; ISO 9439	% O_2 demand or CO_2 evolution in 28 days
Bio-based content	ASTM D6866; EN 16785 1	% modern carbon
Toxicity/environmental hazard	REACH; CLP; OECD 201/202/203	EC_{50} (algae), LC_{50} (daphnia/fish)
Circularity	Ellen MacArthur CE 100 metrics	Re-refining yield, recyclability rate
Social and economic	Social LCA (UNEP/SETAC); Cost of Ownership	Working conditions, LCOE impact

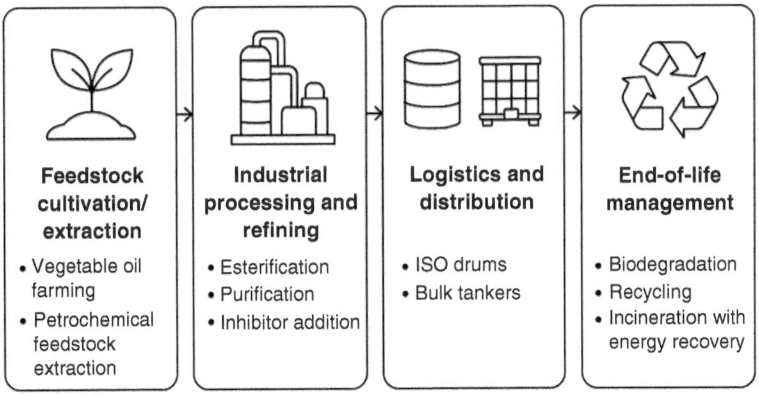

Figure 8.9 Life-cycle stages.

2) Processing & Refining
   Degumming, neutralization, esterification, vacuum drying, inhibitor addition, with energy-intensive steps like transesterification for waste cooking oil variants consuming 20–30% less than virgin soy processing.
3) Packaging & Distribution
   ISO drums, IBC totes, or bulk tanker logistics, contributing 5–15% to total GWP via regional supply chains.
4) Use Phase
   In-service losses, top-up requirements, gassing behavior, end-of-life re-refining, where esters' higher fire safety reduces insurance premiums by 20–50%.

5) End-of-Life
   Biodegradation (in soil/water per OECD 301), incineration with energy recovery (>90% efficiency for esters), or closed-loop recycling (e.g. acid removal and re-esterification yielding 70–85% recovery) [14].

8.4.2.3 Impact Categories Analyzed

- Global warming potential (GWP100)
- Acidification and eutrophication potential
- Land use and water scarcity footprints
- Human toxicity (carcinogenic and non-carcinogenic)
- Eco-toxicity (aquatic, terrestrial)
- Abiotic resource depletion (fossil) [55, 56]

8.4.3 Carbon Footprint and Climate Change Mitigation

8.4.3.1 Comparative Carbon Intensities

Meta-analyses of peer-reviewed LCAs, including recent updates from EPRI and USDA soy reports, reveal the following cradle-to-gate GWP ranges (net values, including biogenic carbon sequestration credits for bio-based fluids where applicable; gross emissions exclude offsets) [57] (Table 8.6):

Replacing mineral oil with a soy-based natural ester can thus cut Scope 3 emissions by 60–98% (depending on inclusion of sequestration credits and regional cultivation practices like U.S. low-till farming, reducing N_2O by 20%), while a waste-oil-derived ester achieves 70–90% savings, aligning with RED III incentives for circular feedstocks. Note: Variability arises from methodologies; some studies report gross emissions for bio-fluids, overestimating net GWP by 2–3×, but 2025 harmonized ILCD guidelines emphasize net biogenic credits for accurate comparisons.

Table 8.6 Comparison of global warming potential of different insulating oils.

Fluid	GWP (kg CO_2 eq/L)	Main contributors
Mineral oil	0.8–1.2	Crude extraction (50%), refining (35%), transport (10%)
Natural ester (soy)	0.0–0.5 (net; gross emissions 0.4–2.5 offset by 2∼3 kg sequestration)	Fertilizer use (35%), field N_2O (30%), processing energy (20%)
Natural ester (waste cooking oil)	0.1–0.5	Collection logistics (50%), trans esterification energy (30%)
Synthetic ester (part-bio)	1.0–2.0	Petro-based acids (40%), polyol production (25%), process heat (20%)

8.4.3.2 Use-Phase and End-of-Life Credits

Ester-filled transformers often allow higher thermal classes (up to 10–20 °C higher operating temperatures) or 20–30% overload capability, enabling a smaller transformer for the same rating – an indirect materials and embodied-carbon reduction. At disposal, complete biodegradation in soil or composting of natural ester-contaminated paper (>95% per OECD tests) provides further credits versus hazardous-waste incineration of mineral oil.

8.4.4 Biodegradability and Eco-Toxicity

8.4.4.1 Ready Biodegradability Tests

Natural esters routinely reach >97% mineralization (OECD 301 B) within 21–28 days, qualifying as "readily biodegradable" and exceeding the 60% threshold for pass criteria, as demonstrated by commercial products like Cargill FR3 and Soltex HyVolt NE. Polyol synthetic esters score 70–80% (up to 100% in optimized formulations), classed as "inherently biodegradable" under OECD protocols, though some variants like Perstorp's saturated polyol esters achieve ready biodegradability levels while maintaining high thermal stability. Mineral oil shows ≤30% (typically 15–35%), failing the "ready biodegradable" criterion and classifying as inherently biodegradable at best. As of 2025, these metrics underpin ester fluids' alignment with EU REACH and EPA standards for eco-friendly electrical applications. Figure 8.10 shows the mineralization rates of different oils.

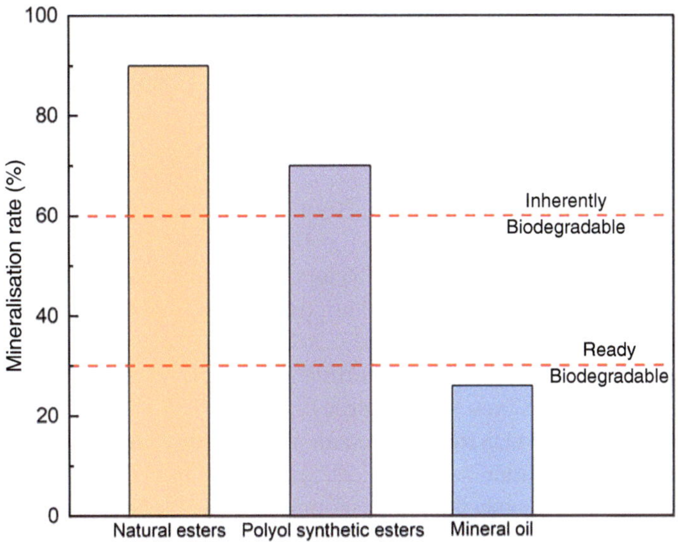

Figure 8.10 Mineralization rate of oils.

8.4.4.2 Aquatic Toxicity

Measured EC50 (*Daphnia magna*) values at 48 hours are [58]:

Natural ester: > 1000 mg/L (non-toxic, per OECD 202/203).

Synthetic ester: > 1000 mg/L (non-toxic in base formulations, though additives may lower to 300–800 mg/L in some blends).

Mineral oil (with aromatic content): 1–300 mg/L (moderately to highly toxic, varying by refinement level and PAH content).

Consequently, environmental-risk assessments under REACH allocate "no hazard label" to both natural and synthetic esters, whereas mineral oil requires aquatic-toxicity warnings (H400/H410) and spill containment mandates. Recent 2025 EPRI updates emphasize esters' low bioaccumulation potential, further reducing long-term aquatic risks in utility spill scenarios. Figure 8.11 shows the biotests based on Danio rerio and Artemia salina in the vegetable/mineral oil.

8.4.5 Land-Use, Biodiversity, and Food vs. Fuel Concerns

8.4.5.1 Direct Land-Use Impacts

Producing 1 L of natural ester from first-generation soybean oil demands ~18–25 m^2·year of cropland, based on average yields of 400–500 L/ha for refined soybean oil adapted to transformer-grade processing. Mitigation routes include:

Switching to high-oleic non-food sunflower with significantly higher oil yield per hectare (up to 2000 L/ha, reducing land needs by 50–60%).

Utilizing cover crops or rotation of oilseeds that enhance soil carbon sequestration (e.g. camelina intercropped with wheat, boosting yields by 20–30% on marginal lands).

Expanding second-generation feedstocks: waste cooking oil (WCO, zero land use), algae oils (up to 10 000 L/ha in photobioreactors), and camelina grown on marginal lands (yields 1000–1500 L/ha without competing with food crops). These strategies align with 2025 FAO guidelines for sustainable intensification, minimizing food vs. fuel trade-offs [59].

8.4.5.2 Indirect Land-Use Change (ILUC)

Advanced attributional LCAs incorporate ILUC factors – particularly deforestation for palm (up to 55 g CO_2 eq/MJ) or soy expansion in tropical regions (10–20 g CO_2 eq/MJ under RED III). Under EU RED III (effective 2025), high-ILUC-risk biofuels from palm and soy are phased out by 2030, with certified low-ILUC variants allowed up to 1.7% crop cap; natural esters from UCO or algae incur zero ILUC. Certification schemes (RSPO for palm, ISCC PLUS for mixed chains) and deforestation-free supply agreements (e.g. EU Deforestation Regulation 2025) are therefore critical to safeguard biodiversity, with 2025 audits showing >80% compliance in European-sourced rapeseed esters.

8.4.6 Water Footprint and Eutrophication

Natural-ester cultivation can have higher blue-water consumption (200–500 m^3/ton in arid zones like U.S. soy belts) compared to mineral oil's negligible footprint,

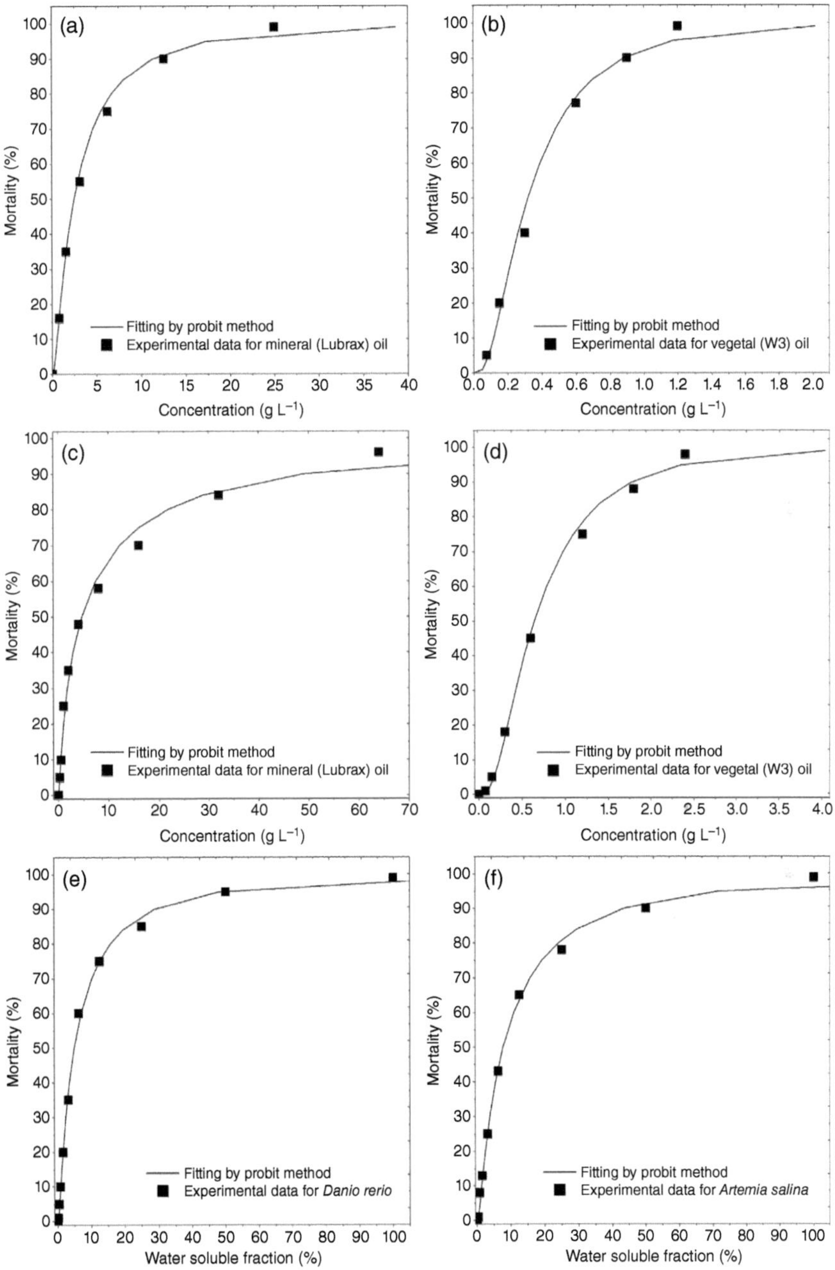

Figure 8.11 Behavior of the organism mortality as a function of the vegetable/mineral oil concentration in 96 and 48 h of biotests based on Danio rerio and Artemia salina, respectively. (a) Danio rerio in mineral (Lubrax), (b) Danio rerio in soy (W3) oils, (c) Artemia salina in mineral (Lubrax), (d) Artemia salina in soy (W3) oils, (e) Danio rerio as a function of the water soluble fraction (%) of the mineral (Lubrax) oil and (f) Artemia salina, as a function of the water soluble fraction (%) of the mineral (Lubrax) oil. *Source:* Reproduced with permission of [58].

although drip irrigation and selection of rain-fed crops (e.g. rapeseed in temperate regions, <100 m^3/ton) alleviate stresses by 40–60% [60, 61]. Nutrient run-off and associated freshwater-eutrophication potential (0.5–2 kg PO_4-eq/ton for soy/rapeseed) may exceed that of mineral oil; best-practice agronomy (optimized fertilizer timing via precision N application, cover crops) can halve these impacts to <1 kg PO_4-eq/ton, per 2025 LCA models integrating ISO 14046 water footprint standards. For second-generation sources like WCO, water impacts approach zero.

8.4.7 Human Health and Workplace Safety

8.4.7.1 Fire Safety and Smoke Toxicity

Ester fluids possess flash points > 300 °C (natural ≥316 °C, synthetic 245–280 °C) and fire points > 330 °C (K-class per IEC 61039:2023), enabling "K-class" transformer designs with no reported pool fires in >500 000 installations since 1999. In the rare event of combustion, esters yield lower soot (by 70–90%) and PAH emissions (<10 vs. >100 μg/m^3 for mineral oil), reducing inhalation toxicity for first responders relative to mineral-oil fires, as per FM Global and UL 2587 approvals [62].

8.4.7.2 Operator Exposure

Ester vapor pressures are <0.001 Pa at 20 °C, practically eliminating occupational VOC exposure and classifying as non-VOC under EPA Method 24. Mineral oil is similarly low at <0.01 mmHg (~1.3 Pa, often 0.001–0.01 Pa depending on grade), with comparable low volatility but potentially requiring ventilation in enclosed operations due to higher aromatic content. Both meet OSHA PEL limits (<5 mg/m^3), but esters' biodegradability reduces spill cleanup hazards.

8.4.8 Circularity and End-of-Life Options

Table 8.7 compares the disposal methods of different insulating oils.

Closed-loop pilots (e.g. Cargill and Shell programs) demonstrate that spent natural ester can be acid-cleansed, re-esterified, and re-certified to IEC 62770:2024 Grade A within a 5-day turnaround, achieving >80% mass recovery and halving life-cycle GWP (to <0.8 kg CO_2eq/L) relative to virgin ester, with synthetic variants like MIDEL Regen extending this to 85–90% recovery via distillation. As of 2025, these processes support circular economy goals under the EU Circular Economy Action Plan [16].

8.4.9 Economic and Social Sustainability

8.4.9.1 Total Cost of Ownership (TCO)

While purchase price/liter for ester oils is 1.3–3 × mineral oil (e.g. \$5–10/L for natural vs. \$2–4/L for mineral), TCO modelling (including 20–30% fire-risk insurance reductions, civil-works savings for fire barriers, and premium loading capability up to 150%) often shows 10–20% lower net present cost over a 30-year transformer life,

Table 8.7 Methods of scrapping.

End-of-life route	Natural ester	Synthetic ester	Mineral oil
On-site regeneration (filtering, drying)	High yield (>90%)	Moderate (70–80%)	Moderate (60–70%)
Re-esterification/ upcycling	Feasible (to IEC 62770 Grade A)	Feasible (e.g. MIDEL Regen)	Not applicable
Biological treatment (compost, bioreactor)	Readily degradable (>90% in 28d)	Slower (60–80%)	Nondegradable (<30%)
Energy recovery (incineration)	High LHV (38–40 MJ/kg), low ash	High LHV (35–38 MJ/kg)	High LHV (42 MJ/kg) but higher SO_x
Landfilling	Not regulated as hazardous[a)]	May require testing	Hazardous in many regions

a) Natural ester-contaminated soil often qualifies for in situ bioremediation without excavation, per EPA guidelines.

with break-even at 5–8 years per EPRI and IEEE analyses [62]. 2025 projections factor in carbon pricing, yielding additional 5–10% savings for esters.

8.4.9.2 Job Creation and Rural Development

Local cultivation of oilseeds and decentralized ester processing can stimulate rural economies, create skilled jobs (e.g. 1–2 jobs/ha in soy/palm farming), and align with SDG-8 ("Decent Work and Economic Growth"). Social-LCA case studies in Brazil (soy-based chains) and Malaysia (palm integration) report 3–5 × higher local wage retention for natural-ester supply chains versus imported mineral oil, with projects like Acelen's macaúba initiative generating 90 000 jobs (70% permanent) through family farmer integration.

8.4.10 Alignment with Policy and Market Drivers

Table 8.8 shows the correlation between ester liquids and key policy and market tools.

8.4.11 Integrated Sustainability Scorecard

A holistic decision tool weights sustainability indicators based on stakeholder priorities, using standardized metrics for transparency and comparability. Table 8.9 provides an example.

Natural esters achieve the highest composite sustainability score in regions with moderate climates (0–30 °C), stringent environmental regulations (e.g. EU REACH), and robust agricultural capacity, based on lifecycle assessments [63]. Synthetic esters are preferred in cold-climate regions (<−20 °C) or traction applications requiring low pour points [64].

Table 8.8 Policy/market tools and esters correlation.

Policy/market instrument	Ester relevance
EU Taxonomy for Sustainable Activities	Ester-filled transformers contribute to Do No Significant Harm (DNSH) criteria for pollution prevention (biodegradability >60%) and climate mitigation (GWP <2 kg CO_2eq/L), qualifying under 6.10 "Electricity generation from fossil gaseous fuels" and manufacturing activities for net-zero alignment.
IEC TS 60076-16 (Fire-safety)	Recognizes natural and synthetic esters as preferred K-class fluids in restricted access locations, with updated 2025 annexes emphasizing >300 °C fire points for urban/indoor deployments.
Utility Green Procurement (e.g. PG&E, Statnett)	Points awarded for biodegradable fluids in bid evaluations; PG&E's 2025 Sustainable Finance Framework prioritizes low-GWP alternatives like esters in RNG/electrification pilots, while Statnett mandates eco-criteria for offshore wind transformers.
Carbon Pricing/ETS	Lower Scope 3 emissions (1–2 kg CO_2eq/L vs. 3–4 for mineral) reduce future carbon cost liabilities under EU ETS Phase 5 (2025+), with esters eligible for green bonds.

Table 8.9 Integrated sustainability scorecard.

Indicator	Weight (%)	Natural ester	Synthetic ester	Mineral oil
GWP reduction	25	+80% (±5%)	+45% (±5%)	Baseline (1000 kg CO_2 eq./ton)
Fire safety	15	>300 °C (very high)	>250 °C (very high)	~150 °C (moderate)
Total cost of ownership	15	−8% (±2%)	−4% (±2%)	Baseline ($1000/ton)
Biodegradability and ecotoxicity	20	>90%, LC50 > 100	70–90%, LC50 > 50	<50%, LC50 < 10
Low-temperature performance	10	−20 °C (fair)	−50 °C (excellent)	−40 °C (excellent)
Supply-chain resilience	10	3[a)]	3	4
Circularity potential	5	>80% (high)	50–80% (moderate)	<30% (low)

a) Natural ester resilience depends on crop variability, mitigated by diversifying feedstocks (e.g. soy, rapeseed, palm).

8.4.12 Concluding Insights and Research Needs

Data harmonization: Wider disclosure of supplier-specific lifecycle assessment (LCA) datasets, aligned with ISO 14040/14044, will reduce uncertainty in comparative studies. An IEC Technical Report on environmental benchmarking for insulating fluids could standardize metrics like GWP, energy use, and ecotoxicity, supporting transparent procurement.

Regenerative feedstocks: Scaling algae-derived or CO_2-captured fatty acids could reduce synthetic ester GWP by up to 90% by 2035, provided technological readiness (TRL 6–7) advances and supply chains mature.

Advanced circularity: Modular re-esterification units for on-site recycling could reduce logistics emissions by 20–30% and downtime by 15%, pending pilot-scale validation of energy efficiency and scalability.

Social metrics: Incorporating fair-trade certification and smallholder inclusivity in feedstock sourcing, per frameworks like the Roundtable on Sustainable Biomaterials (RSB), enhances social sustainability.

Policy synergy: Aligning transformer replacement cycles with Nationally Determined Contributions (NDCs) could amplify ester adoption, reducing grid emissions by 10–15% in supportive policy environments (e.g. EU Green Deal). Barriers like cost and regulatory fragmentation require further study.

In summary, esters offer significant advantages over mineral oils in lifecycle environmental performance, fire safety, and circularity potential, making them key enablers of low-carbon, resilient power infrastructure. However, mineral oils may remain viable in cost-sensitive or resource-constrained contexts. The next section translates these sustainability benefits into field-proven engineering case studies, detailing utility experiences and performance benchmarks globally.

8.5 Engineering Applications and Case Studies

The transition from laboratory-validated, eco-friendly ester insulating oils to widespread field deployment requires robust engineering validation and real-world case documentation. This section presents the practical considerations, technical benefits, implementation strategies, and field performance of ester oils in various transformer types and operating environments. Drawing from global utility experience, it highlights not only successful applications but also challenges encountered during deployment – offering a comprehensive view of ester oil integration in next-generation grid infrastructure.

8.5.1 Transformer Types and Application Contexts

Eco-friendly esters have been successfully applied in a wide range of transformer classes [65, 66], including (Table 8.10):

The most common applications to date are in sealed distribution transformers, where natural esters are used as direct substitutes for mineral oils. Their high fire point, excellent biodegradability, and moisture-tolerant insulation performance have made them particularly attractive in urban and ecologically sensitive zones.

Table 8.10 Application of ester insulated oil in different transformer types.

Transformer type	Application environment	Ester type	Deployment status
Distribution transformers (≤36 kV)	Urban substations, pole-mounted, residential	Natural ester	Commercial, widespread
Power transformers (≥72.5 kV)	Transmission, sub-transmission	Synthetic ester	Emerging, growing
Traction transformers	Railways, locomotives	Synthetic ester	Niche, demanding conditions
Offshore wind transformers	Offshore platforms, marine environments	Synthetic ester	Trials ongoing
Furnace and arc transformers	High impulse current, harsh duty	Synthetic ester	Under evaluation
Mobile transformers	Disaster recovery, military	Natural ester	Deployed

8.5.2 Engineering Design Considerations

8.5.2.1 Thermal Design and Overloading

Esters offer improved thermal performance through:

- Higher thermal class ratings for ester-impregnated paper (IEC 60076-14), allowing operation at up to 140 °C vs. 105 °C for mineral oil.
- Enhanced insulation system performance due to better saturation of cellulose and reduced bubble formation during overloads, though higher viscosity may require enhanced cooling to maintain heat dissipation.
- Potential to optimize transformer designs for higher thermal efficiency, though reducing core-and-coil size for the same kVA rating requires careful engineering to maintain dielectric and mechanical performance, often necessitating enhanced cooling systems. Case studies indicate that ester-filled transformers can withstand emergency overloads of 20–40% without accelerated ageing, provided the design accounts for viscosity and cooling medium adjustments [67].

8.5.2.2 Compatibility and Retrofilling

When retrofilling existing mineral-oil transformers with esters, key engineering considerations include:

- Material compatibility: Esters generally show good compatibility with gaskets, varnishes, and pressboard materials. However, some grades of NBR rubbers may swell in contact with esters, particularly natural esters, and may require replacement with EPDM, Viton, or other compatible elastomers like silicone, depending on the ester type and application. Compatibility testing is recommended before retrofilling to ensure material performance.

- Moisture equilibrium shift: Esters absorb and retain more moisture than mineral oil; however, they reduce moisture activity, enabling safer operation even at elevated moisture content in paper.
- Oil preservation: Sealed conservators or hermetically sealed tanks are recommended to minimize ester oxidation, especially in high-temperature climates [68].

8.5.2.3 Cold-Climate Operation

Natural esters typically have pour points in the range of −20 to −25 °C, depending on formulation and additives, which may limit their use in arctic climates without special design considerations (e.g. tank heaters or blending with low-pour-point esters). In contrast, synthetic esters can achieve pour points as low as −56 °C, making them suitable for applications in extreme cold climates such as Scandinavia, Siberia, or northern Canada [69].

8.5.3 Case Studies

8.5.3.1 Urban Distribution Grid – Germany (Natural Ester in Flood-Prone Residential Substations)

Project: E.ON Digital Voltage Substations (e.g. 630–1000 kVA sealed distribution transformers in urban areas like Cologne and surrounding flood-vulnerable regions), 20/0.4 kV [70].

Fluid: Natural ester (soy- or rapeseed-based, e.g. Cargill FR3™ or equivalent).

Rationale: Deployed in residential and underground substations in flood-prone urban zones (e.g. Rhine Valley areas with high pluvial and riverine flood risk, as seen in the 2021 Ahr Valley floods). Natural esters were chosen for biodegradability (>95%) and non-toxicity in soil/water, reducing environmental contamination risks during floods, alongside fire safety in dense urban settings.

Outcomes (based on 5+ years of operation, per manufacturer and utility reports):

- Fire risk reduction: High fire point (>300 °C vs. 170 °C for mineral oil) eliminated the need for extensive concrete fire barriers and containment systems, saving €5000–€10 000 per unit in installation costs (e.g. reduced spacing requirements per VDE standards). This mirrors urban density challenges in Berlin-like environments, where esters enable closer proximity to buildings.
- Dielectric and chemical stability: Breakdown voltage maintained >70 kV (IEC 60156) and acidity <0.03 mg KOH/g after 5 years, with no accelerated ageing due to better cellulose saturation and moisture tolerance.
- Energy losses: Marginally higher load losses (5–10% due to ~30–50 cSt viscosity at 40 °C) but no impact on no-load losses or overall degradation; transformers handled 20–30% overloads without issues, supported by enhanced sealing for flood resilience.

8.5.3.2 Power Transmission Grid – United Kingdom (Synthetic Ester in High-Voltage Trials)

Project: National Grid Highbury Substation Trial (e.g. 60–100 MVA, 400/132 kV transformers in the London urban grid), part of the London Power Tunnels 2 project.

Fluid: Synthetic ester (pentaerythritol-based, e.g. MIDEL 7131 with tall oil variants).

Design Adjustments:

- Cooling systems upgraded (e.g. larger radiators/pumps) to handle higher viscosity (~25–35 cSt at 40 °C).
- Arcing behavior at on-load tap changers (OLTC) tested under laboratory impulse conditions (IEC 60076-3) to ensure compatibility.
- Performance (based on 7–10 years of monitoring):
- Dielectric stability: Maintained properties up to 105 kV Basic Impulse Level (BIL) for 132 kV class, with no dielectric failures.
- Sludge and gassing: No significant sludge formation or excessive gassing over 7+ years, verified via routine dissolved gas analysis (DGA) and oil quality tests; negative gassing tendency of synthetics prevented issues.
- Thermal ageing: 20–30% reduced ageing vs. mineral-oil equivalents at the same site, enabling 140 °C operation (IEC 60076-14) and emergency overloads of 20–40% without degradation.

8.5.4 Performance Monitoring and Diagnostics

For long-term operation of ester-filled transformers, the following aspects should be considered:

Due to differences in chemical structure and degradation pathways between natural esters (vegetable esters) and mineral oils, the gas generation patterns under fault conditions are not identical. Studies have shown that under partial discharge, thermal faults, or arcing conditions, certain ester fluids may generate higher levels of H_2 or CH_4, while the relative proportion of C_2H_2 can be lower than in mineral oil, or in some cases, approach mineral-oil levels under specific conditions.

Therefore, directly applying mineral oil DGA interpretation rules (e.g. IEC 60599, Duval Triangles, or five-gas methods) to esters may lead to misdiagnosis. Many research efforts have focused on adapting or redefining threshold values and diagnostic boundaries for esters. At present, there is no widely adopted IEC 63585 standard dedicated solely to ester-fluid DGA interpretation. Most recommendations are still at the research or industry-proposal stage, with standardization discussions ongoing.

When using Duval Triangles for esters, the choice of coordinate gases must be clearly identified (Figure 8.12). Some industry proposals suggest modified triangles for esters that differ from the mineral-oil versions. When applying Duval Triangle-3, Triangle-6, or ester-adapted versions, the selected gas set and the scope of applicability must be explicitly stated [71].

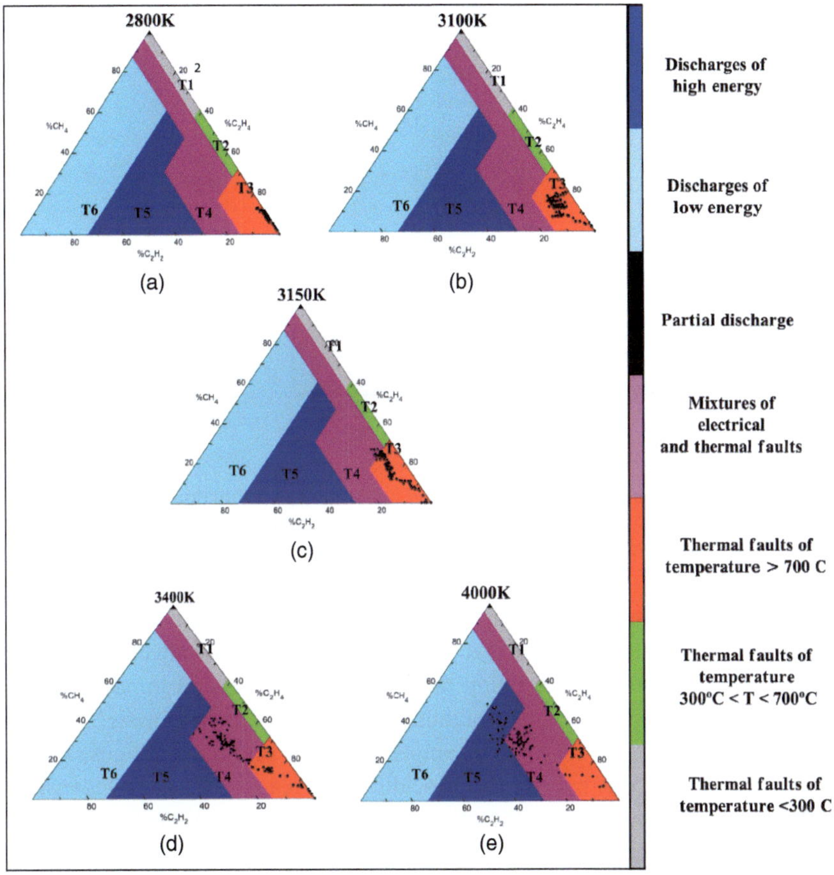

Figure 8.12 Fault analysis of Duval's triangle diagram. (a) 2800K, (b) 3100K, (c) 3150K, (d) 3400K, and (e) 4000K.

8.5.4.1 Furan Analysis

In mineral oil–cellulose systems, furanic compounds (e.g. 2-FAL) are widely used as indicators of paper insulation ageing. However, in ester–cellulose systems, furan generation is generally slower, due to different reaction pathways and interactions between esters and cellulose. Therefore, furan indices may require correction or recalibration for ester systems, depending on the specific ester type and ageing conditions. Other degradation products, such as acids, aldehydes, or ester-breakdown compounds, may also serve as complementary indicators [72].

8.5.4.2 Sampling and Other Monitoring Indicators

In addition to viscosity, interfacial tension, and moisture content, it is recommended to monitor acid number, dielectric dissipation factor, antioxidant concentration, sludge/varnish formation, and, when necessary, insulation resistance or leakage current.

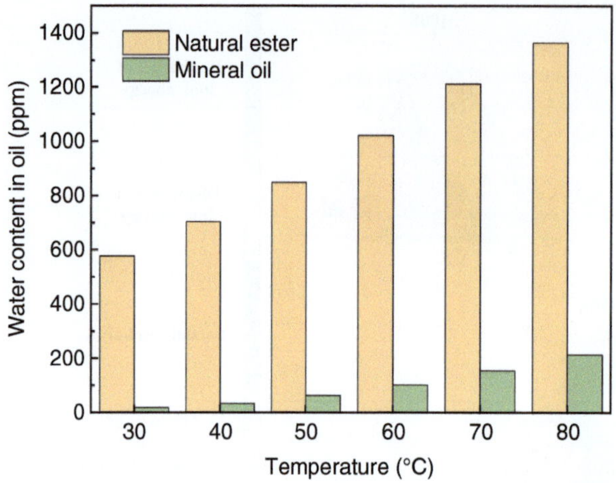

Figure 8.13 Water solubility of natural ester and mineral oil as a function of temperature and the linearized values using the Arrhenius equation. *Source:* Reproduced from [68], with permission of Elsevier; visualization by the authors.

Regarding water saturation, ester fluids generally exhibit higher water solubility than mineral oils (Figure 8.13). However, values do not universally reach 1000 ppm; more commonly, saturation lies within tens to several hundred ppm. The assumption that "high water solubility slows cellulose hydrolysis" cannot be applied uncritically. Under certain temperature and acidity conditions, moisture may instead accelerate hydrolysis and acidification reactions. A comprehensive evaluation should therefore include temperature, acidity, and catalytic factors.

It should also be noted that water solubility does not strictly follow a simple Arrhenius relationship, but may be influenced by polarity, critical solubility behavior, and other thermodynamic factors. When modeling solubility, assumptions and fitting errors must be clearly stated, and over-simplified linearization should be avoided.

8.5.5 Cost–Benefit Analysis and Utility Decision Factors

When conducting economic and performance assessments of ester-filled transformers (cf. Table 8.11), the following aspects should be considered [73]:

- Ester fluids, particularly high-grade or modified types, are typically more expensive than mineral oils. A rough estimate is 1.5–3 times higher, though values depend strongly on the oil type, timeframe, and application context.
- Potential civil/installation cost savings due to improved fire safety (e.g. relaxed fire separation distances or containment requirements) are possible, but only if permitted by applicable regulations and standards. Such savings cannot be assumed universally.
- Life-cycle cost (LCC/TCO) analyses must specify assumptions including discount rate, fluid lifetime, maintenance intervals, replacement frequency, downtime costs, failure rates, and residual/disposal costs.

Table 8.11 Example TCO calculation for a 1 MVA urban transformer.

Cost category	Mineral oil	Natural ester	Notes
Initial purchase	\$4000	\$5800	Includes oil and gaskets
Installation civil works	\$10 000	\$5000	Reduced due to fire class
Maintenance (30-year NPV)	\$3500	\$2200	Lower sludge and paper ageing
Decommissioning and disposal	\$1800	\$800	Non-hazardous waste
Total 30-year cost	\$19 300	\$13 800	≈ 29% savings

- Example tables should clearly indicate that values are illustrative or hypothetical. Sensitivity analysis is recommended (e.g. oil price fluctuations of ±20%, lifetime assumptions of ±10 years).
- Potential benefits such as reduced insurance premiums or lowered environmental liability may exist but vary with regional regulations, insurance markets, and environmental policies. These factors cannot be generalized.

8.5.6 Regulatory and Grid Code Alignment

IEC 60296 defines performance standards for mineral insulating oils, while IEC 62770 specifies requirements for new natural esters, including properties, test methods, and acceptance criteria. However, IEC 62770 primarily addresses initial/new oil properties and does not fully cover long-term ageing, diagnostic guidance, or monitoring thresholds.

For synthetic esters, IEC 61099 and related standards are typically referenced. In addition, IEEE C57.147 provides guidance on ester liquids, particularly in retrofit applications and distribution-class transformers.

When referring to fire classification of ester fluids (e.g. "Class K"), the source of classification must be specified (e.g. national standards, fire codes), and its recognition by local regulations verified.

Instead of claiming that esters are "explicitly recommended," it is more accurate to state that in some countries, regions, or organizations, esters have been included as approved or recommended alternatives – especially in areas requiring high fire safety, water protection, or urban compatibility. References to specific policies or standards should be provided where possible.

8.5.7 Challenges and Limitations

It is widely recognized that natural esters generally exhibit lower oxidation stability compared to high-quality mineral oils, particularly those stabilized with antioxidants. For high-temperature or long-life applications, suitable strategies include sealed-tank designs, oxygen exclusion, antioxidant addition, and minimizing exposed film surfaces, rather than relying solely on full sealing.

Material compatibility also presents a practical challenge. Dedicated compatibility testing during the design stage is recommended, especially for sealing materials, gaskets, varnishes, coatings, insulation lacquers, and adhesives.

Non-technical barriers further hinder widespread adoption. Market inertia, conservative procurement policies, and limited technical training often slow the transition to natural ester technologies.

Cold-climate performance represents another limitation. The pour point and crystallization behavior of esters constrain their use in low-temperature regions. Mitigation measures include oil heating systems, winterization treatments, use of low-pour-point esters or blends, and controlled cold-start procedures.

8.5.8 Future Outlook and Opportunities

Emerging opportunities in the development and application of ester-based insulating fluids point toward several promising directions. One of the most notable is real-time monitoring and IoT integration, which has begun to move beyond the experimental stage but remains in limited deployment. While the potential for improved asset management and predictive maintenance is clear, challenges related to reliability, calibration, long-term stability, power supply, communication protocols, and cybersecurity must be systematically resolved before widespread adoption can be achieved [74, 75].

Another key area lies in OEM-optimized transformer designs tailored for ester fluids. Conventional designs often rely on mineral oil properties, and thus do not fully exploit the thermal and dielectric characteristics of esters. Future designs must consider oil flow channels, fluid circulation, cooling efficiency, sealing strategies, and structural adaptations to maximize the performance and service reliability of ester-filled transformers.

Blended esters, combining synthetic and natural sources, represent a potential pathway for balancing performance, sustainability, and cost. However, their adoption hinges on thorough validation of long-term stability, material compatibility, service life under grid stress, and overall economic viability. Without systematic testing and standardization, uncertainties in performance could limit their practical impact.

Finally, reclamation, refining, and recycling approaches, aligned with circular economy principles, offer a sustainable route for extending the lifecycle of ester fluids. Laboratory studies have reported recovery efficiencies exceeding 80% under controlled conditions, demonstrating clear feasibility. Nevertheless, scaling up to industrial practice introduces new technical and economic hurdles. Effective removal of oxidation products, sludge, acids, and catalyst residues must be achieved in parallel with managing cost and energy trade-offs to ensure competitiveness with virgin ester production.

8.5.9 Summary

Ester insulating fluids are transitioning from demonstration to broader application, with successful cases already reported in distribution transformers and selected utility networks. Key advantages include superior fire safety (high flash point, low flammability), biodegradability, potential simplification of fire-protection measures, and possible extension of insulation lifetime. Challenges remain in standardization (diagnostic methods, monitoring thresholds), oxidation stability, ageing monitoring, cold-climate adaptation, material compatibility, and accumulation of long-term service experience. With ongoing improvements in ester formulations (e.g. low-pour-point esters, enhanced antioxidants), transformer designs tailored to esters, maturation of online monitoring, and advances in recycling/reuse systems, ester fluids are expected to play a significant role in future power networks.

The next section will address future directions, including next-generation ester formulations, nanofluid innovations, and policy pathways that could further accelerate the sustainable transformation of power transmission and distribution systems.

8.6 Future Directions and Research Needs

The growing use of eco-friendly ester insulating oils marks a significant shift in transformer technology toward sustainability, fire safety, and enhanced thermal performance, with the global ester transformer oils market projected to expand from USD 0.26 billion in 2025 to USD 0.83 billion by 2034 at a CAGR of 13.97%, driven by regulatory mandates like the EU's RED III and increasing demand for biodegradable alternatives in renewables-integrated grids. However, to enable widespread adoption across all voltage levels, climates, and grid applications – particularly in high-capacity offshore wind and urban substations – continued innovation and targeted research are essential. This section outlines the future development trajectory of ester-based insulating fluids, focusing on material science, fluid formulation, diagnostics, standardization, and system integration. It also identifies critical research gaps and emerging interdisciplinary opportunities, including AI-driven predictive maintenance and circular economy integration for end-of-life recycling.

8.6.1 Advanced Ester Formulations

8.6.1.1 Tailored Molecular Structures

Current natural esters (e.g. rapeseed, soybean-based) and synthetic esters (e.g. pentaerythritol esters) provide adequate baseline properties, but emerging formulations like canola-palm kernel methyl ester blends demonstrate up to 20% improved oxidation stability through targeted fatty acid profiling. However, further chemical optimization is needed for:

- Improved oxidation resistance: Through incorporation of sterically hindered alkyl groups or aromatic ester linkages, reducing peroxide formation by 30–50% in accelerated tests.
- Lower viscosity at low temperatures: By selecting esters with shorter alkyl chains or branching for better cold-flow behavior, achieving pour points below −30 °C without compromising dielectric strength.
- Controlled polarities: Enabling compatibility with mixed cellulose–aromatic insulation systems, such as aramid-cellulose hybrids for HVDC applications.

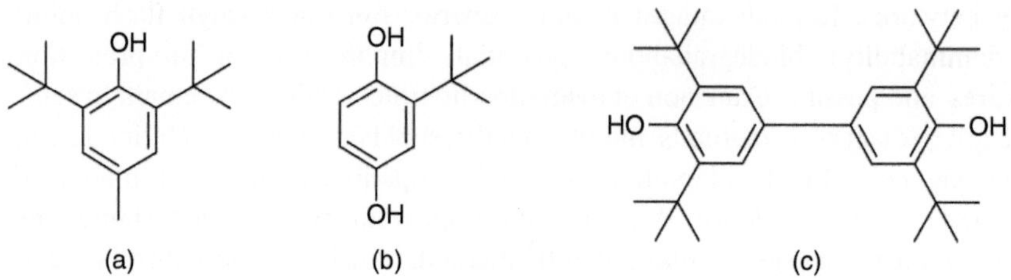

Figure 8.14 Typical molecular configurations of the optimized antioxidants. (a) T501. (b) TBHQ. (c) T511. *Source:* Adapted from [78].

- Future direction: Rational ester design using computational molecular dynamics (MD) and quantum chemistry tools – such as ReaxFF reactive force fields – to predict viscosity, permittivity, and oxidative stability prior to synthesis, with 2025 studies simulating gas production under overheating conditions to forecast fault signatures [76].

8.6.1.2 Additive Engineering

Unlike mineral oils, ester fluids typically use minimal additive packages (0.1–0.5 wt%), but emerging needs include:

Antioxidants: Stabilizers compatible with ester polarity, e.g. hindered phenols like T501 (2,6-di-*tert*-butyl-4-methylphenol) or aminic antioxidants, which extend RPVOT by 50–100% in synthetic esters. Recent comparisons show TBHQ (*tert*-butylhydroquinone) outperforming T501 in natural esters by reducing acid value rise to <0.05 mg KOH/g after 1000 hours at 130 °C, while T511 (2,2′-methylenebis[4-methyl-6-*tert*-butylphenol]) excels in viscosity retention during thermal aging [77]. Figure 8.14 shows the typical molecular configuration of optimized antioxidants.

Anticorrosive agents: For copper and silver-based windings, such as benzotriazoles that suppress TAN increase by 40% in copper-catalyzed environments.

Moisture scavengers: Polar compounds or nanoparticles that reduce moisture activity without compromising dielectric strength, maintaining aw <0.2 in humid conditions.

Research Need: Long-term stability testing of additive-infused esters under high voltage (up to 500 kV), temperature (140–170 °C), and oxidation conditions, including multi-stressor simulations per IEEE C57.154 to validate 20–30 year service life.

8.6.2 Ester-Based Nanofluids

Incorporating nanoparticles into ester oils – forming so-called ester nanofluids – offers promise for improving electrical, thermal, and oxidation performance, with 2025 pilots in 10–50 MVA transformers reporting 15–25% higher loading capacities.

8.6.2.1 Benefits Observed

Studies have shown that dispersing small amounts of nanoparticles – such as TiO_2 (anatase, 20–50 nm), Al_2O^3, SiO_2, or BN (0.01–0.5 wt%) – into esters can significantly

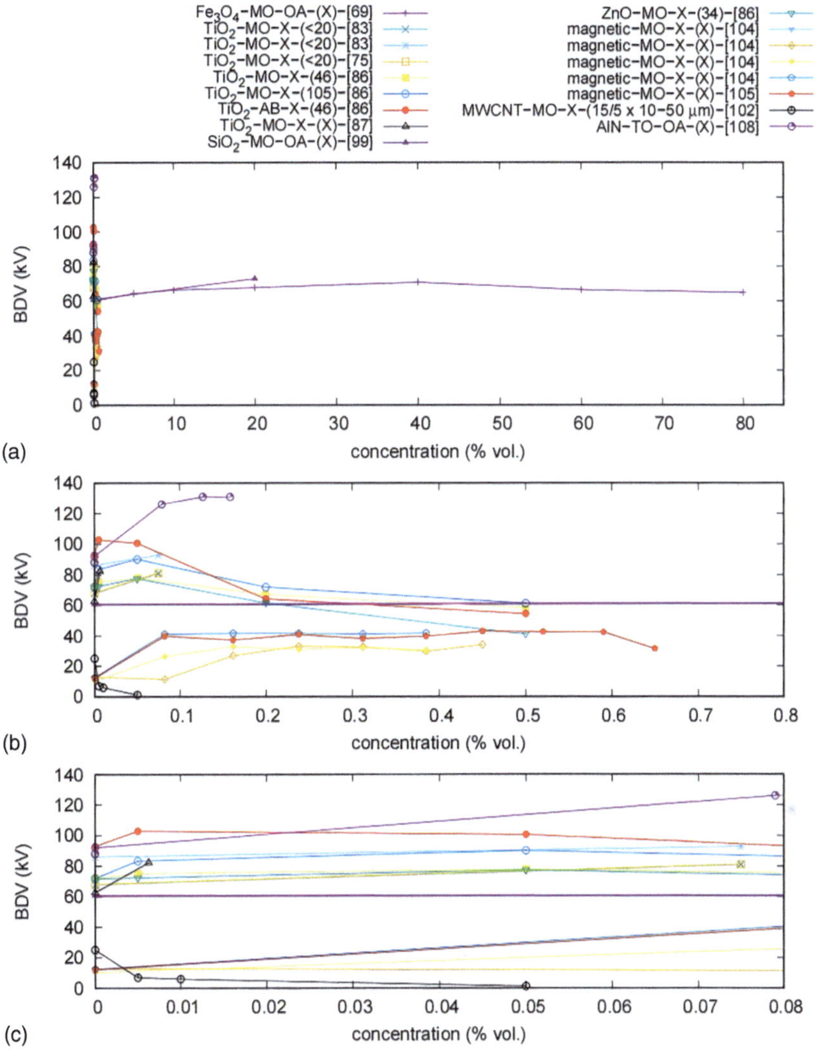

Figure 8.15 Graphical representation of values of AC BDV for various types of oil-based nanoparticle nanofluids in concentrations expressed by % of volume. (A) full range of concentration, (B) magnification in concentration range 0–0.8 vol% and (C) magnification in concentration range 0–0.08 vol%.

enhance their dielectric and thermal performance. Specifically, AC breakdown strength increases by 15–35% (and up to 58% in vegetable ester–TiO_2 systems) due to improved electron trapping and streamer suppression [78]. Thermal conductivity rises by 10–20% (e.g. 12–18% with h-BN in natural esters), enabling more effective heat dissipation in compact designs [79]. Nanoparticles also reduce partial discharges and slow streamer propagation by 20–40% through interfacial polarization, while delaying oxidative ageing via radical scavenging, thereby extending RPVOT by 30–50% in aged fluids. Figure 8.15 shows the relationship between breakdown voltage and mass fraction of nanoparticles in plant insulating oil.

8.6.2.2 Challenges

Nanofluid stability remains a major challenge, as agglomeration and sedimentation of nanoparticles significantly influence long-term performance. Experimental evidence shows that once the zeta potential falls below −30 mV after approximately 5000 hours of operation, the system experiences a measurable decline in dielectric properties, with performance losses in the range of 10–15%. These stability issues highlight the importance of surface chemistry and dispersion strategies in maintaining reliable performance over extended service times.

Another critical aspect is dielectric homogeneity. When nanoparticle concentrations exceed 0.3 wt%, the risk of localized electric field enhancement becomes non-negligible. Such inhomogeneities can serve as initiation sites for partial discharge (PD), undermining the insulating capability of the liquid dielectric [80]. Therefore, careful control of filler content and uniform dispersion is essential to avoid premature breakdown phenomena.

Processability also poses practical limitations. Standard vacuum filling techniques commonly used in insulating oil preparation are not directly transferable to nanoparticle-modified systems. The need for auxiliary treatments such as ultrasonication or surfactant addition can extend processing time by a factor of two to three, creating barriers to large-scale application. Thus, improvements in preparation routes are as important as advances in material design.

Recent research directions have focused on overcoming these limitations. Surface-functionalized nanoparticles, such as silane-coated SiO_2, have demonstrated enhanced compatibility with polar ester matrices, enabling dispersions that remain sedimentation-free for over 10 000 hours. At the same time, two-dimensional nanomaterials – including graphene oxide and hexagonal boron nitride – are being explored to impart multifunctionality, with 2025 studies reporting up to 25% suppression of PD activity. In parallel, multiphysics simulation frameworks that combine molecular dynamics (MD) and computational fluid dynamics (CFD) are emerging as powerful tools to model the aging behavior of these complex fluids, providing predictive insights into their long-term reliability under real grid operating conditions.

8.6.3 Diagnostics and Monitoring Innovation

The move to ester-based systems demands a paradigm shift in condition monitoring techniques, as diagnostic signatures differ from those of mineral oils [81], with recent 2025 studies highlighting the need for ester-specific algorithms to interpret gas evolution and aging byproducts accurately. Ongoing advancements, including AI-driven pattern recognition, are addressing these gaps to enable predictive maintenance in real-time.

8.6.3.1 Redefining Dissolved Gas Analysis (DGA)

Ester-based insulating fluids exhibit distinctive gas generation patterns compared to conventional mineral oils when subjected to electrical and thermal faults. These differences arise primarily from their polar molecular structure and higher oxygen

Table 8.12 Dominant gases generated under fault conditions in mineral oils and esters.

Fault type	Dominant gases in mineral oil	Dominant gases in esters
Partial discharge	H_2, CH_4	H_2, trace CO, lower CH_4
Arcing	C_2H_2, H_2	Lower C_2H_2, more CO_2, trace dimethyl disulfide
Thermal (>700 °C)	CH_4, C_2H_4	CO, CO_2, ethane as primary stray gas

reactivity, which influence both the type and concentration of decomposition products. Table 8.12 summarizes the dominant gases produced under typical fault conditions in mineral oil versus ester fluids. While mineral oils commonly generate hydrocarbon gases such as hydrogen (H_2), methane (CH_4), ethylene (C_2H_4), and acetylene (C_2H_2), ester fluids tend to produce more oxygenated species, particularly carbon monoxide (CO) and carbon dioxide (CO_2).

As shown in the table, partial discharges in esters still produce hydrogen but yield significantly less methane, with trace amounts of carbon monoxide as a distinguishing marker. Arcing events are associated with lower acetylene formation compared to mineral oils, while ester fluids generate more carbon dioxide and occasionally sulfur-containing compounds such as dimethyl disulfide. Under severe thermal faults (above 700 °C), esters primarily produce carbon monoxide and carbon dioxide, accompanied by ethane as a characteristic stray gas, in contrast to the hydrocarbon-rich gas profile of mineral oils.

These fundamental differences have important implications for dissolved gas analysis (DGA). Conventional IEC and IEEE interpretation tools, which were originally developed for mineral oils, cannot be directly applied to ester-based fluids without adjustment. Comparative studies conducted in 2025 on rapeseed- and soybean-derived natural esters confirmed this limitation, reporting hydrogen generation reduced by as much as 50–80% under thermal fault conditions relative to mineral oils. Consequently, new diagnostic ratios and interpretation algorithms tailored to ester fluids are needed to ensure reliable fault identification in ester-filled transformers.

Future Work:

A key direction for future research is the development of ester-specific dissolved gas analysis (DGA) codes. Current diagnostic frameworks, largely standardized under IEC TC 10, were designed for mineral oils and remain insufficiently adapted to the behavior of alternative fluids such as synthetic and natural esters. Ongoing efforts within IEC and related CIGRE working groups aim to establish qualification criteria and fault interpretation guidelines for esters, with particular relevance to their application in on-load tap changers (OLTCs) and bushings.

Another promising avenue involves the integration of machine learning techniques into DGA interpretation. Recent pilot studies conducted in 2025 demonstrated that advanced classification algorithms can achieve fault detection accuracies in the range of 90–95%. Such approaches not only enable more reliable identification of incipient faults but also support automated diagnostics by recognizing complex gas-evolution patterns that conventional ratio methods may overlook.

Finally, progress is being made in the development of real-time DGA sensors optimized for ester-based insulating liquids. Commercial solutions, such as Vaisala's online monitors originally designed for natural esters, are being adapted to capture ester-specific gas profiles with high sensitivity. These technologies promise early-stage fault identification in distribution transformers, thereby extending asset life and improving grid reliability through continuous, condition-based monitoring.

8.6.3.2 Ageing and Moisture Models

Because esters interact differently with cellulose and moisture [82], conventional ageing markers such as furan compounds (notably 2-FAL) are less reliable in natural ester systems. Compared with mineral oil–paper insulation, ester–paper systems exhibit both faster diffusion and lower formation rates of 2-FAL, reducing its effectiveness as a diagnostic indicator. Recent research (2025) has therefore proposed alternative markers, including methanol and levoglucosan, which show stronger correlations with the degree of polymerization (DP) of paper.

Moisture equilibrium in cellulose is also altered significantly in ester environments. While paper moisture typically stabilizes at 1–2% in mineral oil, it shifts to 0.3–0.5% in ester systems due to the 20–30 times higher moisture solubility of esters. This redistribution of water content influences both dielectric performance and degradation pathways. Under wet conditions, hydrolytic degradation behaves differently: natural esters exhibit acid formation rates 2–3 times slower than mineral oils, yet simultaneously generate larger quantities of glycerol byproducts.

These differences highlight a clear research need. Future diagnostic models must incorporate the coupled kinetics of ester–paper–moisture interactions, accounting for their unique transport and reaction behaviors. Validation will require long-term accelerated ageing studies – such as tests conducted at 130 °C for more than 5000 hours – together with the evaluation of alternative chemical markers and the integration of AI-based predictions of DP loss.

8.6.4 Grid-Level Integration Strategies

8.6.4.1 Modular, Ester-Optimized Transformer Designs

Instead of adapting mineral-oil transformers to esters, future designs will incorporate compact windings utilizing higher thermal classes (up to 140 °C hot-spot per IEEE C57.91), as demonstrated in ABB's eco-friendly ester-filled units up to 500 kVA. Hybrid cooling systems (oil–air–water) enhance efficiency by 15–20% in high-load scenarios, while intelligent diagnostics ports compatible with ester sensors support IoT integration for remote monitoring. Design Trend: Modular transformer platforms adaptable for mineral, ester, or synthetic biofluids based on regional grid needs, such as Virginia Transformer's 150 MVA natural ester prototypes and Trench Group's modular ester CTs for simplified retrofits [83].

8.6.4.2 Ester Use in HVDC and FACTS Devices

Ester fluids are being considered for advanced applications such as HVDC converter transformers – subject to high-frequency transients (up to 10 kHz) – with Hitachi Energy's 2025 successful testing of a 765 kV natural ester-filled unit showing 2× higher flash point and self-extinguishing properties. Static synchronous compensators (STATCOMs) benefit from minimized dielectric losses (<0.05% $\tan\delta$), while offshore floating substations prioritize biodegradability for spill mitigation in marine environments. Research Challenge: Validating ester performance under high switching frequencies (1–5 kHz), harmonic voltages (THD >5%), and power electronic interface stresses, with 2025 studies confirming 10–15% lower space charge accumulation in ester-HVDC insulation compared to mineral oil.

8.6.5 Policy and Standardization Gaps

8.6.5.1 Global Fluid Standards Alignment

Current ester-related standards have been progressively refined to support transformer applications. Key documents include IEC 62770:2024, which covers natural esters for transformers with updated requirements for unused fluids and enhanced oxidation stability tests; IEC 61099:2010, which specifies synthetic esters and is under revision to incorporate hybrid blends; and IEEE C57.147-2018, an application guide currently supplemented by 2025 draft updates addressing maintenance thresholds [84–86].

Despite these advances, significant discrepancies remain across regions and standards. Differences exist in test methodologies – such as adaptations of the Rotating Pressure Vessel Oxidation Test (RPVOT) – in fluid classification systems, including K-class definitions per IEC 61039 and Class C fire safety requirements, and in the interpretation of dielectric performance under DC stress. These inconsistencies highlight the need for stronger alignment of global fluid standards.

Future priorities in standardization include the harmonization of ester ageing indicators, with methanol proposed as a practical alternative to furan derivatives; the development of ester-specific dissolved gas analysis (DGA) protocols, potentially through Annex updates to IEC 60599; and the formalization of definitions for "hybrid esters," typically comprising 15–40% synthetic blends, alongside clear thresholds for bio-based content (e.g. >60% as specified in the EU RED III framework).

8.6.5.2 Sustainability Metrics and Carbon Accounting

The sustainability profile of ester insulating fluids varies considerably depending on feedstock, production process, and end-of-life management. Natural esters derived from rapeseed oil typically show a lifecycle carbon intensity of 1.1–1.6 kg CO_2 eq/L, whereas palm-derived esters and conventional mineral oils fall within 3.5–4.0 kg CO_2 eq/L. Bio-based manufacturing routes can further reduce emissions by 20–40% compared to petrochemical processes. Transport logistics and fluid recyclability also play critical roles: recovery rates of up to 90% through re-use and re-refining can cut net global warming potential (GWP) by as much as 50%.

From a policy perspective, there is a clear need for ISO-compatible models for carbon footprinting of transformer fluids, particularly cradle-to-gate LCAs aligned with ISO 14067. Incentive mechanisms, such as EU Emissions Trading System (ETS) credits, could accelerate the adoption of low-carbon ester fluids by utilities. Moreover, embedding ester fluid selection criteria within ESG frameworks and green finance instruments would support wider decarbonization strategies. Projections for 2025 indicate that ester-filled grids could deliver GWP reductions of 30–50% compared with mineral-oil-based systems.

8.6.6 Circular Economy and Recycling Innovation

End-of-life handling of ester fluids has the potential to significantly enhance grid sustainability. By enabling closed-loop re-refining processes, the demand for virgin feedstock can be reduced by 70–80%, contributing to both environmental and economic gains.

8.6.6.1 Re-Refining Technologies

Recent advances in re-refining technologies have demonstrated that 80–90% of used ester fluids can be recovered and restored. Techniques such as solvent extraction with *N*-methyl-2-pyrrolidone and hydroprocessing effectively remove moisture, acids, and polar degradation products, yielding regenerated fluids with dielectric strength above 50 kV and oxidation stability (RPVOT) exceeding 300 hours.

Looking ahead, a research frontier is the development of in-situ regeneration modules that can be integrated directly within transformers. These systems combine catalytic filtration, for instance, using alumina nanocarriers, with additive replenishment strategies [87]. Early 2025 pilot projects for natural esters report regeneration efficiencies of approximately 85%, highlighting the feasibility of real-time life extension technologies for insulating fluids.

8.6.6.2 Biodegradation Pathway Studies

Ester-based insulating oils are known for their rapid biodegradability, typically exceeding 90% within 28 days under aerobic conditions, as defined by OECD 301B protocols. However, their degradation dynamics under different environmental conditions require further investigation. In particular, anaerobic biodegradation in soils, especially at temperatures below 10 °C, progresses at rates 40–60% slower than aerobic breakdown due to limited microbial activity and oxygen availability. Understanding these differences is crucial for accurate environmental risk assessment and remediation planning. Figure 8.16 shows the biodegradability of different insulating oils.

Several important research questions remain open. One critical area is the identification of byproducts formed during the degradation of long-chain esters in soil. For example, short-chain fatty acids and glycerol are commonly observed intermediates under anaerobic conditions, yet their persistence and environmental impacts are not fully understood. Another key issue concerns the role of microbial populations. Strains such as Pseudomonas and Candida have been shown to

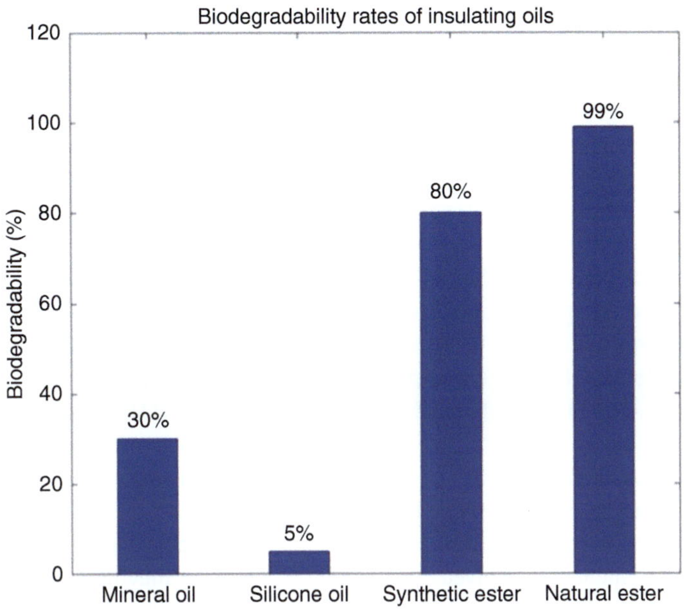

Figure 8.16 Biodegradability of different insulating oils. *Source:* Reproduced from [88], with permission of John Wiley & Sons.

influence degradation kinetics significantly, with recent isolates (2025) achieving up to 70% removal in vegetable oil analog systems. Finally, the potential of biotechnological interventions is increasingly relevant: genetically engineered microbes, optimized for esterase overexpression, could enhance ester degradation rates by twofold to threefold in oxygen-limited environments, offering a promising pathway for accelerated remediation following accidental spills.

8.6.7 Interdisciplinary Collaboration Opportunities

The transformation of transformer insulation fluids should no longer be regarded as an isolated challenge within electrical engineering. Instead, it represents a truly interdisciplinary field that spans multiple domains. In green chemistry, novel ester synthesis routes are emerging, including CO_2 electroreduction pathways, which align with global carbon-neutrality goals. Environmental toxicology plays a vital role in assessing spill risks, with studies showing that ester spills generally result in less than 1% long-term toxicity persistence. Data science contributes by advancing diagnostics, with predictive models now achieving fault identification accuracies of up to 95%. Mechanical engineering offers insights into fluid dynamics optimization for hybrid cooling systems, while public policy establishes roadmaps for sustainable grid operation, aligned with net-zero targets for 2050.

Addressing such a multifaceted challenge requires collective action. A strategic priority is the establishment of global consortia that bring together utilities, OEMs, fluid producers, and academic institutions. Examples include initiatives under CIGRE's Joint Working Group A2/D1 on modern insulating liquids, IEC TC 10, and the IEEE PES Transformers Committee. The 2025 update of CIGRE Technical Brochure 436 illustrates the value of such collaborations, consolidating service experiences and providing guidance for the safe and effective adoption of ester-based fluids worldwide.

8.6.8 Summary

The future of ester insulating oils lies at the intersection of material science, diagnostics, environmental policy, and grid innovation. To fully unlock their potential, several priority areas must be advanced. These include the development of customized ester formulations with improved cold-temperature performance (pour points down to −56 °C) and enhanced oxidation stability (RPVOT exceeding 600 hours through optimized antioxidant packages). Nanotechnology offers opportunities for next-generation performance improvements, such as the use of 0.1 wt% hexagonal BN to increase breakdown strength by up to 30%.

Equally important are diagnostic techniques tailored to ester-specific degradation signatures, with AI-enabled dissolved gas analysis now achieving accuracies approaching 90%. On the regulatory side, harmonized international standards – such as the upcoming IEC 62770:2024 extensions – will help ensure consistent quality and safety across global markets. Furthermore, circular economy strategies, enabling ester fluid recovery rates of 80–90% and lifecycle carbon reductions of 30–50%, highlight the broader environmental benefits of ester technologies.

By advancing these directions, ester fluids can achieve technical viability across the full voltage spectrum, including applications up to 765 kV. At the same time, they will contribute to a more sustainable, resilient, and low-carbon power grid infrastructure. With the ester market projected to grow to USD 300–600 million by 2032, their role in the next generation of insulating liquids is set to expand significantly.

8.7 Conclusion

The transition toward low-carbon and environmentally sustainable electrical systems is one of the most defining technological imperatives of the 21st century, accelerated by global net-zero targets and regulations like the EU's Renewable Energy Directive III (RED III), which prioritizes low-ILUC biofuels and insulating fluids. In this context, the development and deployment of environmentally friendly ester insulation oils represents both a necessary and promising evolution in transformer and switchgear technologies, with the global biodegradable transformer oil market valued at USD 5.895 billion in 2025 and projected to reach USD 7.958 billion by 2032 at a CAGR of 5.4%. Over the past two decades, natural and synthetic esters have matured from niche alternatives into credible mainstream dielectric fluids – offering not only safety and performance benefits but also aligning with global climate and environmental goals, as evidenced by their integration into over 500 000 ester-filled transformers worldwide by mid-2025.

This section has comprehensively reviewed the scientific basis, technological progress, and application prospects of ester-based insulating oils. The discussion spans their chemistry, feedstocks, physicochemical and electrical performance, environmental behavior, standardization, and research gaps. The following key conclusions can be drawn.

8.7.1 Technical Viability and Performance

Ester fluids, especially natural esters derived from vegetable oils and synthetic esters engineered for performance, have demonstrated excellent dielectric strength (50–65 kV AC breakdown for 2.5 mm gaps), high fire points (>300 °C for natural esters), and strong moisture-handling capabilities (up to 1200 ppm saturation). Their polar nature makes them particularly suitable for insulation-paper impregnation and long-term cellulose preservation, which translates to extended transformer life expectancy – up to 5–8 times longer than mineral oil systems under equivalent conditions. Despite some limitations such as high pour points (for natural esters, −15 to 0 °C unmodified) or higher cost (synthetic esters at 20–35% premium), the balance of performance and safety provided by esters is increasingly attractive – particularly in Ref. [88]:

Distribution transformers in urban and indoor settings, where K-class fire safety reduces insurance premiums by 20–30%.
Renewable energy installations with fire safety demands, such as offshore wind farms supporting 150% overload capacities.
Environments sensitive to environmental contamination (e.g. water protection zones, marine/offshore), where rapid biodegradation minimizes spill risks.

Moreover, recent innovations in formulation chemistry, additive systems (e.g. benzotriazole passivators), and nanotechnology integration – such as 0.02–0.1 wt% SiO_2 nanoparticles boosting breakdown strength by 20–30% – are paving the way for customized esters that can serve across medium and high voltage levels up to 500 kV.

8.7.2 Environmental and Sustainability Benefits

The sustainability advantages of ester fluids are perhaps their most distinguishing feature. Their high biodegradability (>97% within 28 days under OECD 301B for natural esters), non-toxicity, and renewable sourcing potential make them well aligned with circular economy principles and environmental regulations such as RED III, which restricts high-ILUC feedstocks like palm oil [89]. In contrast to mineral oils, esters pose a much lower ecological risk in the event of leaks or accidental release, as they typically degrade within weeks under aerobic conditions, reducing environmental persistence by 90–95%.

From a life cycle assessment (LCA) perspective, ester fluids – particularly those derived from waste feedstocks or certified sustainable agriculture – achieve significantly lower carbon footprints than conventional mineral oils. For example, natural esters such as FR3 emit up to 56 times less greenhouse gases on a cradle-to-gate

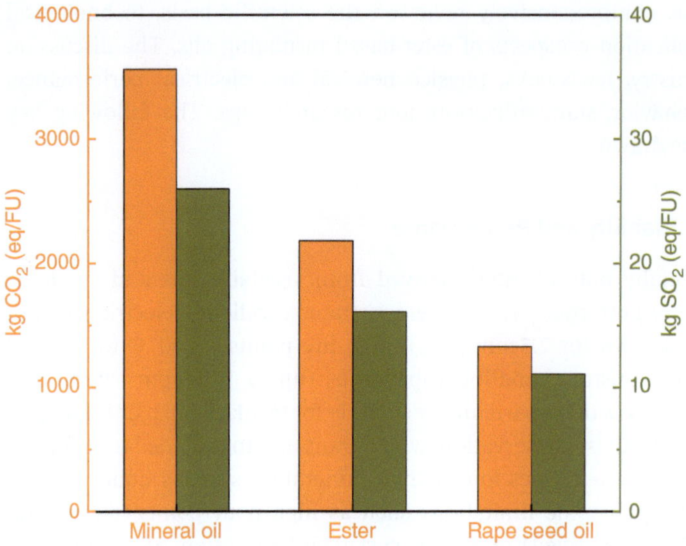

Figure 8.17 The contribution of (a) global warming potential and (b) total acidification potential of the base fluid. *Source:* Reproduced from [91], with permission of Springer Nature; visualization by the authors.

basis. Figure 8.17 shows contribution of different base fluid to global warming and acidification. Beyond direct emissions, ester fluids can also contribute to indirect reductions through several mechanisms [90]: improved cooling efficiency due to their 10–20% higher thermal conductivity (saving 5–10% in operational energy use), lower transformer failure rates due to better aging resistance (enhancing grid reliability by 15–25%), and extended equipment lifetimes that reduce replacement needs and cut lifecycle emissions by 20–40%.

Despite these benefits, wider adoption will depend on stronger feedstock sustainability tracking systems (e.g. ISCC PLUS certification), standardized carbon accounting methodologies such as ISO 14067, and robust end-of-life handling infrastructure. Recycling and regeneration efforts, in particular, must scale up toward ambitious targets, with recycling rates aiming for 80% by 2030.

8.7.3 Standardization and Diagnostics Gaps

The widespread deployment of ester insulation oils requires harmonization across international standards. At present, inconsistencies exist between IEC, IEEE, and various regional frameworks regarding test methods, classification schemes (e.g. K-class fluids under EN IEC 61039:2025), and diagnostic procedures such as dissolved gas analysis (DGA). For example, IEC 62770:2024 specifies natural ester limits, while IEC 61099:2016 (currently under review) governs synthetics.

Because ester fluids respond differently to thermal and electrical stress compared with mineral oils, diagnostic protocols must be adapted [91]. Ester-filled transformers show distinct DGA signatures, including higher CO_2/CO ratios (10–50) and lower H_2/methane concentrations during thermal faults, which deviate from

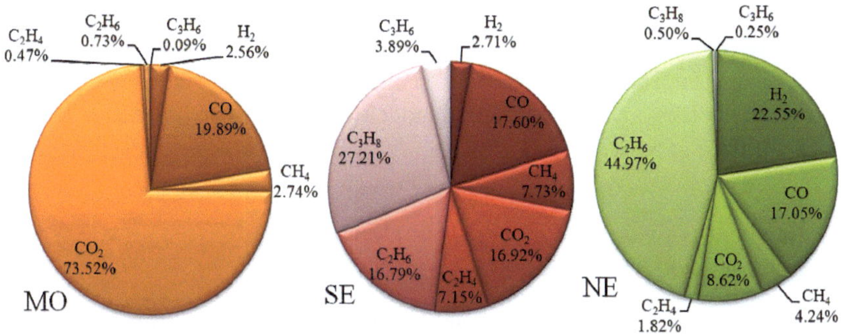

Figure 8.18 The percentage share of analyzed gases in the sum of gases generated in mineral oil (MO), synthetic ester (SE) and natural ester (NE) at 350 °C.

conventional fault gas ratios. Figure 8.18 shows the total dissolved gas ratio in different insulating oils during arc discharge. Recent studies on stray gassing in retrofilled units confirm these unique patterns [92]. Likewise, the use of furan compounds as aging markers is problematic, as esters suppress furan formation by about 50% but introduce new degradation products such as ethyl methyl carbonate.

Addressing these challenges will require the development of new diagnostic models, data-driven algorithms (e.g. AI-enhanced adaptations of the Duval Triangle), and ester-specific standards. Progress is underway with documents like IEEE C57.154-2022, while ongoing IEC task forces are targeting full harmonization of ester diagnostics and classification by 2027.

8.7.4 Research Needs and Future Development

Future progress in ester insulation oils hinges on both technological refinement and system-level integration. Several critical research directions stand out.

Formulation chemistry must advance toward esters with tailored properties suited to diverse climates and voltage classes. Hybrid blends, for example, may achieve pour points of −30 °C without compromising biodegradability. Nano-enhanced esters also show promise, with surface-modified SiO_2 or h-BN nanoparticles boosting dielectric strength by up to 30% and thermal conductivity by 12%, as demonstrated in 2025 Manchester University pilot studies.

Another important priority is aging behavior modeling under coupled thermal, electrical, and moisture stresses. This includes tackling knowledge gaps in long-gap breakdown (>30 mm) and hydrolysis kinetics. From a sustainability perspective, integration into ESG frameworks, carbon credit systems, and certification schemes such as RSB will become essential, with LCAs emphasizing the advantages of waste-derived feedstocks. In parallel, recycling and regeneration technologies are needed to close the loop, with filtration-based reclamation systems targeting 70–90% efficiency.

Collaborative initiatives are crucial to accelerate these developments. Platforms like CIGRE Working Group D1.52 illustrate how utilities, manufacturers, research institutes, and policymakers can coordinate to address technical, regulatory, and sustainability challenges. Ultimately, the future of ester research will depend on interdisciplinary innovation that combines electrical engineering, materials science, environmental studies, and computational modeling. EU Horizon programs, which in 2025 prioritized nano-ester scalability, exemplify the type of system-level support needed.

8.7.5 Outlook and Strategic Importance

Within the broader shift toward sustainable energy infrastructure, ester insulating oils represent far more than a simple replacement for mineral oils. They act as enablers of transformation. Market forecasts project a compound annual growth rate (CAGR) of 6.8% for ester-based transformer oils through 2032, driven by the expansion of smart grid systems. As the grid becomes more decentralized, digitized, and decarbonized, demand for environmentally responsible and high-performance insulation fluids will intensify.

Ester fluids are uniquely suited to meet these requirements. Their high flash points – two to three times greater than mineral oils – provide intrinsic fire safety that supports deployment in urban areas without the need for barriers. Their environmental compatibility, including life-cycle greenhouse gas emissions that are up to 56 times lower, makes them an essential tool for net-zero roadmaps by 2050. Their thermal efficiency (10–15% improved cooling) and superior dielectric performance under moist conditions (20–30% higher than mineral oils) further reinforce their role as next-generation insulating fluids.

Moreover, ester compatibility with innovative transformer architectures – ranging from offshore wind prototypes (e.g. 10 MVA units enhanced with nanofluids) to EV charging networks and containerized substations – positions them as critical elements in the future smart and green grid. Market penetration of 20–30% by 2030 appears feasible under current trajectories.

Ultimately, ester insulating oils illustrate a compelling case where performance and sustainability reinforce one another rather than compete. With continued advances in science, diagnostics, and policy, esters are poised to become the default insulating fluid for future power grids, directly contributing to global agendas on resilience, net-zero emissions, and environmental stewardship.

References

1 Martin, R., Athanassatou, H., Duart, J.C. et al. (2010). Experiences in service with new insulating liquids. Cigré Technical Brochure, 436.

2 Tenbohlen, S. and Koch, M. (2010). Aging performance and moisture solubility of vegetable oils for power transformers. *IEEE Transactions on Power Delivery* 25 (2): 825–830.

3 IEC (2020). *Fluids for Electrotechnical Applications – Mineral Insulating Oils for Electrical Equipment (IEC 60296:2020).* International Electrotechnical Commission.

4 Koch, M., Krueger, M., and Tenbohlen, S. (2010, April). On-site methods for reliable moisture determination in power transformers. In: *IEEE PES T & D 2010*, 1–6. IEEE.

5 Rafiq, M., Shafique, M., Ateeq, M. et al. (2024). Natural esters as sustainable alternating dielectric liquids for transformer insulation system: analyzing the state of the art. *Clean Technologies and Environmental Policy* 26 (3): 623–659.

6 Chronis, I., Kalogeropoulou, S., and Psomopoulos, C.S. (2021). A review on the requirements for environmentally friendly insulating oils used in high-voltage equipment under the eco design framework. *Environmental Science and Pollution Research* 28 (26): 33828–33836.

7 Organization for Economic Co-operation and Development (1992). Ready biodegradability: manometric respirometry test. In: . OECD Guideline for Testing of Chemicals. No. 301F.

8 Moody, G. and Stephenson, G. (2024, January). How can esters improve the sustainability of both intrinsic. *24th International Colloquium Tribology*, 245. Expert Verlag GmbH.

9 IEC (2013). Fluids for electrotechnical applications – unused natural esters for transformers and similar electrical equipment (IEC 62770:2013). International Electrotechnical Commission.

10 International Energy Agency (2023). *Electricity Market Report 2023*. IEA.

11 Zou, M., Ke, J., Tao, Q. et al. (2025, May). Dielectric and gas generation characteristics of synthetic ester insulation oil under accelerated thermal aging. *2025 IEEE 8th International Electrical and Energy Conference (CIEEC)*, 3396–3401. IEEE.

12 IEEE Std C57.147-2018 *Guide for Acceptance and Maintenance of Natural Ester Fluids in Transformers*. IEEE Power and Energy Society.

13 Thiel, C., Amillo, A.G., Tansini, A. et al. (2022). Impact of climatic conditions on prospects for integrated photovoltaics in electric vehicles. *Renewable and Sustainable Energy Reviews* 158: 112109.

14 Shen, Z., Wang, F., Wang, Z., and Li, J. (2021). A critical review of plant-based insulating fluids for transformer: 30-year development. *Renewable and Sustainable Energy Reviews* 141: 110783.

15 IEC (2010). *Insulating Liquids – Specifications for Unused Synthetic Organic Esters for Electrical Purposes (IEC 61099:2010)*. International Electrotechnical Commission.

16 Oparanti, S.O., Obebe, E.O., Fofana, I., and Jafari, R. (2025). A state-of-the-art review on the potential of waste cooking oil as a sustainable insulating liquid for green transformers. *Applied Sciences* 15 (14): 7631.

17 Chairul, I.S., Bakar, N.A., Othman, M.N. et al. (2018). Development of waste cooking oil methyl ester as potential electrical insulating fluid for power transformer. *ARPN Journal of Engineering and Applied Sciences* 13 (20): 8154–8162.

18 Gunstone, F.D. (2019). *Individual Commodity Oils and Fats*. AOCS.

19 Karthik, M. and Narmadhai, N. (2021). A survey on natural esters based insulating fluid medium for transformer applications. *Materials Today: Proceedings* 45: 2022–2028.

20 Deraman, M.N., Bakar, N.A., Ab Aziz, N.H., and Chairul, I.S. (2020). The experimental study on the potential of waste cooking oil as a new transformer insulating oil. *Journal of Advanced Research in Fluid Mechanics and Thermal Sciences* 69 (1): 74–84.

21 Abdelmalik, A.A. (2012). The feasibility of using a vegetable oil-based fluid as electrical insulating oil. Doctoral dissertation. University of Leicester.

22 Raof, N.A., Hamid, H.A., Mohamad Aziz, N.A., and Yunus, R. (2022). Prospects of plant-based trimethylolpropane esters in the biolubricant formulation for various applications: a review. *Frontiers in Mechanical Engineering* 8: 833438.

23 Tao, Y., Chen, B., Liu, L., and Tan, T. (2012). Synthesis of trimethylolpropane esters with immobilized lipase from Candida sp. 99–125. *Journal of Molecular Catalysis B: Enzymatic* 74 (3–4): 151–155.

24 Cilliyuz, Y., Bicen, Y., Aras, F., and Aydugan, G. (2021). Measurements and performance evaluations of natural ester and mineral oil-immersed identical transformers. *International Journal of Electrical Power & Energy Systems* 125: 106517.

25 Wang, X. (2011). *Partial Discharge Behaviours and Breakdown Mechanisms of Ester Transformer Liquids Under ac Stress.* United Kingdom: The University of Manchester.

26 Jayyid, U.L., Fadillah, A., Kurniawan, M. et al. (2024, November). Comparative study of mineral oil, synthetic ester, and natural Ester as liquid insulation in transformers. *2024 6th International Conference on Power Engineering and Renewable Energy (ICPERE)*, 1–6. IEEE.

27 Malde, J., Daghrah, M., and Gyore, A. (2020). Natural and synthetic ester liquids-how they differ what they deliver. *Transformer Technology* 5: 56–67.

28 Bandara, K., Ekanayake, C., and Saha, T.K. (2015, July). Compare the performance of natural ester with synthetic ester as transformer insulating oil. *2015 IEEE 11th International Conference on the Properties and Applications of Dielectric Materials (ICPADM)*, 975–978. IEEE.

29 POLICY UPDATE (2012). *Proposed Amendments to Eu Fuel Quality and Renewable Energy Directives.* POLICY UPDATE.

30 Business Research (2025). *Ester Transformer Oils Market Size, Share, Growth, and Industry Analysis, By Type (Natural Ester Transformer Fluid, Synthetic Ester Transformer Oil), By Application (Ordinary Transformer, EHV Transformer), Regional Insights and Forecast from 2025 to 2034.* Business Research.

31 Yu, J., Ma, Y., Sun, J. et al. (2017). Microalgal oil from *Schizochytrium* sp. prevents HFD-induced abdominal fat accumulation in mice. *Journal of the American College of Nutrition* 36 (5): 347–356.

32 Mao, J., Wang, Y., Zhang, B. et al. (2024). Advances in electrocarboxylation reactions with CO_2. *Green Carbon* 2 (1): 45–56.

33 Zhang, H., Li, H., Xu, C.C., and Yang, S. (2019). Heterogeneously chemo/enzyme-functionalized porous polymeric catalysts of high-performance for efficient biodiesel production. *ACS Catalysis* 9 (12): 10990–11029.

34 Fofana, I. (2013). 50 years in the development of insulating liquids. *IEEE Electrical Insulation Magazine* 29 (5): 13–25.

35 CIGRÉ Working Group D1.70 (2021). Dielectric performance of insulating liquids for transformers (Technical Brochure 856). CIGRÉ.

36 Dang, V.H., Beroual, A., and Perrier, C. (2010, October). Comparative study of streamer phenomena in mineral, synthetic and natural ester oils under lightning impulse voltage. *2010 International Conference on High Voltage Engineering and Application*, 560–563. IEEE.

37 Skytt, T. (2012). Considerations for the design, manufacture, and retro–filling of power transformers with high fire point, biodegradable Ester fluids. *Cigré 2012 Paris Session.*

38 Li, S., Wang, F., Rao, U.M. et al. (2024). Hints to lightning breakdown enhancement for synthetic ester insulating oil: unraveling the impact of molecular chain length on electron structural parameters under electric field stress. *Journal of Molecular Liquids* 408: 125364.

39 Shu, Y., Wang, C., Huang, Z., et al. (2024, August). Enhanced oxidation stability of natural ester transformer oil by a natural antioxidant. *2024 IEEE International Conference on High Voltage Engineering and Applications (ICHVE)*, 1–5. IEEE.

40 Li, S., Wang, F., Rozga, P. et al. (2024). The hidden vulnerabilities of pentaerythritol synthetic ester insulating oil for eco-friendly transformers: a dive into molecular behavior. *IEEE Transactions on Dielectrics and Electrical Insulation* 32 (3): 1413–1421.

41 Kurzweil, P., Schell, C., Haller, R. et al. (2021). Environmental impact and aging properties of natural and synthetic transformer oils under electrical stress conditions. *Advanced Sustainable Systems* 5 (8): 2100079.

42 Wang, F., Ouyang, L., Song, C. et al. (2023). Enhancing dielectric and thermal performances of synthetic-ester insulating oil via blending with natural ester. *IEEE Transactions on Dielectrics and Electrical Insulation* 30 (3): 1115–1124.

43 Lyutikova, M., Korobeynikov, S., and Konovalov, A. (2021). Evaluation of the properties of mixtures of aromatic mineral oil and synthetic ester for high-voltage equipment. *IEEE Transactions on Dielectrics and Electrical Insulation* 28 (4): 1282–1290.

44 Hamdi, A., Fofana, I., and Djillali, M. (2017). Stability of mineral oil and oil–ester mixtures under thermal ageing and electrical discharges. *IET Generation Transmission and Distribution* 11 (9): 2384–2392.

45 Huang, Z., Cheng, C., Zhuo, R. et al. (2025, August). Divergent voltage dependence of frequency dielectric response in natural ester and mineral oil-impregnated press-board. *2025 IEEE 5th International Conference on Electrical Materials and Power Equipment (ICEMPE)*, 1–4. IEEE.

46 Kumar, A. (2022, August). Natural ester oil a sustainable alternative insulation oil to mineral oil for power transformer industries and electric power system. *ICSET: International Conference on Sustainable Engineering and Technology* (Vol. 1, No. 1, pp. 119–125).

47 Carvalhosa, S., Leite, H., Soares, M. et al. (2022, March). Ester-based dielectric fluid for power transformers: design and test experience under the GreenEst project. In: *Journal of Physics: Conference Series* (Vol. 2213, No. 1, p. 012026). IOP Publishing.

48 Rao, U.M., Fofana, I., Jaya, T., et al. (2019). Preliminary study of ester-based fluid for application in transformers serving in cold climatic regions. *Proc. 14th CIGRE Canada Expo Power Syst*, 1–6.

49 Li, S., Wang, F., Ouyang, L. et al. (2024). Differential low-temperature AC breakdown between synthetic ester and mineral oils: insights from both molecular dynamics and quantum mechanics. *IEEE Transactions on Dielectrics and Electrical Insulation* 31 (3): 1296–1304.

50 Obebe, E.O., Hadjadj, Y., Oparanti, S.O., and Fofana, I. (2025). Enhancing the performance of natural ester insulating liquids in power transformers: a comprehensive review on antioxidant additives for improved oxidation stability. *Energies* 18 (7): 1690.

51 Wang, F., Zhou, J., Dan, L. et al. (2024). Advancing thermal stability in natural ester oil-paper insulation systems via precision nanostructuring with parylene films: experimental and molecular-level comprehensive assessment. *Polymer Testing* 140: 108592.

52 Hamed, G.R. (1992). Materials and compounds. In: *Engineering with Rubber, How to Design Rubber Components* (ed. A.N. Gent), 11–31. Hanser Publishers.

53 Daris, I., Ahmad, M., and Nurhakim, H.R. (2021, October). Comparative study of accelerated thermal aging of papers in mineral oil, natural ester, and gas-to-liquid. *2021 3rd International Conference on High Voltage Engineering and Power Systems (ICHVEPS)*, 323–328. IEEE.

54 Piotrowski, T. and Markowska, D. (2025). Carbon footprint of power transformers evaluated through life cycle analysis. *Energies* 18 (6): 1373.

55 Kong, M., Liu, F., Zhang, W. et al. (2021). High resolution investigation of pH on control of mobile and organic phosphorus and associated release kinetics following lanthanum modified bentonite (LMB) amendment at the water-sediment interface. *Journal of Cleaner Production* 295: 126148.

56 Wang, H., Zhang, Z., Li, J. et al. (2022). Study on bubble formation of different moisture content oil-immersed cardboards in natural ester oil. *Energies* 15 (3): 957.

57 Liu, Y. (2022). Sustainability Evaluation of Transformer Insulating Liquid against UN SDGs.

58 Módenes, A.N., Sanderson, K., Trigueros, D.E.G. et al. (2018). Insights on the criteria of selection of vegetable and mineral dielectric fluids used in power transformers on the basis of their biodegradability and toxicity assessments. *Chemosphere* 199: 312–319.

59 Taheripour, F., Mueller, S., Emery, I. et al. (2024). Biofuels induced land use change emissions: the role of implemented land use emission factors. *Sustainability* 16 (7): 2729.

60 Deepa, R., Anandhi, A., and Alhashim, R. (2021). Volumetric and impact-oriented water footprint of agricultural crops: a review. *Ecological Indicators* 130: 108093.

61 Luna Juncal, M.J., Masino, P., Bertone, E., and Stewart, R.A. (2023). Towards nutrient neutrality: a review of agricultural runoff mitigation strategies and the development of a decision-making framework. *Science of the Total Environment* 874: 162408.

62 Reid, J. (2011). The use of ester-based transformer liquids for reduced fire risk and lower costs. *Hazards 31 Conference*, IChemE.

63 Nogueira, T., Carvalho, J., and Magano, J. (2022). Eco-friendly ester fluid for power transformers versus mineral oil: design considerations. *Energies* 15 (15): 5418.

64 Oparanti, S.O., Rao, U.M., and Fofana, I. (2023). Natural esters for green transformers: challenges and keys for improved serviceability. *Energies* 16 (1): 61.

65 Wang, Y., Wang, R., Pan, K. et al. (2022). Detailed procedures of retrofilling transformers with FR3 natural ester and residual mineral oil content testing. *IET Generation, Transmission and Distribution* 16 (10): 1901–1913.

66 Düzkaya, H. (2024). Applications of natural esters alternative to mineral oils in power transformers. *Mugla Journal of Science and Technology* 10 (1): 8–16.

67 Villarroel, R., García de Burgos, B., and García, D.F. (2021). Moisture dynamics in natural-ester filled transformers. *International Journal of Electrical Power & Energy Systems* 124: 106172.

68 Deb, S. and Srikant, R. (2001). Congestion control for fair resource allocation in networks with multicast flows. *Proceedings of the 40th IEEE Conference on Decision and Control, Orlando, FL, USA*, 2, 1911–1916.

69 Gehani, A., Hanz, D., Rushby, J. et al. (2011). On the (f)utility of untrusted data sanitization. *MILCOM 2011 Military Communications Conference*, Baltimore, MD, USA, 1261–1266.

70 Ahmad, A. (2022). *Sustainable Transformer Technologies for an Evolving Grid*. Berlin: CWIEME.

71 Huang, Y., Lu, W., Qi, W., and Zhang, J. (2025). Molecular dynamics simulation of the decomposition and gas generation characteristics of natural ester insulating oils under overheated conditions. *IET Nanodielectrics* 8 (1): e70016.

72 Zhang, Y., Zhang, L., Zhang, Y. et al. (2021). Effect of cellulose particle and moisture content on insulation characteristics of natural ester. *IEEE Transactions on Dielectrics and Electrical Insulation* 28 (3): 1061–1069.

73 Villante, C., Gallucci, K., Di Giuliano, A. et al. (2025). Liquid e-fuels for a sustainable future: a comprehensive review of production, regulation, and technological innovation. *Renewable and Sustainable Energy Reviews* 175: 113067.

74 Guo, D., Xue, Z., Yin, Y. et al. (2025). Nano-modifiers enhance the performance of natural ester transformer oils: challenges and future directions. *Frontiers of Chemical Science and Engineering* 19 (45).

75 Yang, Z., Wang, F., Wang, C. et al. (2023). Optimizing molecular structure for trimethylolpropane ester-insulating oil: achieving high fluidity and stability. *IEEE Transactions on Dielectrics and Electrical Insulation* 30 (6): 1432–1440.

76 Dai, W., Yan, J., Zou, D. et al. (2025). Molecular simulation on a new type of two-element mixed natural ester insulating oil with upgraded fluidity and oxidative stability. *ACS Omega* 10 (16): 16633–16641.

77 Zhang, Y., Wang, F., Li, S. et al. (2025). Comparison of phenolic antioxidants' impact on thermal oxidation stability of pentaerythritol ester insulating oil. *IEEE Transactions on Dielectrics and Electrical Insulation* 32 (3): 1518–1527.

78 Fal, J., Mahian, O., and Żyła, G. (2018). Nanofluids in the service of high voltage transformers: breakdown properties of transformer oils with nanoparticles, a review. *Energies* 11 (11): 2942.

79 Huang, Z., Li, J., Yao, W. et al. (2019). Electrical and thermal properties of insulating oil-based nanofluids: a comprehensive overview. *IET Nanodielectrics* 2 (1): 27–40.

80 Chen, J., Sun, P., Sima, W. et al. (2019). A promising nano-insulating-oil for industrial application: electrical properties and modification mechanism. *Nanomaterials* 9 (5): 788.

81 Bustamante, S., Manana, M., Arroyo, A. et al. (2019). Dissolved gas analysis equipment for online monitoring of transformer oil: a review. *Sensors* 19 (19): 4057.

82 Li, J., Zhang, Z., Grzybowski, S., and Liu, Y. (2012). Characteristics of moisture diffusion in vegetable oil-paper insulation. *IEEE Transactions on Dielectrics and Electrical Insulation* 19 (5): 1650–1656.

83 Tenbohlen, S., Koch, M., Vukovic, D. et al. (2008, August). Application of vegetable oil-based insulating fluids to hermetically sealed power transformers. *Cigre Session*, 42, 24–29.

84 International Electrotechnical Commission (2021). *IEC 62770: Fluids for Electrotechnical Applications – Unused Natural Esters for Transformers and Similar Electrical Equipment*. Geneva: IEC.

85 International Electrotechnical Commission (2011). *IEC 61099: Insulating Liquids – Specifications for Unused Synthetic Organic Esters for Electrical Purposes*. Geneva: IEC.

86 Institute of Electrical and Electronics Engineers (2008). *IEEE C57.147–2008: IEEE Guide for the Acceptance and Maintenance of Natural Ester Fluids in Transformers*. Piscataway, NJ: IEEE.

87 Jacob, J., Preetha, P., and Krishnan, S.T. (2020). Review on natural ester and nanofluids as an environmental friendly alternative to transformer mineral oil. *IET Nanodielectrics* 3: 33–43.

88 Rozga, P., Beroual, A., Przybylek, P. et al. (2020). A review on synthetic ester liquids for transformer applications. *Energies* 13 (23): 6429.

89 Asano, R. and Page, S.A. (2014). Reducing environmental impact and improving safety and performance of power transformers with natural ester dielectric insulating fluids. *IEEE Transactions on Industry Applications* 50 (1): 134–141.
90 Våg, C., Marby, A., Kopp, M. et al. (2002). A comparative life cycle assessment of the manufacture of base fluids for lubricants. *Journal of Synthetic Lubrication* 19 (1): 39–57.
91 Hussein, A.R., Dakhil, A.M., Rashed, J.R., and Othman, M.F. (2022). Intelligent expert system for diagnosing faults and assessing quality of power transformer insulation oil by DGA method. *Misan Journal of Engineering Sciences* 1 (1): 47–57.
92 Przybylek, P. and Gielniak, J. (2019). Analysis of gas generated in mineral oil, synthetic ester, and natural ester as a consequence of thermal faults. *IEEE Access* 7: 65040–65051.

9

Multi-Dimensional Evaluation, Optimization, and Operating Condition Evolution Characteristics of Eco-Friendly Insulating Gas Performance

Song Xiao, Yi Li, and Shengyao Shi

Wuhan University, State Key Laboratory of Power Grid Environmental Protection, School of Electrical Engineering and Automation, Bayi Road 299#, Hubei 430072, China

9.1 Introduction

Gas-insulated equipment (GIE) in the power industry has the advantages of a small footprint, high reliability, and a long dimensional cycle, playing an important role in the transmission and distribution systems at various voltage levels. GIE mainly includes gas-insulated switchgear (GIS), gas-insulated line (GIL), gas-insulated transformer (GIT), and gas-insulated cabinet (GIC), which are mostly filled with high-pressure (0.13–0.6 MPa) sulfur hexafluoride (SF_6) gas as an excellent insulation or arc extinguishing medium [1].

However, the global warming potential (GWP) of SF_6 is 24 300 times that of CO_2, far exceeding the other five greenhouse gases limited in the Kyoto Protocol, and is currently the gas with the strongest greenhouse effect [2]. At the same time, SF_6 is not easy to degrade due to its very stable physical and chemical properties, and its atmospheric life can reach more than 3200 years [3]. Therefore, even if only a small amount of SF_6 leaks into the atmosphere from electrical equipment, it will have a serious impact on the global greenhouse effect after a long period of accumulation. Figure 9.1 shows the trend of changes in the global SF_6 atmospheric content. According to monitoring, the global SF_6 atmospheric content is 1.163 times that of 5 years ago, and the global temperature rise caused by SF_6 in the atmosphere has reached 0.004°C [5, 6]. Therefore, strengthening the control of SF_6 gas use and reducing SF_6 gas emissions are of great significance for countries around the world to achieve greenhouse gas emission reduction targets and achieve "net zero".

Gases such as perfluoroisobutyronitrile (C_4F_7N), perfluoro(3-methyl-2-butanone) ($C_5F_{10}O$), perfluoro(2-methyl-3-pentanone) ($C_6F_{12}O$), 1,3,3,3-tetrafluoropropene (HFO-1234ze(E)), perfluorocyclobutane(c-C_4F_8), and trifluoroiodomethane (CF_3I) have attracted widespread attention due to their great eco-friendly and insulating properties. To comprehensively evaluate the application prospects of the above SF_6 eco-friendly insulating alternative gases, three basic requirements must be met: environmental, health and safety, and technical requirements [7–9]. This chapter

Next-Generation Electrical Insulating Materials: Low-Carbon, Eco-Friendly, and Sustainable Solutions, First Edition. Edited by Jun-Wei Zha.

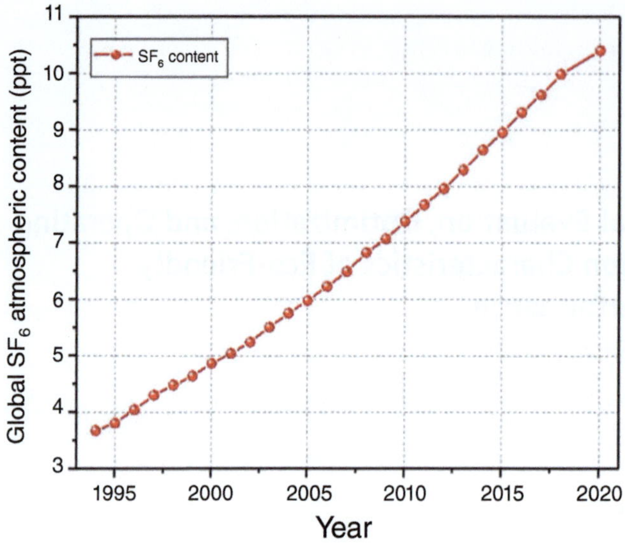

Figure 9.1 Trends in global SF_6 atmospheric content. *Source:* Adapted from [4].

will focus on the above-mentioned basic requirements, based on the previous research results, will conduct a multi-dimensional evaluation of various SF_6 eco-friendly insulating alternative gases, and comprehensively propose reasonable multi-dimensional performance optimization suggestions based on the state evolution characteristics of eco-friendly insulating gases under the background of working conditions.

9.2 Multi-Dimensional Evaluation of Eco-Friendly Insulating Gases

At present, there is no comprehensive evaluation system for the performance of eco-friendly gas-insulating media. Due to the wide variety of potential gas-insulating media, systematic research on each type of gas will consume a lot of manpower and financial resources. Therefore, it is necessary to comprehensively evaluate the common types of gas-insulating media from multiple perspectives, select one or two potential SF_6 substitutes with excellent comprehensive performance, and conduct in-depth systematic research in the future.

As a substitute for SF_6, it must meet the following basic requirements:

1) Environmental requirements: As a new type of gas-insulating medium, it should first have excellent environmental protection characteristics. That is, it needs to have a sufficiently low greenhouse effect relative to SF_6, and at the same time, it does not damage the ozone layer and has no ecological toxicity. The original intention of the development of eco-friendly gases is to solve the greenhouse

effect problem of SF_6 gas, making it necessary to investigate the environmental protection characteristics of various gas-insulating media.

2) Health and safety requirements: Since the pressure inside gas-insulated electrical equipment is generally 0.15–0.5 MPa, which is higher than one atmosphere, the gas inside the equipment will flow into the atmosphere around the equipment through diffusion or leakage due to equipment failure; at the same time, maintenance personnel often need to open the equipment safety valve to take gas or change air during routine maintenance, that is, the operation and maintenance personnel will inevitably come into contact with the gas inside the equipment. Therefore, it is required that the new gas-insulating medium should meet the requirements of nontoxicity or sufficiently low toxicity; in addition, the new gas-insulating medium should meet the requirements of stable chemical properties and nonflammable and nonexplosive, otherwise combustion or explosion under conditions such as electric sparks will cause irreversible damage to the equipment and personal safety.
3) Technical requirements: On the basis of meeting environmental protection and safety, the new insulating medium should also have a low liquefaction temperature, good insulation strength, and excellent material compatibility. Since the internal gas pressure of GIE is high, the gas-insulating medium should have a high saturated vapor pressure to ensure that it does not liquefy at a certain pressure and temperature; in addition, the gas-insulating medium itself should have excellent insulation properties. The insulation performance of the new gas determines whether it can completely replace SF_6 gas in actual engineering applications. If the insulation performance of the new gas does not reach the level of pure SF_6 under the same conditions, it is necessary to improve the insulation performance through equipment pressure or volume, which is not economical for equipment manufacturing and application. In addition, there are various insulating materials and metal materials inside the equipment. The new gas-insulating medium should meet the requirement of not chemically reacting with various materials, otherwise it will seriously affect the service life of the equipment and the insulation performance of the gas-insulating medium itself.

Based on the above three requirements, this chapter intends to determine the following evaluation angles:

1) Physical and chemical properties: This item focuses on the GWP value and ozone depletion potential (ODP) value of potential SF_6 alternative gases to measure their environmental performance; in addition, it focuses on the liquefaction temperature (saturated vapor pressure characteristics) of the eco-friendly gas-insulation medium itself to evaluate the pressure restrictions of its later engineering applications.
2) Safety performance: This item focuses on the toxicity of potential alternative gases and the types and toxicity of important decomposition products, mainly using acute toxicity LC50 as a quantitative standard to evaluate its application safety; at the same time, it focuses on its own flammability and corrosiveness to determine its safety under high voltage or discharge conditions.

3) Insulation performance: This item focuses on the insulation performance of potential alternative gases, using the critical breakdown field strength relative to SF_6 as a quantitative standard to evaluate the insulation performance of gas-insulation media.

From the perspective of physical and chemical properties, as shown in Table 9.1, traditional gases such as CO_2 and N_2 perform outstandingly, with the lowest liquefaction temperature among all potential gases and a relatively outstanding GWP value. Fluorocarbon gases such as CF_4, C_2F_6, C_3F_8, and c-C_4F_8 have high GWP values and unsatisfactory environmental protection characteristics. As the number of carbon atoms in the molecule increases, the liquefaction temperature of the gas molecule also increases. For example, the liquefaction temperature of c-C_4F_8 gas is only −8°C. The liquefaction temperature and GWP value of CF_3I gas are outstanding; the GWP values of $C_5F_{10}O$ and $C_6F_{12}O$ are only 1, but due to their large molecular weight, their liquefaction temperatures are relatively high, reaching 27 and 49°C [10]. The liquefaction temperature and GWP of C_4F_7N gas are at a medium level. From the scoring results of the comprehensive physical and chemical properties, it can be seen that the physical and chemical properties of traditional gases CO_2, N_2, and CF_3I are more outstanding.

From the perspective of insulation performance, the insulation performance of traditional gases such as CO_2 and N_2 is poor, which is about 0.3 times that of pure SF_6 gas. Among fluorocarbon gases such as CF_4, C_2F_6, C_3F_8, and c-C_4F_8, the insulation performance of CF_4, C_2F_6, and C_3F_8 is weaker than that of pure SF_6 gas. As the number of carbon atoms in the molecule increases, the insulation performance of the gas improves. The insulation performance of pure c-C_4F_8 and CF_3I gases exceeds

Table 9.1 Comparison of main parameters of SF_6 alternative gases.

Gas	GWP	ODP	Liquefaction temperature/°C	Flammability	LC50 (ppmv)	Main toxic decomposition products	Relative SF_6 electrical strength
N_2	<1	0	−196	Nonflammable	Nontoxic	None	0.3
CO_2	1	0	−78	Nonflammable	Nontoxic	None	0.3
CF_4	6630	0	−128	Nonflammable	895 000	HF, etc.	0.41
C_2F_6	12200	0	−78	Nonflammable	Nontoxic	HF, etc.	0.80
C_3F_8	8900	0	−38	Nonflammable	Nontoxic	HF, etc.	0.94
c-C_4F_8	8700	0	−8	Nonflammable	Nontoxic	HF, etc.	1.25
CF_3I	<5	0	−22.5	Nonflammable	May induce mutations in organisms	HF, etc.	1.2
$C_5F_{10}O$	<1	0	26.5	Nonflammable	20 000	COF_2, etc	≈2
$C_6F_{12}O$	1	0	49	Nonflammable	20 000	COF_2, etc	2.7
C_4F_7N	2090	0	−4.7	Nonflammable	12 000–15 000	Cyanide, HF, etc.	≈2

that of SF_6 gas, reaching about 1.2 times. The insulation performance of $C_5F_{10}O$ and C_4F_7N is similar, reaching more than twice that of pure SF_6 gas. The insulation performance of $C_6F_{12}O$ is the most outstanding among all gases, reaching 2.7 times that of pure SF_6 gas. Overall, $C_5F_{10}O$, C_4F_7N, and $C_6F_{12}O$ have the best insulation performance among all potential alternative gases.

From the perspective of safety, CO_2, N_2, CF_4, C_2F_6, C_3F_8, and c-C_4F_8 gases are actually nontoxic substances, among which CO_2, CF_4, C_2F_6, C_3F_8, and c-C_4F_8 gases are asphyxiating under high concentration conditions. At present, most of the above gases are widely used in industrial fields, such as plasma etching. Therefore, it is safe and reliable to use them as insulating media. $C_6F_{12}O$ was used as a new type of fire extinguishing agent in the early days [11], and its LC50 value and that of $C_5F_{10}O$ are also high [12]. In addition, considering its extremely high liquefaction temperature, it needs to be mixed with a buffer gas with a low liquefaction temperature in actual engineering applications, which can significantly reduce its toxicity to a certain extent; C_4F_7N is highly toxic, and its LC50 value is between 10 000 and 15 000 ppm, but similar to $C_5F_{10}O$ and $C_6F_{12}O$, its proportion in actual applications is within 10%, so it is relatively safe.

In addition, it should be pointed out that solid substances are precipitated from CF_3I and c-C_4F_8 gases after discharge. Figure 9.2a,b shows the iodine precipitation of CF_3I at the moment of discharge and after discharge. It can be seen that a large amount of purple iodine vapor is generated at the moment of discharge, and a thick layer of solid substance (iodine) is attached to the surface of the ball electrode after

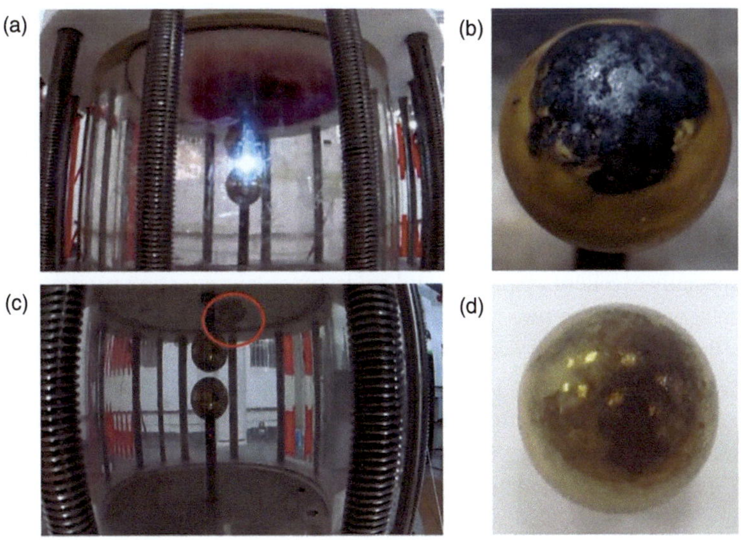

Figure 9.2 Iodine precipitation and carbon deposition from CF_3I and c-C_4F_8 at the moment of discharge and after discharge. *Source:* Xiao et al. [13] / IOP Publishing; (a) iodine is released at the moment of discharge; (b) electrode after discharge; (c) carbon deposits are released at the moment of discharge; (d) electrode after discharge. *Source:* [16] / with permission of IOP Publishing.

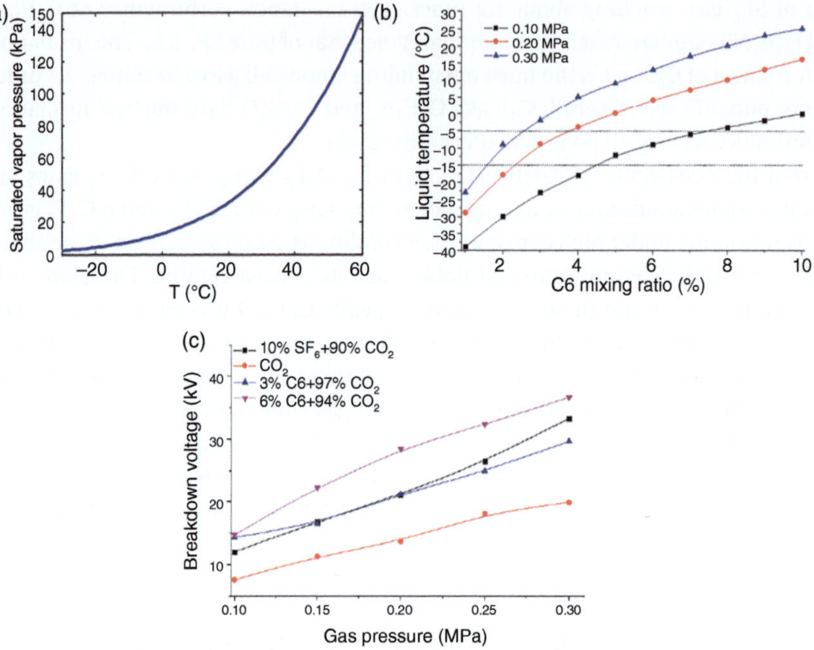

Figure 9.3 Saturated vapor pressure curve of $C_6F_{12}O$ and the relationship between liquefaction temperature and mixing ratio; (a) saturated vapor pressure curve; (b) relationship between liquefaction temperature and mixing ratio; (c) insulation properties of $C_6F_{12}O$-CO_2 mixed gas. *Source:* [19] / with permission of John Wiley & Sons.

the discharge, which seriously limits its engineering application [14]. Figure 9.3c,d shows the electrode changes of c-C_4F_8 before and after multiple discharges. It can be seen that after multiple discharges, the surface of the ball electrode is covered with a layer of black powdery solid (carbon) [16, 17].

In general, the three gases $C_6F_{12}O$, $C_5F_{10}O$, and C_4F_7N have relatively balanced overall performance in terms of insulation energy, physical and chemical properties, and safety, especially excellent insulation and environmental protection properties. Since its liquefaction temperature is inferior to that of other gases, it needs to be mixed with a buffer gas with a lower liquefaction temperature. The main alternatives are CF_4, N_2, and CO_2. Since CF_4 itself is a serious greenhouse gas, it is not suitable for use as a buffer gas; compared with N_2, existing studies have shown that CO_2 has better arc extinguishing performance, and it is recommended to choose it as a buffer gas when the insulating gas needs to interrupt the current. Therefore, we believe that $C_6F_{12}O$, $C_5F_{10}O$, and C_4F_7N mixed with N_2 or CO_2 have the potential for engineering applications. In fact, large foreign equipment companies have already achieved commercial testing of $C_5F_{10}O$ and C_4F_7N insulating media, proving that they have the ability to be used in actual engineering applications. For example, in 2017, ABB launched an eco-friendly medium-voltage GIE (Air-plus) in China with $C_5F_{10}O$ as the main insulating medium [18, 19]. Therefore, this project believes that $C_6F_{12}O$ has the potential to be used in medium-voltage equipment such as ring network switchgear (considering its high liquefaction temperaturc). On the one hand, the development of $C_6F_{12}O$ mixed gas equipment will break the monopoly of international electrical giants on eco-friendly GIE and form products with independent intellectual property rights. On the other hand, the price of $C_6F_{12}O$ is lower than that of $C_5F_{10}O$, and China can currently produce it independently, so its cost is lower and the threshold for later application and promotion is lower.

Figure 9.3a,b shows the saturated vapor pressure curve of $C_6F_{12}O$ and the relationship between liquefaction temperature and mixing ratio. It can be seen that for medium and low voltage equipment, the operating temperature range is mostly between −5 and 40°C. In order to ensure that the mixed insulating gas does not liquefy, the content of $C_6F_{12}O$ should be kept below 10%. When the air pressure is 0.10 MPa and the equipment is operating at −5°C, the mixing ratio of $C_6F_{12}O$ should be less than 7%. When the minimum operating temperature is −15°C, the mixing ratio of $C_6F_{12}O$ should be less than 5%.

The insulation performance of $C_6F_{12}O$-CO_2 mixed gas was tested, and the test results are shown in Figure 9.3c. It can be seen that the breakdown voltage of $C_6F_{12}O$ mixed gas is significantly greater than the breakdown voltage of pure CO_2. When the gas pressure is less than or equal to 0.15 MPa, the breakdown voltage of 3% $C_6F_{12}O/CO_2$ mixed gas is greater than the breakdown voltage of 10% SF_6/CO_2; when the gas pressure is greater than 0.20 MPa, the breakdown voltage of 10% SF_6/CO_2 is greater than the breakdown voltage of 3% $C_6F_{12}O/CO_2$ mixed gas, but less than the breakdown voltage of 6% $C_6F_{12}O/CO_2$ mixed gas. Therefore, the use of 3% $C_6F_{12}O$ mixed gas in low-voltage equipment can reduce the use of SF_6 while ensuring insulation performance, taking into account environmental protection principles [20, 21].

In addition, the decomposition characteristics of $C_6F_{12}O$ mixed gas were investigated. The experiment found that the main decomposition products of the mixed gas after 50 breakdowns were CF_4, C_2F_6, C_3F_8, C_4F_{10}, etc., and no solid matter (carbon) was precipitated on the electrode surface after multiple discharges. In fact, since the content of $C_6F_{12}O$ in the mixed gas is low (generally below 5%) and the molecular structure of $C_6F_{12}O$ does not contain C=C (CF_2 group), the fragments produced by its bond breaking are mainly CF_3, and CF_3 further ionization to produce C requires higher energy, so the mixed gas is not easy to precipitate solid matter after multiple breakdowns. In addition, considering that CO_2 gas has good arc extinguishing characteristics, $C_6F_{12}O$ and CO_2 mixed gas has a certain ability to break small currents [22, 23].

In summary, existing research has confirmed that $C_6F_{12}O$ mixed gas has excellent environmental protection characteristics and insulation properties. Due to its low saturated vapor pressure, it cannot be used in high-voltage equipment, but it has the potential to be used in low-voltage equipment.

In addition, screening found that the liquefaction temperature of C_4F_7N is much lower than that of $C_5F_{10}O$ and $C_6F_{12}O$, but its insulation performance is also twice that of SF_6 gas. Relevant research at this stage also pointed out that the insulation performance of C_4F_7N-CO_2 mixed gas containing 20–18% C_4F_7N

reaches that of pure SF_6, and the GWP value of the mixed gas is only 2% of that of pure SF_6 [24].

This is an experimental study on the power frequency breakdown characteristics, partial discharge characteristics, and lightning impulse withstand characteristics of C_4F_7N/CO_2 mixed gas. The test found that the power frequency breakdown voltage of SF_6 and C_4F_7N/CO_2 mixed gas under quasi-uniform electric field showed a linear growth trend with gas pressure, and a linear saturation growth trend under extremely nonuniform field conditions; the power frequency breakdown voltage and partial discharge starting voltage of the mixed gas under different electric field conditions showed a linear saturation growth trend with the increase of the mixing ratio. The power frequency breakdown voltage of the mixed gas containing 2% C_4F_7N under a quasi-uniform electric field can reach 50% of pure SF_6 under the same conditions, and the insulation performance of the mixed gas containing 8% C_4F_7N can reach 75% of pure SF_6.

At the same time, the formation mechanism of discharge decomposition components of C_4F_7N and micro-water conditions was calculated using density functional theory, and the energy, free energy, and activation energy required for different paths were explored. The ionization parameters of C_4F_7N and various decomposition products were calculated and analyzed from the perspective of molecular structure, and the decomposition characteristics of C_4F_7N and micro-water environments were evaluated in combination with toxicity. The study found that C_4F_7N would break chemical bonds under high-energy electric fields or local overheating, and the main products produced by various free radical recombination were C_2F_5CN, CF_3CN, CF_4, C_2F_6, C_3F_8, C_4F_{10}, etc. Among them, C_2F_5CN, CF_3CN, C_3F_8 and C_4F_{10} all had good electronegativity and insulating properties. In the micro-water environment, the presence of H· and OH· would accelerate the consumption of free radical particles produced by the decomposition of C_4F_7N in the system to form new products such as CF_3H, CF_2O, C_3F_7H, C_3F_7OH, CF_3CFHCN, and CH_2FCN, and the ionization parameters of the above products were poor. Among the various decomposition products, C_2F_5CN and CF_3CN are toxic substances, and fluorocarbons such as CF_4, C_2F_6, and C_3F_8 have no effect on human safety; CF_2O, C_3F_7H, CF_3CH_2CN, CF_2HCN, CH_2FCN, HF, and HCN produced in a micro-water environment are all highly toxic and some products are highly corrosive, which will cause potential damage to the internal structure of the equipment and seriously threaten the safety of maintenance personnel. In addition, the discharge decomposition characteristics of the C_4F_7N/N_2 mixed gas were tested using a gas chromatography-mass spectrometer. The test found that with the increase in the number of experiments, the breakdown voltage of the C_4F_7N/N_2 mixed gas did not decrease significantly, indicating that the insulation self-recovery characteristics of the mixed gas are excellent. In addition, after the discharge, the mixed gas produced decomposition products such as CF_4, C_2F_6, C_3F_8, CF_3CN, C_2F_4, C_3F_6, and C_2F_5CN, which are consistent with the simulation results.

9.3 Eco-Friendly Gas-Insulation Synergy Mechanism and Formulation Optimization

9.3.1 Experimental Study on Power Frequency Breakdown Characteristics of C_4F_7N/CO_2 Mixed Gas

This section conducts relevant experimental analysis on the power frequency breakdown characteristics of C_4F_7N/CO_2 mixed gas under quasi-uniform electric field and extremely nonuniform electric field conditions. It mainly discusses the influence mechanism of gas pressure, mixing ratio, and electric field utilization coefficient on the insulation characteristics of C_4F_7N/CO_2 mixed gas. By comparing and analyzing with pure SF_6 gas, the most suitable working gas pressure and mixing ratio are obtained, and the feasibility of using it as an insulating medium to replace SF_6 gas is analyzed.

9.3.1.1 Power Frequency Breakdown Characteristics of C_4F_7N/CO_2 Mixed Gas Under Quasi-Uniform Field

Ball electrodes are used to simulate the quasi-uniform electric field environment, and the electrode spacing is 2 mm. The experiment is completed under the condition of an ambient temperature of 20 C. Gas pressure and mixing ratio are two important factors affecting the insulation performance of mixed gas. To avoid the liquefaction of the C_4F_7N/CO_2 mixed gas, considering the working environment under actual engineering conditions [25], this experiment selects C_4F_7N in C_4F_7N/CO_2 mixed gas of 2%, 4%, 6%, 8%, 10%, and 12%. The gas pressure in the gas chamber of the device is set to 0.1 MPa, 0.2 MPa, 0.3 MPa, 0.4 MPa, 0.5 MPa, and 0.6 MPa. This section studies the insulation characteristics of C_4F_7N/CO_2 mixed gas under a quasi-uniform field through power frequency test, and analyzes the influence of gas pressure and mixing ratio on the insulation performance of mixed gas. Finally, the influence mechanism of mixing ratio and gas pressure on the insulation performance of C_4F_7N/CO_2 mixed gas under a quasi-uniform electric field is discussed, and the insulation characteristics of pure SF_6 gas under the same conditions are compared and analyzed.

1) Effect of gas pressure on power frequency breakdown characteristics of C_4F_7N/CO_2 mixed gas under quasi-uniform field

Based on the data in Table 9.2, the effect of gas pressure on the power frequency breakdown value of C_4F_7N/CO_2 mixed gas and pure SF_6 gas is plotted, as shown in Figure 9.5a.

As can be seen from Figure 9.5a, within the range of 0–0.6 MPa, the breakdown voltage of C_4F_7N/CO_2 mixed gas and pure SF_6 gas at various mixing ratios increases with the increase of pressure, and the breakdown voltage of C_4F_7N/CO_2 mixed gas in this test is lower than that of pure SF_6 gas. In a quasi-uniform electric field environment, the breakdown voltage of 8% C_4F_7N/92% CO_2 mixed gas at a pressure of

Table 9.2 Breakdown voltage of C_4F_7N/CO_2 mixed gas under quasi-uniform electric field conditions.

	Breakdown voltage (kV)					
	Air pressure (MPa)					
Mixing ratio k (%)	**0.1**	**0.2**	**0.3**	**0.4**	**0.5**	**0.6**
0	11.5	13.9	15.5	19.2	21.9	25.9
2	12.5	17.0	22.7	26.3	33.5	40.4
4	12.9	18.9	25.2	30.3	38.2	45.1
6	13.5	20.9	28.1	34.9	43.2	50.9
8	13.6	22.1	31.5	38.8	47.6	54.6
10	14.3	25.4	33.4	40.4	51.2	59.3
12	15.6	27.6	36.3	44.3	55.4	61.9
SF_6	17.2	31.2	43.7	56.7	68.6	73.7

0.3 MPa can reach the level of 0.2 MPa pure SF_6 gas; the breakdown voltage of 8% C_4F_7N/92% CO_2 mixed gas at a pressure of 0.6 MPa is close to the level of 0.4 MPa pure SF_6 gas; the highest breakdown voltage of C_4F_7N/CO_2 mixed gas in this test is 12% C_4F_7N/88% CO_2 mixed gas at a pressure of 0.6 MPa, which is equivalent to the breakdown strength of pure SF_6 gas at a pressure of 0.4–0.5 MPa.

At the same time, in a quasi-uniform electric field environment, C_4F_7N/CO_2 mixed gas with different mixing ratios showed an increasing trend with pressure in the pressure range of 0–0.6 MPa, while pure SF_6 gas showed an increasing trend in the pressure range of 0–0.5 MPa, and the growth rate decreased in the pressure range of 0.5–0.6 MPa, indicating that pure SF_6 gas showed a saturated growth trend in the pressure range of 0–0.6 MPa. And under the pressure of 0–0.5 MPa, the growth rate of C_4F_7N/CO_2 mixed gas was lower than that of pure SF_6 gas with the change of pressure. For example, when the pressure increased from 0.3 MPa to 0.4 MPa, the breakdown voltage of 12% C_4F_7N/88% CO_2 mixed gas increased by 8 kV, and that of pure SF_6 gas increased by 13 kV. At 0.5–0.6 MPa, the breakdown voltage of C_4F_7N/CO_2 mixed gas increases faster than pure SF_6 gas with pressure changes. When the pressure increases from 0.5 MPa to 0.6 MPa, the breakdown voltage of 12% C_4F_7N/88% CO_2 mixed gas increases by 6.5 kV, while that of pure SF_6 gas increases by 5.1 kV. In summary, in a quasi-uniform electric field, when the pressure is in the range of 0.1–0.6 MPa, the effect of increasing the pressure on the breakdown characteristics of C_4F_7N/CO_2 mixed gas is lower than that of SF_6 gas, that is, the breakdown voltage of SF_6 can increase significantly with pressure, while the breakdown voltage of C_4F_7N/CO_2 mixed gas increases at a lower rate with pressure.

2) Effect of mixing ratio on power frequency breakdown characteristics of C_4F_7N/CO_2 mixed gas under quasi-uniform field

Figure 9.5b shows the breakdown voltage curve of C_4F_7N/CO_2 mixed gas with the mixing ratio under different gas pressures. As shown in the figure, as the mixing ratio increases, the breakdown voltage of the C_4F_7N/CO_2 mixed gas also increases. This shows that the higher the C_4F_7N content, the stronger its insulation properties. When comparing pure CO_2 gas with a mixed gas containing 2% C_4F_7N, we can see that when the gas pressure is 0.6 MPa and the mixing ratio increases from 0 to 2%, the breakdown voltage increases from 25.9 to 40.4 kV, an increase of 14.5 kV. After the mixing ratio exceeds 2%, the breakdown voltage of C_4F_7N/CO_2 mixed gas continues to increase with the increase of the mixing ratio, but the rate of increase slows down. The results show that the addition of C_4F_7N gas to CO_2 gas can effectively improve its insulation strength. The higher the C_4F_7N content at high pressure, the more obvious the enhancement effect; the increase in the C_4F_7N content in the C_4F_7N/CO_2 mixed gas can improve its breakdown strength.

Figure 9.5c shows the change of ratio of the breakdown voltage of C_4F_7N/CO_2 mixed gas to pure SF_6 gas (defined as the relative insulation strength of C_4F_7N/CO_2 mixed gas) with the change of C_4F_7N content under different gas pressures. When the gas pressure is 0.1 MPa, the relative breakdown voltage is the largest, and it shows a slow upward trend with the increase of C_4F_7N content; when the gas pressure is greater than or equal to 0.2 MPa, the breakdown voltage of pure CO_2 gas relative to pure SF_6 gas is very low, with the lowest value close to 0.31; with the increase of C_4F_7N content added to CO_2 gas, the breakdown voltage value relative to pure SF_6 gas also increases significantly. The increase of C_4F_7N content in C_4F_7N/CO_2 mixed gas can effectively improve the insulation strength of the gas. Under the same mixing ratio conditions, higher the gas pressure, slightly lower the breakdown voltage of the C_4F_7N/CO_2 mixed gas relative to pure SF_6 gas. The conclusion from the previous section shows that the increase in gas pressure has little effect on improving the insulation capacity of the C_4F_7N/CO_2 mixed gas compared to SF_6.

3) Synergistic effect of C_4F_7N/CO_2 mixed gas under quasi-uniform field

At present, all electronegative gases, when two gases are mixed, can be classified into four categories according to their breakdown of electric fields, namely linear relationship type, synergistic effect type, positive synergistic effect type, and negative synergistic effect type. Linear relationship type: When an electronegative gas is mixed with another gas, the breakdown field strength of the mixed gas is the sum of the breakdown field strengths of the two gases. The breakdown field strengths of most mixed gases with N_2 as a buffer gas have a typical linear relationship. Synergistic effect type: When a strong electronegative gas is mixed with a weak electronegative gas, its actual breakdown field strength is less than the sum of the field strengths of two gases, such as currently commonly used SF_6/N_2 mixed gas, which no longer satisfies the linear relationship and presents a synergistic effect. Positive synergistic effect: When there is a gas in the mixed gas whose electron attachment process is related to the gas pressure, and the other gas has a higher electron attachment rate and can produce more stable negative ions, and the breakdown field strengths of two gases are not much different, then within a certain mixing ratio range, the breakdown field strength is higher than any of the two gases, such as SF_6/C_3F_8 mixed gas.

Negative synergistic effect: Contrary to the positive synergistic effect, the breakdown electric field strength of the mixed gas is less than that of the two gases in the mixed gas. According to the theory of negative synergistic effect, it can be concluded that mixing two electronegative gases does not increase the breakdown field strength, and the opposite result may also occur.

The principle of synergistic effect can be explained by the electron attachment coefficient, which is expressed as [26]:

$$\eta = \frac{N}{\overline{v_d}} \int_0^{\infty} \delta_a(v) v f(v) dv \tag{9.1}$$

where N is the number of molecules per unit volume, $\overline{v_d}$ is the average electron moving speed, δ_a is gas molecule attachment interface, v is electron speed, and $f(v)$ is electron speed distribution function. It can be seen from the formula that increasing the gas molecule attachment interface value δ_a or moving the electron speed distribution function $f(v)$ to a smaller value v as much as possible can effectively improve the electron attachment coefficient.

Combined with the above analysis results, when using mixed gas as a gas-insulating medium, it is necessary to avoid the occurrence of the negative synergistic effect phenomenon, and make full use of the synergistic effect or positive synergistic effect of the two gases to improve the insulation properties of the mixed gas. For example, the currently commonly used 20% SF_6/80% N_2 mixed gas has a good synergistic effect, and its insulation strength is 70% of pure SF_6 gas. If the content of SF_6 gas in the mixed gas continues to increase, the insulation properties of the mixed gas will still increase, but the effect is not obvious; if the gas pressure is increased, the liquefaction temperature of the mixed gas will rise, which cannot meet normal working applications. Therefore, for the use of mixed gases, their synergistic effect should be fully analyzed to achieve the best combination of mixed gases.

It can be seen from Figures 9.4a–c that the power frequency breakdown value of C_4F_7N/CO_2 mixed gas under a quasi-uniform electric field does not change linearly with gas pressure and mixing ratio. This nonlinear change is the synergistic effect of electronegative gas and buffer gas. The synergistic effect expression of electronegative gas and buffer gas [29]:

$$V_m = V_2 + \frac{k(V_1 - V_2)}{k + (1-k)C}, V_1 > V_2 \tag{9.2}$$

where V_1 and V_2 are defined as the power frequency breakdown voltage values when the C_4F_7N content in the C_4F_7N/CO_2 mixed gas is 12% and 0%, respectively. V_m is the power frequency breakdown voltage when the C_4F_7N content in the C_4F_7N/CO_2 mixed gas is 2%, 4%, 6%, 8%, and 10%; k is the mixing ratio; C is the synergy coefficient. Then the synergy coefficient C value of the C_4F_7N/CO_2 mixed gas with the C_4F_7N content in the range of 0–10% is obtained. When $C = 1$, V_m increases linearly from V_2 to V_1 with the increase of k; when $0 < C < 1$, V_m increases nonlinearly from V_2 to V_1. Therefore, the value of C can be used to represent the nonlinearity of the two, that is, the size of the synergistic effect: the smaller C is, the greater the nonlinearity is, and the more obvious the synergistic effect is.

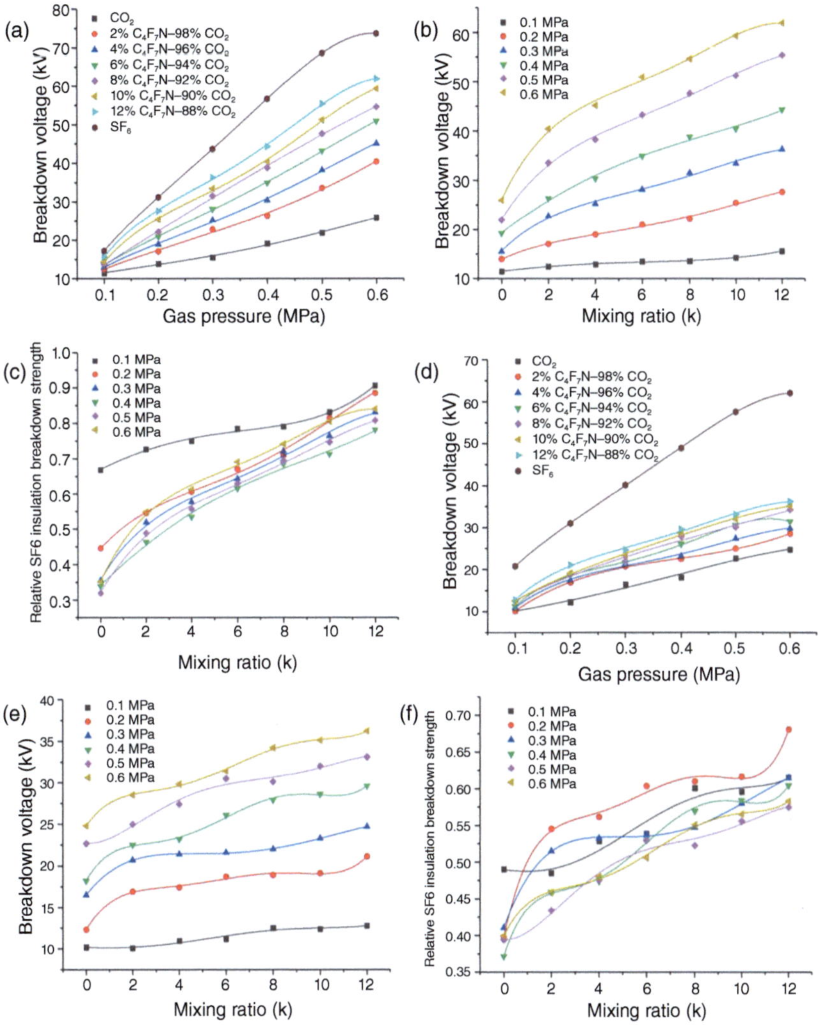

Figure 9.4 Effect of mixing ratio on C_4F_7N/CO_2 mixed gas under extremely nonuniform electric field and quasi-uniform electric field conditions: (a) breakdown voltage curve of C_4F_7N/CO_2 mixed gas and pure SF_6 gas under quasi-uniform electric field conditions versus gas pressure; (b) variation of the breakdown voltage of C_4F_7N/CO_2 gas mixture with mixing ratio at different pressures in a quasi-uniform electric field; (c) variation of the breakdown voltage of C_4F_7N/CO_2 mixed gas relative to pure SF_6 gas with the mixing ratio in a quasi-uniform electric field; (d) breakdown voltage curve of C_4F_7N/CO_2 mixed gas and pure SF_6 gas under extremely nonuniform electric field conditions versus gas pressure; (e) variation of the breakdown voltage of C_4F_7N/CO_2 gas mixture with mixing ratio at different pressures in extremely nonuniform electric field; (f) variation of the breakdown voltage of C_4F_7N/CO_2 mixed gas relative to pure SF_6 gas with the mixing ratio in extremely nonuniform electric field. *Source:* Adapted from [29, 30].

Table 9.3 Synergistic effect coefficient C of breakdown voltage of C_4F_7N/CO_2 mixed gas under quasi-uniform electric field conditions.

Gas pressure/MPa	Synergy coefficient C				
	$k = 2\%$	$k = 4\%$	$k = 6\%$	$k = 8\%$	$k = 10\%$
0.10	0.063	0.080	0.067	0.083	0.052
0.20	0.070	0.073	0.061	0.058	0.021
0.30	0.039	0.048	0.042	0.026	0.018
0.40	0.052	0.053	0.038	0.024	0.020
0.50	0.039	0.044	0.037	0.026	0.016
0.60	0.030	0.036	0.028	0.022	0.009

According to the experimental measured mixed gas power frequency breakdown data, the values of the synergistic effect coefficient C under different gas pressures are calculated (as shown in Table 9.3). The table gives the synergistic effect value C of the breakdown voltage value of C_4F_7N/CO_2 mixed gas under different gas pressures and mixing ratios in a quasi-uniform electric field. It can be seen that the power frequency breakdown synergistic effect coefficient C of C_4F_7N gas and CO_2 gas is in the range of 0.009–0.083. Overall, the same effect coefficient C values of C_4F_7N/CO_2 mixed gas under different gas pressures and mixing ratios are relatively close to 0, indicating that C_4F_7N/CO_2 mixed gas has a good synergistic effect relationship. From the comparison of the synergistic effect coefficients in the table, it can be seen that under the same mixing ratio conditions, the increase in gas pressure makes the synergistic effect coefficient C smaller for the breakdown voltage of C_4F_7N/CO_2 mixed gas; and under the same gas pressure, as the mixing ratio increases, that is, the higher the content of C_4F_7N in the C_4F_7N/CO_2 mixed gas, the smaller the synergistic effect coefficient C value.

In general, a comparative analysis of the synergistic effect of the breakdown voltage of the C_4F_7N/CO_2 mixed gas under a quasi-uniform electric field environment revealed that the C_4F_7N/CO_2 mixed gas has a good synergistic effect. At the same time, increasing the content of C_4F_7N gas or increasing the gas pressure can effectively improve the synergistic effect relationship of the mixed gas.

4) Comparative analysis of the effects of gas pressure and mixing ratio on the power frequency breakdown characteristics of C_4F_7N/CO_2 mixed gas under quasi-uniform field

According to the above results, it can be seen that the power frequency breakdown voltage of C_4F_7N/CO_2 mixed gas can be increased by increasing the gas pressure and the mixing ratio. However, there are certain differences in the mechanism between the two.

For gas-insulating media, as the gas pressure increases, the density of the mixed gas increases, and the mean free path of the free electrons in the gap shortens, which reduces the kinetic energy accumulated by the electrons in the secondary collision, causing the ionization process to weaken and the insulation performance of the mixed gas to be improved, so the breakdown voltage will increase with the increase of gas pressure. However, it does not mean that the breakdown voltage will continue to increase with the increase in gas pressure. This is because the gas pressure is not the only factor that affects the numerical value of the breakdown voltage. When the gas pressure rises to a certain level, other factors will become the main factors affecting the breakdown voltage, such as electrode size and electrode spacing. Therefore, after the gas pressure exceeds a certain value, the breakdown voltage will be in a saturated growth state with the increase of gas pressure.

Under quasi-uniform electric field conditions, according to Baschen's law:

$$U = \frac{Bps}{ln\left[\frac{Aps}{ln\left(1+\frac{1}{\gamma}\right)}\right]} = f(ps) \tag{9.3}$$

where U is the breakdown voltage; A and B are constants related to the properties of the gas; γ is the positive ion surface free coefficient; p is the working gas pressure; s is the electrode spacing. It can be seen that under the quasi-uniform electric field, the gas breakdown voltage is not only related to the type of gas, but also to the gas state. When the pressure is constant, the larger the pressure, the higher the breakdown voltage.

The test results of the effective ionization coefficient of the C_4F_7N/CO_2 mixed gas show that the rate at which the effective ionization coefficient of the mixed gas containing 6.7% C_4F_7N increases with the electric field strength is close to that of pure SF_6 gas [30]. This is the reason why the local discharge voltage intensity of the C_4F_7N/CO_2 mixed gas shows a linear saturation growth trend with the gas pressure under extremely nonuniform electric field conditions. C_4F_7N is a strongly electronegative gas that easily adsorbs free electrons to form negative ions. The average free path length of ions is shorter than that of electrons, so the accumulated kinetic energy is also less, and ions are very easy to lose the accumulated kinetic energy during the collision process. Therefore, after the electrons become negative ions, the development process of gas discharge will be hindered. Secondly, the molecular volume of C_4F_7N is relatively large, and the average free path of free electrons in the C_4F_7N mixed gas will be shortened. The higher the content of C_4F_7N in the mixed gas, the more difficult it will be for the free electrons to accelerate and accumulate enough kinetic energy, thereby weakening the ability of collision electricity. At the same time, the speed of various ions in the mixed gas will also decrease, and the mutual recombination of these ions will reduce the number of charged particles, thereby greatly improving the insulation strength of the gap.

The free electrons generated by various factors in the gap cannot increase infinitely. When the content of C_4F_7N in the C_4F_7N/CO_2 mixed gas further increases, most of the free electrons in the gap have been adsorbed by C_4F_7N to form negative ions, and the breakdown voltage will no longer be significantly improved. With the increase of the mixing ratio k, the breakdown voltage of the mixed gas will reach saturation. Therefore, by increasing the gas pressure or mixing ratio, the insulation performance of the mixed gas can eventually reach or even exceed that of pure SF_6 gas under some conditions.

9.3.1.2 Power Frequency Breakdown Characteristics of C_4F_7N/CO_2 Mixed Gas Under Extremely Nonuniform Field

This section uses needle-plate electrodes to simulate an extremely nonuniform electric field environment. The electrode spacing is 5 mm. The external test environment is the same as quasi-uniform field. The mixing ratio and test pressure of the C_4F_7N/CO_2 mixed gas are the same as those in quasi-uniform field test. This section mainly studies the influence mechanism of the mixing ratio and gas pressure on the power frequency breakdown characteristics of the C_4F_7N/CO_2 mixed gas under extremely nonuniform electric field conditions, and compares and analyzes the power frequency breakdown characteristics of pure SF_6 and CO_2 gases under the same electric field conditions. It also investigates the influence mechanism of insulation performance under an extremely nonuniform electric field and compares and analyzes the insulation characteristics of SF_6 under the same conditions.

Table 9.4 gives the breakdown voltage values of C_4F_7N/CO_2 mixed gas and pure SF_6 gas under extremely nonuniform electric field conditions with an electrode spacing of 5 mm.

1) Influence of gas pressure on power frequency breakdown characteristics of C_4F_7N/CO_2 mixed gas under extremely nonuniform field

According to the data in Table 9.4, the effect of gas pressure on the power frequency breakdown value of C_4F_7N/CO_2 mixed gas and pure SF_6 gas is plotted. As shown in Figure 9.4d, in an extremely nonuniform electric field environment, the breakdown voltage values of C_4F_7N/CO_2 mixed gas under various mixing ratio conditions are generally lower than those of pure SF_6 gas under the same gas pressure, and when

Table 9.4 Breakdown voltage of C_4F_7N/CO_2 gas mixture under extremely nonuniform electric field conditions.

	Breakdown voltage (kV)					
	Gas pressure (MPa)					
Mixing ratio k (%)	0.1	0.2	0.3	0.4	0.5	0.6
0	10.2	12.3	16.5	18.2	22.7	24.8
2	10.1	16.9	20.7	22.5	25.0	28.5
4	11.0	17.4	21.4	23.2	27.4	29.8
6	11.2	18.7	21.6	26.1	30.5	31.4
8	12.5	18.9	22.0	27.9	30.1	34.2
10	12.4	19.1	23.3	28.6	32.0	35.1
12	12.8	21.1	24.7	29.6	33.1	36.2
SF_6	20.8	31.0	40.2	49.0	57.6	62.1

the mixing ratio is constant, the breakdown voltage values of C_4F_7N/CO_2 mixed gas and pure SF_6 gas show an increasing trend with the increase of test gas pressure. The power frequency breakdown voltage strength of the 2% C_4F_7N/98% CO_2 mixed gas at a pressure of 0.3 MPa can reach the level of pure SF_6 gas at a pressure of 0.1 MPa; the power frequency breakdown voltage strength of the 6% C_4F_7N/94% CO_2 mixed gas at a pressure of 0.6 MPa can reach the level of pure SF_6 gas at a pressure of 0.2 MPa; in the test, under the conditions that the maximum breakdown voltage of the C_4F_7N/CO_2 mixed gas is 0.6 MPa and the C_4F_7N content is 12%, its power frequency breakdown voltage value is within the level of the pure SF_6 gas pressure range of 0.2–0.3 MPa.

In an extremely nonuniform electric field environment, the breakdown voltage curves of C_4F_7N/CO_2 mixed gases with different C_4F_7N contents are similar as they change with gas pressure (as shown in Figure 9.4d). For example, when the gas pressure rises from 0.2 to 0.3 MPa, the breakdown voltage of 2% C_4F_7N/98% CO_2 mixed gas increases by 3.8 kV, and the breakdown voltage of 12% C_4F_7N/88% CO_2 mixed gas increases by 3.6 kV, but the breakdown voltage of pure SF_6 gas increases by 9.2 kV; when the gas pressure rises from 0.5 to 0.6 MPa, the breakdown voltage of 2% C_4F_7N/98% CO_2 mixed gas increases by 3.5 kV, and the breakdown voltage of 12% C_4F_7N/88% CO_2 mixed gas increases by 3.1 kV, but the breakdown voltage of pure SF_6 gas increases by 4.5 kV. Combining the specific data and the curve changes in the figure, it is analyzed that the rate at which the breakdown voltage of the C_4F_7N/CO_2 mixed gas increases with gas pressure is lower than that of pure SF_6 gas, indicating that the C_4F_7N/CO_2 mixed gas is less affected by gas pressure than SF_6 gas [27].

2) Effect of mixing ratio on power frequency breakdown characteristics of C_4F_7N/CO_2 mixed gas under an extremely inhomogeneous field

According to the data in Table 9.4, the influence of the mixing ratio on the power frequency breakdown value of C_4F_7N/CO_2 mixed gas and pure SF_6 gas is plotted, as shown in Figure 9.4.

As can be seen from Figure 9.4e, when the gas pressure is 0.1 MPa, the breakdown voltage value only increases slightly with the increase of C_4F_7N content in the C_4F_7N/CO_2 mixed gas. For example, when the C_4F_7N content increases from 0 to 12%, the breakdown voltage value only increases by 2.6 kV. The test results show that under extremely nonuniform electric field conditions, the mixing ratio has little effect on the breakdown voltage of the C_4F_7N/CO_2 mixed gas at normal pressure. When the gas pressure is greater than 0.2 MPa, adding a small amount of C_4F_7N gas to CO_2 gas can effectively increase its breakdown voltage.

For example, at 0.2 MPa pressure, when the C_4F_7N content increases from 0 to 2%, the breakdown voltage value increases by 4.6 kV. At 0.6 MPa pressure, when the C_4F_7N content increases from 0 to 2%, the breakdown voltage value increases by 4.3 kV. In fact, when the pressure is greater than 0.2 MPa, the rate of increase of the breakdown voltage when the C_4F_7N content increases from 0 to 2% is significantly higher than the changes of other mixing ratios, which once again proves that C_4F_7N has significant insulation properties. From the overall change of the breakdown voltage curve in Figure 9.5e, under various pressure conditions in an extremely

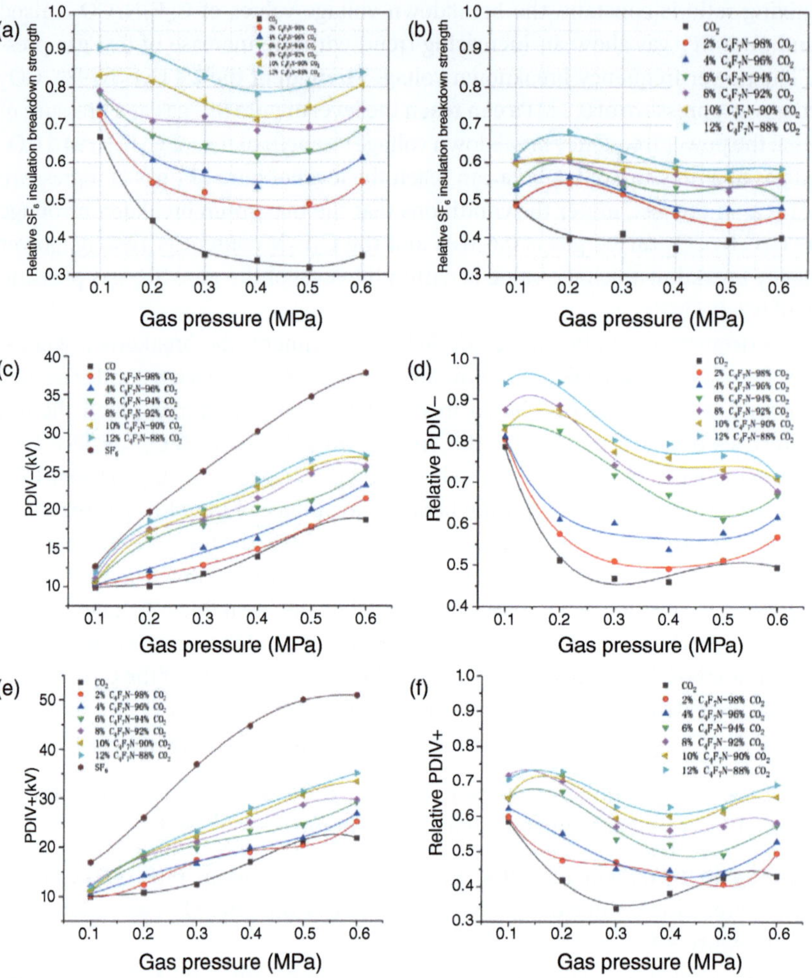

Figure 9.5 Power frequency breakdown strength and PDIV+, PDIV− and relative PDIV+, PDIV− values of C_4F_7N/CO_2 mixed gas at various mixing ratios as a function of gas pressure; (a) quasi-uniform electric field; (b) extremely inhomogeneous electric field; (c) PDIV− of C_4F_7N/CO_2 mixed gas and pure SF_6 gas-changes with gas pressure; (d) PDIV− of C_4F_7N/CO_2 mixed gas relative to pure SF_6 gas; (e) changes of PDIV+ of C_4F_7N/CO_2 mixed gas and pure SF_6 gas with gas pressure; (f) C_4F_7N/CO_2 mixed gas relative to pure SF_6 gas PDIV+. *Source:* Adapted from [30, 33].

nonuniform electric field environment, the increase of C_4F_7N content can increase the breakdown voltage of the C_4F_7N/CO_2 mixed gas.

Figure 9.5f shows the ratio of the breakdown voltage of C_4F_7N/CO_2 mixed gas to the breakdown voltage of pure SF_6 gas under the same pressure conditions. It can be seen from the figure that when the pressure is 0.1 MPa, the C_4F_7N content in the C_4F_7N/CO_2 mixed gas increases from 0 to 2%, and the relative breakdown voltage decreases slightly; when the C_4F_7N content exceeds 2%, the relative breakdown voltage shows a saturated growth trend with the increase of the mixing ratio. Under the conditions of 0.2 and 0.3 MPa, the C_4F_7N content in the C_4F_7N/CO_2 mixed gas increases from 0 to 2%, and the relative breakdown voltage increases significantly, especially under the pressure of 0.2 MPa, the relative breakdown voltage increases from 0.39 to 0.54; when the C_4F_7N content increases from 2% to 8%, the relative breakdown voltage remains basically stable, with only a slight increase; when the C_4F_7N content exceeds 8%, the relative breakdown voltage growth rate begins to rise again. Under the condition of gas pressure of 0.4–0.6 MPa, the breakdown voltage of C_4F_7N/CO_2 mixed gas shows a saturated growth trend with the increase of C_4F_7N content. It has been proven that the C_4F_7N/CO_2 mixed gas is less affected by gas pressure than SF_6 gas, and the breakdown voltage strength of C_4F_7N/CO_2 mixed gas relative to pure SF_6 gas under higher gas pressure is lower than that under a lower gas pressure environment.

3) Synergistic effect of C_4F_7N/CO_2 mixed gas under an extremely inhomogeneous field

According to Equation (9.3), V1 represents the breakdown voltage value of 12% C_4F_7N/88% CO_2 mixed gas, and V2 represents the breakdown voltage value of pure CO_2 gas. Substitute the breakdown voltage data of C_4F_7N/CO_2 mixed gas obtained from the experiment into the expression, and calculate the synergistic effect coefficient C value as shown in Table 9.5 (under extremely nonuniform electric field environment).

Compare the synergistic effect coefficient C of the breakdown voltage of C_4F_7N/CO_2 mixed gas, as shown in Table 9.5. From the data in the table, we can see that when the gas pressure is 0.1 MPa, the value of the synergistic effect coefficient C of the 2% C_4F_7N/98% CO_2 mixed gas is −0.55. Because the gas pressure is low at this time and the C_4F_7N content is relatively small, its breakdown voltage value is not significantly different from that of pure CO_2 gas under 0.1 MPa gas pressure. From the overall calculation results, the synergistic effect coefficient C value of the C_4F_7N/CO_2 mixed gas with a C_4F_7N content of 2–10% is in the range of 0.011–0.102, indicating that the C_4F_7N/CO_2 mixed gas has a good synergistic effect. Under the

Table 9.5 Synergistic effect coefficient C of breakdown voltage of C_4F_7N/CO_2 mixed gas under extremely nonuniform electric field conditions.

Gas pressure/MPa	Synergy coefficient C				
	$k = 2\%$	$k = 4\%$	$k = 6\%$	$k = 8\%$	$k = 10\%$
0.10	−0.55	0.094	0.102	0.011	0.020
0.20	0.018	0.030	0.024	0.029	0.033
0.30	0.019	0.028	0.039	0.043	0.023
0.40	0.033	0.053	0.028	0.015	0.011
0.50	0.072	0.051	0.021	0.035	0.013
0.60	0.042	0.053	0.046	0.019	0.012

same gas pressure conditions, with the increase of the mixing ratio, the synergistic effect coefficient C becomes smaller, especially under high pressure conditions. The change is more obvious, indicating that with the increase of C_4F_7N content, its synergistic effect is better. Under the same mixing ratio, the synergistic effect coefficient of C_4F_7N/CO_2 mixed gas does not change regularly with the change of gas pressure [32], indicating that gas pressure has no regular effect on the synergistic effect of C_4F_7N/CO_2 mixed gas.

According to the synergistic effect coefficient C value of the breakdown voltage of C_4F_7N/CO_2 mixed gas, we can judge that in an extremely nonuniform electric field environment, C_4F_7N and CO_2 have a good synergistic effect, and the increase of the mixing ratio can enhance their synergistic effect.

4) Comparative analysis of the effects of gas pressure and mixing ratio on the power frequency breakdown characteristics of C_4F_7N/CO_2 mixed gas under an extremely nonuniform field

In the same quasi-uniform electric field environment, C_4F_7N has strong electronegative gas characteristics, which makes it easier to adsorb electrons to form negative ions, thereby hindering the gas discharge process; the volume of the C_4F_7N molecule itself is large, and the particle mobility formed by its dissociation is small, resulting in a dense space charge near the needle electrode, and it is not easy to form a uniform space charge layer that can improve the electric field distribution near the electrode. Therefore, in an extremely nonuniform electric field, the increase of C_4F_7N in the C_4F_7N/CO_2 mixed gas makes the breakdown voltage show a saturated growth trend.

The reasons for the abnormal phenomenon of SF_6 gas breakdown in the extremely nonuniform electric field gap are very complicated. The existing research results can only show that the abnormal phenomenon is related to the movement of space charges in the gap. In an extremely nonuniform electric field, increasing the gas pressure has a smaller effect on increasing the breakdown voltage of the mixed gas gap.

9.3.1.3 Analysis of the Influence of Electric Field Uniformity on Power Frequency Breakdown of C_4F_7N/CO_2 Mixed Gas

This section compares the breakdown voltage strength ratio of pure CO_2 gas and C_4F_7N/CO_2 mixed gas in a quasi-uniform electric field and an extremely nonuniform electric field with pure SF_6 gas under the same pressure conditions. Analyze the sensitivity of C_4F_7N/CO_2 mixed gas to electric field uniformity.

Figure 9.5a shows the power frequency breakdown strength of C_4F_7N/CO_2 mixed gas relative to pure SF_6 gas in a quasi-uniform electric field environment; Figure 9.5b shows the power frequency breakdown strength of C_4F_7N/CO_2 mixed gas relative to pure SF_6 gas in an extremely nonuniform electric field environment. It can be seen that the power frequency breakdown strength of pure CO_2 gas is less affected by the uniformity of electric field, and the power frequency breakdown strength relative to pure SF_6 gas is even slightly lower in a quasi-uniform electric field than in an extremely nonuniform electric field, indicating that pure SF_6 gas is more affected by the uniformity of electric field than pure CO_2 gas. When a small amount of C_4F_7N gas is added to CO_2 gas, its power frequency breakdown strength relative to pure SF_6 gas is significantly improved. The power frequency breakdown strength of C_4F_7N/CO_2 mixed gas relative to pure SF_6 gas in a quasi-uniform electric field is generally higher than that in an extremely nonuniform electric field. Among them, the breakdown strength of C_4F_7N/CO_2 mixed gas with a C_4F_7N content of 2%, relative to pure SF_6 gas in a quasi-uniform electric field, is the highest at 0.73 and the lowest at 0.42; in an extremely nonuniform electric field, the highest is 0.54 and the lowest is 0.43. The breakdown strength of C_4F_7N/CO_2 mixed gas with a C_4F_7N content of 12%, relative to pure SF_6 gas in a quasi-uniform electric field, is the highest at 0.91 and the lowest at 0.78; in an extremely nonuniform electric field, the highest is 0.68 and the lowest is 0.57. The results show that the higher the C_4F_7N content in the C_4F_7N/CO_2 mixed gas, the more uniform the electric field, and the higher breakdown voltage strength relative to pure SF_6 gas [33].

The degree to which the insulating performance of a gas in an inhomogeneous electric field decreases compared to a uniform electric field is defined as the sensitivity of the gas to the electric field. The addition of C_4F_7N to CO_2 gas increases the sensitivity to the electric field, and as the C_4F_7N content in the C_4F_7N/CO_2 mixed gas increases, the sensitivity of the mixed gas to the uniformity of the electric field increases accordingly, indicating that C_4F_7N gas is particularly sensitive to the uniformity of the electric field. Therefore, in electrical equipment using C_4F_7N/CO_2 mixed gas as a gas-insulating medium, it is necessary to optimize the design to minimize the unevenness of the electric field in the gas chamber.

9.3.2 Experimental Study on Power Frequency Breakdown Characteristics of C_4F_7N/CO_2 Mixed Gas

In the previous section, the power frequency breakdown characteristics of C_4F_7N/CO_2 mixed gas in a quasi-uniform electric field and an extremely nonuniform electric field environment were studied, and the power frequency breakdown characteristics of pure SF_6 gas under the same conditions were compared and analyzed. The results show that in a quasi-uniform electric field and an extremely nonuniform electric field environment, C_4F_7N/CO_2 mixed gas has good insulation characteristics and is suitable for being considered as a gas-insulation medium to replace SF_6 gas in electrical equipment.

This section mainly conducts relevant experimental analysis on the PD (Partial discharge) insulation characteristics of C_4F_7N/CO_2 mixed gas under extremely nonuniform electric field conditions, and mainly discusses the influence mechanism of gas pressure and mixing ratio on the PD insulation characteristics of C_4F_7N/CO_2 mixed gas. The insulation characteristics of pure SF_6 gas under the same conditions are compared and analyzed to obtain the appropriate working gas pressure and mixing ratio, and the feasibility analysis of it as an insulation medium to replace SF_6 gas is carried out, laying a good foundation for further research on SF_6 alternative gas.

9.3.2.1 PD Characteristics of C_4F_7N/CO_2 Mixed Gas Under Extremely Inhomogeneous Field

This section uses the power frequency experimental platform to study the PD insulation characteristics of C_4F_7N/CO_2 mixed gas under power frequency voltage under extremely nonuniform electric field conditions. The spacing between the needle-plate electrodes in this chapter's experiment to simulate the extremely nonuniform electric field is set to 5 mm, and the C_4F_7N content in the C_4F_7N/CO_2 mixed gas selected for the experiment is 2%, 4%, 6%, 8%, 10%, and 12%; the gas pressure in the gas chamber of the device is selected to be 0.1 MPa, 0.2 MPa, 0.3 MPa, 0.4 MPa, 0.5 MPa, and 0.6 MPa. Based on the test results, the influence mechanism of the mixing ratio and gas pressure on the PD insulation characteristics of the C_4F_7N/CO_2 mixed gas under an extremely nonuniform electric field is analyzed, and compared with the insulation characteristics of pure SF_6 and CO_2 gases under the same conditions.

1) PDIV− and PDIV+ of C_4F_7N/CO_2 gas mixture under extremely nonuniform electric field

Tables 9.6 and 9.7 give the PDIV− and PDIV+ values of C_4F_7N/CO_2 mixed gas and pure SF_6 gas. It can be seen that the addition of C_4F_7N can significantly increase the PDIV− and PDIV+ values of CO_2 gas. Under the condition of 0.3 MPa pressure, the PDIV− value of the 6% C_4F_7N/94% CO_2 mixed gas is about 1.53 times that of pure CO_2 gas, and the PDIV+ value is about 1.58 times that of pure CO_2 gas. When the C_4F_7N content increases from 2% to 4%, the partial discharge starting voltage of the mixed gas increases significantly, and when the C_4F_7N content is in the range of 4% to 12%, the PDIV− and PDIV+ at each pressure increase, but the changes are relatively small.

2) Effect of gas pressure on PD characteristics of C_4F_7N/CO_2 mixed gas

Table 9.6 PDIV values of C_4F_7N/CO_2 mixed gas and pure SF_6 gas.

	PDIV− (kV)					
	Gas pressure (MPa)					
Mixing ratio k (%)	0.1	0.2	0.3	0.4	0.5	0.6
0	9.9	10.1	11.7	13.9	17.7	18.7
2	10.1	11.3	12.7	14.8	17.7	21.4
4	10.2	12.0	15.0	16.2	20.0	23.2
6	10.5	16.2	17.9	20.2	21.1	25.3
8	11.0	17.4	18.5	21.5	24.7	25.6
10	10.4	17.2	19.3	22.9	25.3	26.7
12	11.8	18.5	20.0	23.9	26.5	27.0
SF_6	12.6	19.7	25.0	30.2	34.7	37.8

Table 9.7 PDIV+ values of C_4F_7N/CO_2 mixed gas and pure SF_6 gas.

	PDIV+ (kV)					
	Gas pressure (MPa)					
Mixing ratio k (%)	0.1	0.2	0.3	0.4	0.5	0.6
0	9.9	10.9	12.5	17.1	21.3	21.9
2	10.1	12.3	17.3	18.9	20.3	25.1
4	10.5	14.3	16.6	19.9	21.8	26.8
6	11.0	17.4	19.7	23.2	24.5	29.2
8	12.1	18.2	21.0	25.0	28.5	29.6
10	11.0	18.6	21.9	26.8	30.5	33.3
12	11.9	18.9	23.1	28.0	31.2	35.1
SF_6	16.9	26.0	36.9	44.7	50.0	50.9

According to the data in Table 9.7, the PDIV− and relative PDIV− curves of C_4F_7N/CO_2 mixed gas with pressure are plotted, as shown in Figure 9.5. Figure 9.5c shows the change of PDIV− value of C_4F_7N/CO_2 mixed gas with pressure at different mixing ratios, and Figure 9.5d shows the change of PDIV− value of C_4F_7N/CO_2 mixed gas relative to pure SF_6 gas with pressure at different mixing ratios.

When the electrode spacing is fixed and the mixing ratio k value is constant, the PDIV− value of C_4F_7N/CO_2 mixed gas and pure SF_6 gas increases with the increase of pressure. When the pressure is 0.1 MPa, the difference between the PDIV− value of CO_2 gas and SF_6 gas is about 5 kV, and when pressure is 0.6 MPa, the difference between the PDIV− value of CO_2 gas and SF_6 gas is about 20 kV. When C_4F_7N is added to CO_2 gas, the PDIV− of C_4F_7N/CO_2 mixed gas with a C_4F_7N content of 2% to 12% is lower than that of pure SF_6 gas under various pressure conditions. The PDIV− of 12% C_4F_7N/88% CO_2 mixed gas at 0.3 MPa can reach the level of pure SF_6 gas at 0.2 MPa, and the PDIV− of 8% C_4F_7N/92% CO_2 mixed gas at 0.6 MPa can reach the level of pure SF_6 gas at 0.3 MPa. With the increase of pressure, the PDIV− value of C_4F_7N/CO_2 mixed gas shows a saturated growth trend, but its growth rate is lower than that of pure SF_6 gas, indicating that the C_4F_7N/CO_2 mixed gas is less affected by pressure than SF_6 gas.

For the ratio of PDIV− of C_4F_7N/CO_2 mixed gas to pure SF_6 gas. The ratio of the insulation performance of C_4F_7N/CO_2 mixed gas to pure SF_6 gas under the same pressure is now defined as relative PDIV−. From the figure, it can be seen that the relative PDIV− of C_4F_7N/CO_2 mixed gas at different mixing ratios is affected differently by pressure.

When the C_4F_7N content in the C_4F_7N/CO_2 mixed gas is less than 4%, the relative PDIV− value shows a rapid downward trend as the pressure increases from 0.1 MPa to 0.2 MPa; from 0.2 to 0.4 MPa, the relative PDIV− still shows a downward trend, but the rate of decline slows down; from 0.4 to 0.6 MPa, the relative PDIV− begins to show a slow upward trend. When the C_4F_7N content in the C_4F_7N/CO_2 mixed gas exceeds 4%, the relative PDIV− value remains basically unchanged as the pressure increases from 0.1 to 0.2 MPa. When the pressure exceeds 0.2 MPa, the relative PDIV− shows a slow downward trend and eventually stabilizes. Under high pressure (0.5–0.6 MPa), the PDIV− of 12% C_4F_7N/88% CO_2 mixed gas relative to pure SF_6 gas is stable at 70–75%, indicating that under high pressure and high mixing ratio conditions, C_4F_7N/CO_2 mixed gas has good relative insulation properties.

According to the data in Table 1.3.6, the PDIV+ and relative PDIV+ curves of C_4F_7N/CO_2 mixed gas with pressure are plotted. Figure 9.5e shows the change of PDIV+ value of C_4F_7N/CO_2 mixed gas with pressure at different mixing ratios, and Figure 9.5f shows the change of PDIV+ value of C_4F_7N/CO_2 mixed gas relative to pure SF_6 gas with pressure at different mixing ratios.

As the gas pressure increases, the PDIV+ value of the C_4F_7N/CO_2 mixed gas also increases, but the PDIV+ value under each condition is lower than that of pure SF_6 gas under the same conditions. For example, the PDIV+ value of the C_4F_7N/CO_2 mixed gas shows an approximately linear growth trend as the gas pressure increases. The PDIV+ of the 4% C_4F_7N/96% CO_2 mixed gas at a pressure of 0.6 MPa is the same as that of pure SF_6 gas at a pressure of 0.2 MPa; the PDIV+ of the 12% C_4F_7N/88% CO_2 mixed gas at a pressure of 0.6 MPa is the same as that of pure SF_6 gas at a pressure of 0.3 MPa.

As can be seen from Figure 9.5f, the relative PDIV+ of the C_4F_7N/CO_2 mixed gas does not change much with the change of air pressure, and is relatively stable, and the curve change law is also similar to that of PDIV−. When the C_4F_7N content in the C_4F_7N/CO_2 mixed gas is lower than 4%, the relative PDIV− of the mixed gas decreases with the increase of air pressure at an air pressure of 0.1–0.3 MPa, and the rate of decrease is smaller than that of PDIV−; at an air pressure of 0.3–0.5 MPa, PDIV+ begins to rise, but the change is not large; when the air pressure is in the range of 0.5–0.6 MPa, the growth rate increases. When the C_4F_7N content exceeds 4%, PDIV+ basically remains stable under the pressure of 0.1–0.2 MPa; when the pressure is in the range of 0.2–0.3 MPa, PDIV+ shows a downward trend. When the pressure is in the range of 0.3–0.5 MPa, PDIV+ tends to be stable again; when the pressure is in the range of 0.5–0.6 MPa, PDIV+ begins to show an upward trend.

Comparing the PDIV− and PDIV+ values of C_4F_7N/CO_2 mixed gas with pure SF_6 gas, it is verified again that the more C_4F_7N content, the higher the insulation strength of C_4F_7N/CO_2 mixed gas. And Figure 9.5d,f both show that the relative value of 12% C_4F_7N/88% CO_2 mixed gas is the highest at 0.2 MPa pressure, and it shows an upward trend at 0.5–0.6 MPa pressure. The relative PDIV− of 12% C_4F_7N/88% CO_2 mixed gas at 0.6 MPa pressure reaches 0.71, and the relative PDIV+ reaches 0.69. Therefore, under high mixing ratio conditions, the relative PDIV− and PDIV+ of C_4F_7N/CO_2 mixed gas are relatively high.

3) Effect of mixing ratio on PD characteristics of C_4F_7N/CO_2 mixed gas

According to the data in Table 9.7, the PDIV− and relative PDIV− curves of C_4F_7N/CO_2 mixed gas are plotted as a function of gas pressure. Figure 9.6a shows

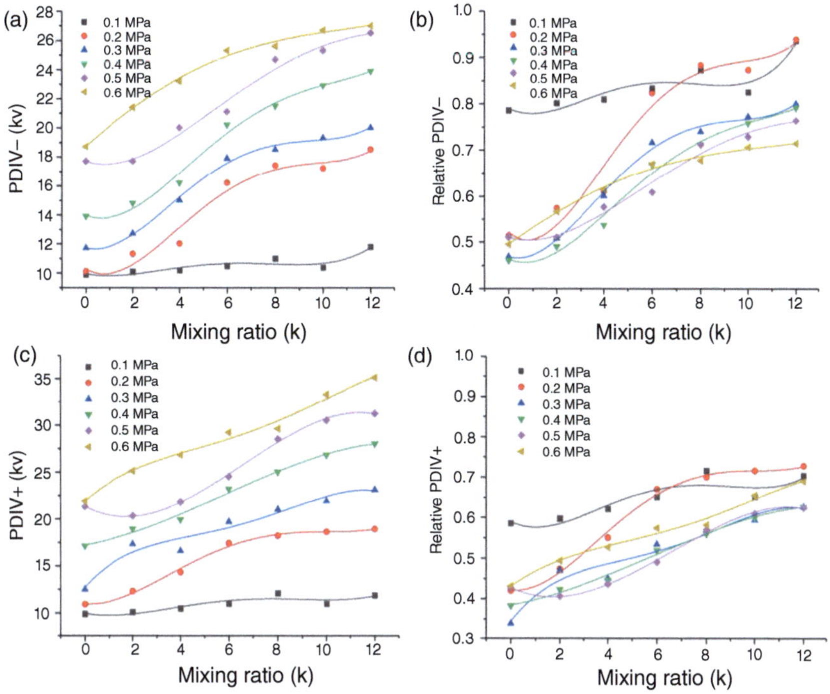

Figure 9.6 PDIV+, PDIV– and relative PDIV+, PDIV– values of C_4F_7N/CO_2 mixed gas at different pressures as a function of mixing ratio; (a) PDIV– of C_4F_7N/CO_2 mixed gas under different pressures-varies with mixing ratio; (b) PDIV– of C_4F_7N/CO_2 mixed gas relative to pure SF_6 gas; (c) PDIV+ of C_4F_7N/CO_2 mixed gas changes with mixing ratio under different pressures; (d) C_4F_7N/CO_2 mixed gas relative to pure SF_6 gas PDIV+. *Source:* Adapted from [30, 33].

the change of PDIV– value of C_4F_7N/CO_2 mixed gas with the mixing ratio under different mixing ratios.

Figure 9.6a shows the effect of different mixing ratios on the PDIV– value of C_4F_7N/CO_2 mixed gas in the gas pressure range of 0.1–0.6 MPa. It can be seen from the figure that under various gas pressure conditions, when the mixing ratio is 0–4%, the PDIV– value of C_4F_7N/CO_2 mixed gas increases slowly with the increase of the mixing ratio; when the mixing ratio is in the range of 4–6%, the PDIV– value of C_4F_7N/CO_2 mixed gas increases fastest with the mixing ratio. When the mixing ratio exceeds 6%, its growth rate begins to slow down and shows a saturated growth trend. The results show that the increase of C_4F_7N gas can effectively improve the PD characteristics of C_4F_7N/CO_2 mixed gas.

Figure 9.6b shows the variation of PDIV– of C_4F_7N/CO_2 mixed gas relative to pure SF_6 gas with the mixing ratio. It can be seen from the figure that with the increase of C_4F_7N content, the relative PDIV– of C_4F_7N/CO_2 mixed gas shows a linear saturated growth trend, and the relative PDIV– of C_4F_7N/CO_2 mixed gas under low pressure conditions is slightly higher than that under high pressure conditions.

Under 0.2 MPa, the relative PDIV− value of 2% C_4F_7N/98% CO_2 mixed gas is about 80.1%, while the relative PDIV− value of 12% C_4F_7N/88% CO_2 mixed gas is about 93.7%; in 0.3 MPa pressure environment, the relative PDIV− value of 12% C_4F_7N/88% CO_2 mixed gas is about 80.0%. In 0.6 MPa pressure environment, the relative PDIV− value of 2% C_4F_7N/98% CO_2 mixed gas is about 56.6%, while the relative PDIV− value of 12% C_4F_7N/88% CO_2 mixed gas is about 71.4%, indicating that the insulation performance of C_4F_7N/CO_2 mixed gas in low pressure environment is slightly better than that of pure SF_6 gas under the same pressure in high pressure environment.

According to the data in Table 1.3.6, the PDIV+ and relative PDIV+ curves of C_4F_7N/CO_2 mixed gas are plotted for the mixing ratio. Figure 9.6c shows the changes in PDIV+ value of C_4F_7N/CO_2 mixed gas with pressure at different mixing ratios, and Figure 9.6d shows the changes of PDIV+ value of C_4F_7N/CO_2 mixed gas relative to pure SF_6 gas with pressure at different mixing ratios.

According to Figure 9.6c, PDIV+ of C_4F_7N/CO_2 mixed gas increases with the increase of mixing ratio. When the mixing ratio is higher than 6%, the growth rate of PDIV+ value of C_4F_7N/CO_2 mixed gas with the mixing ratio is relatively low, and the mixing ratio (less than 6%) decreases slightly. The relative PDIV+ of C_4F_7N/CO_2 mixed gas is similar to PDIV−, and also shows a linear growth trend, but its growth rate is lower than the relative PDIV− value. As can be seen from Figure 9.6d, the relative PDIV+ of C_4F_7N/CO_2 mixed gas is not greatly affected by changes in gas pressure. For example, the relative PDIV+ of C_4F_7N/CO_2 mixed gas with a mixing ratio in the range of 3% to 5% is almost the same.

9.3.2.2 Synergistic Effect of C_4F_7N/CO_2 Mixed Gas

According to the synergistic effect expression in the previous section, V1 represents the PDIV value of the 12% C_4F_7N/88% CO_2 mixed gas, and V2 represents the PDIV value of pure CO_2 gas. Substitute the PDIV data of the C_4F_7N/CO_2 mixed gas obtained in the previous experiment into the expression, and calculate the synergistic effect coefficient C value as shown in Table 9.8 and Table 9.9.

According to Table 9.8, it can be found that the local discharge synergistic effect coefficient C of C_4F_7N gas and CO_2 gas is in the range of 0.004–0.206. Overall,

Table 9.8 Synergistic effect coefficient C of C_4F_7N/CO_2 mixed gas PDIV− value.

Gas pressure/MPa	Synergy coefficient C				
	k = 2%	k = 4%	k = 6%	k = 8%	k = 10%
0.10	0.173	0.222	0.138	0.062	0.311
0.20	0.122	0.143	0.024	0.013	0.020
0.30	0.149	0.063	0.021	0.019	0.010
0.40	0.206	0.139	0.037	0.027	0.012
0.50	0.201	0.117	0.101	0.022	0.018
0.60	0.042	0.035	0.016	0.018	0.004

Table 9.9 Synergistic effect coefficient C of C_4F_7N/CO_2 mixed gas PDIV+ value.

Gas pressure/MPa	Synergy coefficient C				
	$k = 2\%$	$k = 4\%$	$k = 6\%$	$k = 8\%$	$k = 10\%$
0.10	0.184	0.097	0.052	−0.008	0.091
0.20	0.096	0.056	0.014	0.008	0.004
0.30	0.025	0.066	0.030	0.021	0.014
0.40	0.103	0.120	0.050	0.033	0.014
0.50	−0.222	0.783	0.134	0.033	0.008
0.60	0.064	0.071	0.052	0.062	0.018

the same effect coefficient C values of C_4F_7N/CO_2 mixed gas under different gas pressure and mixing ratio conditions are relatively close to 0, indicating that C_4F_7N/CO_2 mixed gas has a good synergistic effect relationship. From the comparison of synergistic effect coefficients in the table, it can be seen that under the same mixing ratio conditions, the change in gas pressure has little effect on the synergistic effect coefficient C of C_4F_7N/CO_2 mixed gas PDIV−. With the increase of the mixing ratio, that is, the higher the content of C_4F_7N in C_4F_7N/CO_2 mixed gas, the smaller the synergistic effect coefficient C value. Comparing the changes in the synergistic effect coefficient C of C_4F_7N/CO_2 mixed gas PDIV+ with gas pressure and mixing ratio in Table 9.9, its changes are similar to those of PDIV−. The synergistic effect of C_4F_7N/CO_2 mixed gas can be further improved under high mixing ratio conditions. In addition, the synergistic effect coefficient C in the table appears negative under some conditions, but it is concentrated in low mixing ratios or low-pressure environments. This is due to the small difference in the insulation performance of the mixed gas in low mixing ratios and low-pressure environments. At the same time, when comparing the specific values of PDIV− and PDIV+ of the C_4F_7N/CO_2 mixed gas corresponding to the negative synergistic effect coefficient C, we will find that its values are basically equal to the PDIV− and PDIV+ of pure CO_2 gas under the same conditions. This once again shows that the appearance of a negative value of the synergistic effect coefficient C does not mean that the mixed gas has a negative synergistic effect.

In general, a comparative analysis of the PDIV− and PDIV+ synergistic effects of the C_4F_7N/CO_2 mixed gas revealed that C_4F_7N and CO_2 have good synergistic effects, and increasing the content of C_4F_7N gas can effectively improve the synergistic effect relationship of the mixed gas.

9.3.3 Eco-Friendly Gas Biosafety Assessment and Industrial Protection Technology

This section uses a multifunctional dynamic gas distributor (as shown in Figure 9.7), (model: GC_400) to accurately distribute the test gas. The distributor has four channels. The gas flows out of the compressed steel cylinder and enters each channel

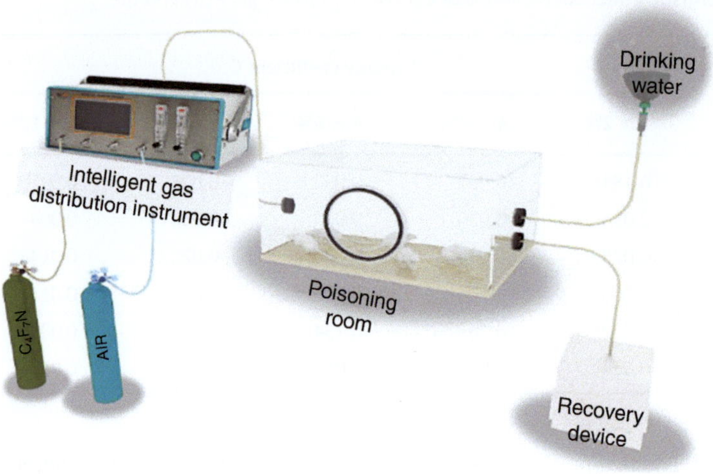

Figure 9.7 C_4F_7N gas toxicity test platform.

of the distributor through the pressure-reducing valve. The distributor uses the principle of mass flow mixing to mix multiple high-concentration gases according to the set ratio to obtain the test gas of the required concentration. At the same time, the instrument can control the flow of the prepared gas and output it at the set flow rate; the length, width, and height of the poison chamber are 40, 30, and 20 cm, respectively, and it is composed of a polytetrafluoroethylene plate with a thickness of 1 cm. There are three holes on both sides of the poison chamber, which are used for air intake, water intake and air outlet respectively; a round window with a radius of 7.5 cm is opened on one side for taking in and placing experimental rats and food. The gas outlet is connected to a beaker filled with NaOH solution as an exhaust gas filter.

During the test, C_4F_7N and dry air were introduced into the two channels of the dynamic gas distributor, respectively, and then introduced into the left inlet of the poisoning room through the right outlet after being configured according to the set mixing ratio. Three monitoring cameras were installed around the entire test platform, two of which were used to monitor the activity status of animals in the poisoning room, and one was used to monitor key parameters such as the real-time output flow of the gas distributor control panel.

Considering the asphyxiation of CO_2 gas, this project mixed C_4F_7N with dry air to study the toxicity of C_4F_7N gas. In terms of the selection of the mixing ratio, in order to take into account both insulation performance and liquefaction temperature in actual engineering applications, the concentration of C_4F_7N in the C_4F_7N mixed gas is generally in the range of 4%–10%, but the actual leakage after the equipment fails depends on the severity of the failure, the equipment volume, the equipment inflation pressure, etc., and the specific concentration cannot be accurately estimated. Therefore, this project is mainly based on the data disclosed in existing literature, that is, the LC50 of C_4F_7N is in the range of 15 000 ppm–20 000 ppm, and

Table 9.10 C_4F_7N gas toxicity test conditions.

Group	Animal combination	Mixing ratio	Exposure time
1	2 males, 2 females	100% dry air	24 h
2	2 males, 2 females	1.5% C_4F_7N/98.5% air	24 h
3	2 males, 2 females	1% C_4F_7N/99% air	24 h
4	2 males, 2 females	0.5% C_4F_7N/99.5% air	24 h
5	2 males, 2 females	2% C_4F_7N/98% air	4 h
6	2 males, 2 females	1.5% C_4F_7N/98.5% air	4 h
7	2 males, 2 females	1% C_4F_7N/99% air	4 h

the mixed gas containing 0.5–2% C_4F_7N is tested for toxicity. Table 9.10 gives the relevant test conditions. Among them, 1 control group and 6 test groups were set up to test the 4-hour acute toxicity and 24-hour acute toxicity of C_4F_7N gas, respectively (the 24-hour test was carried out according to DL/T 921-2005).

Before the test, the air tightness of the poisoning room was first checked. Use a precision digital barometer to connect the air outlet of the poisoning room, and pass compressed dry air into the poisoning room. When the air pressure reaches 0.13 MPa, stop ventilation and observe whether the precision air pressure indication number will change within 15 minutes. It is defined that the device passes the air tightness test when the air pressure indication number decreases by less than 1% within 15 minutes.

Then, according to the test conditions in Table 9.10, the test animal samples were numbered and placed in the poisoning room, and mouse food and drinking water were provided at the same time; the prepared mixed gas was passed into the poisoning room, and the flow rate was set to 3000 ml/min, that is, the gas flowing into the poisoning room per minute was not less than one-eighth of the total volume of the poisoning room, which met the relevant test requirements in DL/T921-2005. After the poisoning started, the monitoring camera was turned on to monitor the vital signs and gas distribution parameters of the rats in the poisoning room, so as to trace the relevant data after the test.

After the poisoning test, the surviving test samples were returned to the original feeding container for further observation. If the test samples were active during the poisoning time and the following 72 hours, the gas was nontoxic. The ambient temperature was controlled at 25 °C during the test.

For the biological toxicity test results of the new insulating gas, the control group was first tested for 24 hours by passing 100% dry air. During the entire observation period, the vital signs of the four rats were normal, and they were able to eat and drink water independently. After the 24-hour control test, the vital signs of the above four rats were also normal during the 72-hour observation period. The above results confirm that the dry air used in the test is nontoxic and harmless, and on the other hand, they also confirm that the key parameters such as flow rate in the test plan of this article are reasonable.

Four rats were exposed to 1.5% C_4F_7N/98.5% air for 24 hours, and the related poisoning symptoms are summarized as follows:

Before poisoning, the four rats were able to move relatively calmly in the poisoning room; 2 minutes after the start of poisoning, the four rats changed from calm to overexcited, specifically climbing on the inner wall of the poisoning room, trying to break free from the confined space, indicating that the rats began to feel uncomfortable 2 minutes after poisoning. As the poisoning time continued, the rats became restless 7 minutes after poisoning, and repeatedly rubbed their faces (nose) with their forefoots, and then several other test samples also showed similar symptoms. This behavior may be caused by the discomfort caused by C_4F_7N gas irritating the rat's nasal mucosa; about 10 minutes after poisoning, the first rat showed a poor mental state, stopped moving and curled up in the corner of the poisoning room. After 23 minutes, all four rats stopped moving, curled up in the corner of the poisoning room, and slightly closed their eyes. At the same time, the rats began to have difficulty breathing, accompanied by obvious arrhythmia and convulsions during breathing, indicating that the gas containing 1.5% C_4F_7N has caused serious damage to the rats' respiratory system. After 60 minutes of poisoning, all rats completely stopped moving and eating, and their vital signs were weak.

Continue to conduct poisoning tests on samples and observe vital signs. After 10 hours and 17 minutes, the first rat was found dead; after 10 hours and 54 minutes, the second rat died; after 11 hours and 51 minutes, the third rat died; after 14 hours and 03 minutes, the fourth rat died. Therefore, the mixed gas of 1.5% C_4F_7N/98.5% air is toxic.

In addition, we collected blood samples from normal rats and rats that died after exposure to poison and conducted blood cell tests and analyses (Table 9.11 gives the test results). The results of the blood cell analysis show that the white blood cell, red blood cell and platelet counts of the control rats were within the reference value range, indicating that the experimental rats were healthy; the red blood cell count of the rats that died after exposure to poison was slightly lower than the reference value, while the white blood cell count increased significantly, reaching more than 10 times the reference maximum value. Since an increase in white blood cells is common in inflammatory infections, bleeding, poisoning, etc., it was confirmed that rats were severely poisoned after contact with 1.5% C_4F_7N/98.5% air. In addition, the platelet

Table 9.11 Analysis of blood cells before and after 24-hour exposure.

No.	White blood cells (10^9/L)	Red blood cells (10^{12}/L)	Platelets (10^9/L)
Control group	4.50	5.11	151
Test group	163.50	4.33	1360
Reference value	2.90–15.30	5.60–7.89	100–1610

Table 9.12 Results of 24-hour toxicity test of C_4F_7N gas.

Contents	1.5% C_4F_7N/98.5% Air	1% C_4F_7N/99% Air	0.5% C_4F_7N/99.5% Air
Complete inactivity time	60 min	75 min	86 min
First death time	10 h 17 min	13 h 48 min	>24 h
Second death time	10 h 54 min	14 h 04 min	>24 h
Third death time	11 h 51 min	15 h 39 min	>24 h
Fourth death time	14 h 03 min	18 h 35 min	>24 h

count of rats after exposure increased, but was within the reference value range, suggesting that C_4F_7N gas has the potential to affect the coagulation system of rats.

The following is an analysis and discussion of the results of the 24-hour exposure test of C_4F_7N mixed gas with different concentrations. Based on the main characteristic symptoms found in the above tests, Table 9.12 gives the 24-hour toxicity test results of C_4F_7N-air mixed gas with C_4F_7N content of 0.5–1.5%.

After 2–3 minutes of exposure to different concentrations of C_4F_7N gas, rats all showed an overexcited state and tried to break free from the poisoning room environment; then they rubbed their faces (nose) with their front paws. After a period of continued poisoning, all rats completely stopped moving or even died.

As the C_4F_7N content increased, the time for complete cessation of activity gradually shortened. For example, rats stopped moving after 86 minutes of exposure to a gas environment with a C_4F_7N content of 0.5%, and stopped moving after 60 minutes of exposure to a gas environment with a C_4F_7N content of 1.5%. Rats completely died after being exposed to an environment with a C_4F_7N content of 1% or more for 24 hours. After being exposed to an environment with a C_4F_7N content of 0.5% for 24 hours, they still had weak breathing and heartbeats, but were in an extremely weak state and could not move or eat normally. The four rats that had been poisoned in the 0.5% C_4F_7N environment were moved to a breeding environment for further observation, and all four rats died within 72 hours. According to the LC50 test standard, after the poisoning, the surviving samples need to be kept in a well-ventilated animal room for observation for 14 days, and the LC50 lethal concentration is determined based on the number of survivors. Therefore, the LC50 (rat, 24h) of C_4F_7N gas is less than 5000 ppm.

In order to further test the acute toxicity of C_4F_7N gas, this paper conducted a 4-hour poisoning test on gases with C_4F_7N content of 1%, 1.5% and 2%.

Similar to the previous test, rats showed slightly closed eyes and stopped moving after being poisoned in the C_4F_7N gas environment for a period of time. After four hours of poisoning, the rats under each test condition did not die, but they were weak, their activities decreased or even stopped. All test samples were transferred to the breeding environment for further observation, and no deaths occurred after 72 hours.

Table 9.13 Analysis of blood cells after 4-hour exposure.

No.	White blood cells (10^9/L)	Red blood cells (10^{12}/L)	Platelets (10^9/L)
Sample 1	0.80	4.48	716
Sample 2	1.30	3.84	862
Reference value	2.90–15.30	5.60–7.89	100–1610

Table 9.13 shows the results of blood cell analysis of rats exposed to 2% C_4F_7N gas environment for 4 hours. It can be seen that the white blood cell and red blood cell counts of rats decreased significantly after 4 hours of poisoning, which were lower than the reference value; the platelet count increased compared with the control group. The results of red blood cell and platelet counts after 4 hours of poisoning were basically consistent with the results of death after 24 hours of poisoning. The white blood cells of rats after poisoning showed a trend of rapid decrease in a short period of time (4 hours) and then a large increase (24 hours), indicating that the rats consumed the existing white blood cells with protective functions in the body after poisoning, and then produced a stress response to regenerate a large number of white blood cells to fight against toxic gases.

In addition, the rat samples exposed to the gas environment with a C_4F_7N content of 1.5% were still in a weak state after 24 hours and could not move independently; the rat samples exposed to the gas environment with a C_4F_7N content of 1% could crawl independently after 19 hours, but the muscle strength of the limbs decreased and it was difficult to support the body. After 52 hours, the rat's motor system recovered and could crawl independently. All test samples recovered their motor function after 96 hours. After 14 days of observation, no dead rats were found in the rats exposed to the environment with C_4F_7N content of 1% and 1.5%. However, two rats exposed to the environment with C_4F_7N content of 2% showed symptoms such as slowed movement and slightly closed eyes on the 13th day, and eventually died. Therefore, the LC50 (rat, 4H) of C_4F_7N gas is higher than 15 000 ppm but lower than 20 000 ppm.

Overall, the gas with C_4F_7N content of 1–2% will not cause the death of rats within 4 hours, but it will cause great damage to their respiratory and motor systems. The breathing of the poisoned samples slowed down, the autonomous motor function gradually weakened, and the whole body was paralyzed. After being poisoned for 4 hours in an environment with C_4F_7N content below 1.5%, the rats were able to gradually return to normal, that is, C_4F_7N did not cause irreversible damage to the respiratory and motor systems of rats.

According to the test results of this project, the LC50 (rat, 24H) of C_4F_7N is less than 5000 ppm, and the LC50 (rat, 4h) is greater than 15 000 ppm, which is basically consistent with the existing literature results (as shown in Table 9.14). According to the classification standard for acute toxicity of chemicals in the Globally Harmonized System of Classification and Labelling of Chemicals (GHS), C_4F_7N is classified

Table 9.14 Toxicity of C_4F_7N and its mixed gases.

Contents	LC50 (ATE)	TWA
C_4F_7N	10 000–15 000 ppm (Rats, 4h)	65 ppm
CO_2	>30%	—
4% C_4F_7N-96% CO_2	160 000–212 000 ppm (Rats, 4h)	—
10% C_4F_7N-90% CO_2	95 500–100 000 ppm (Rats, 4h)	—
57.7% C_4F_7N-42.3% Air	<173 000 (Rats, 4h)	—

Table 9.15 Toxicity of main decomposition products of C_4F_7N gas [40, 41].

Contents	LC50 (ppm)
CF_3CN	500 (1 h, Rats)
C_4F_6	82 (1 h, Rats)
CO	1847 (4 h, Rats)
HF	483 (4 h, Rats)
C_4F_8	0.5 (4 h, Rats)

as Class IV; the occupational exposure limit value (TWA, 8 hours) of C_4F_7N given by 3MTM is 65 ppm, which is also lower than SF_6's 1000 ppm, so the human body should avoid inhaling this gas.

At the same time, combined with the changes in physiological characteristics and behaviors of rats after exposure, it was found that C_4F_7N was irritating to the nasal mucosa of rats. After a period of exposure, the rats' respiratory rate decreased, their motor ability was gradually lost, and their blood cell counts became abnormal. Therefore, C_4F_7N may cause certain damage to the respiratory and motor systems of the human body. Therefore, for the control of contact with C_4F_7N, attention should be paid to eye, skin, and respiratory protection. We communicated with 3MTM China. For scientific research and operation and maintenance personnel, it is recommended to use indirect ventilation goggles to protect the eyes and face; it is recommended to use nitrile gloves to protect the hands; it is recommended to use air-supplied half masks or full-face respirators for respiratory protection to avoid inhalation or contact with C_4F_7N gas. The hidden dangers and hazards to one's own safety are avoided.

In actual engineering applications, considering the limiting factors such as liquefaction temperature, gas pressure, and mixing ratio, the content of C_4F_7N in the C_4F_7N mixed gas inside the GIE is generally in the range of 4–10%. At present, the existing literature has tested the LC50 of C_4F_7N/CO_2 mixed gas, and found that the LC50 (4 hours, ppm) of 4% C_4F_7N-96% CO_2 and 10% C_4F_7N-90% CO_2 mixed gas is in the range of 160 000–212 000 ppm and 95 500–100 000 ppm, that is, the toxicity of C_4F_7N mixed gas is lower than that of pure C_4F_7N gas.

Although it is inevitable for scientific researchers and equipment operation and maintenance personnel to contact C_4F_7N mixed gas, and the C_4F_7N content in the environment is closely related to factors such as the volume of the equipment, inflation pressure, leakage severity, and inflation mixing ratio, the probability of a serious leakage accident in real equipment is extremely low, and the C_4F_7N content of the leaked gas will also be greatly reduced after diffusion, so the C_4F_7N mixed gas is still safe for application. However, it is recommended to establish a C_4F_7N gas concentration monitoring and early warning system in the test area and equipment operation area to avoid safety hazards caused by large-scale leakage of C_4F_7N gas due to factors such as malfunctions.

Gas-insulating media will inevitably decompose under partial discharge or overheating conditions. Existing studies have confirmed that C_4F_7N gas will produce toxic decomposition products such as C_2F_5CN, CF_3CN, COF_2, C_3F_6, C_2F_4, HF, and CO under discharge faults and overheating conditions. Table 1.3.14 summarizes the LC50 of the above decomposition products. It can be seen that the toxicity of the decomposition products is generally stronger than that of C_4F_7N gas. Therefore, in actual scientific research or engineering applications, more attention should be paid to avoid contact with C_4F_7N mixed gas inside the equipment after testing or running for a period of time.

In terms of protection recommendations, C_4F_7N mixed gas has some similarities with SF_6. Due to its small environmental impact and relatively low toxicity, its protection requirements can be relatively relaxed in some aspects, but similar protection measures as SF_6 are still required in other aspects.

1) Gas leakage detection and monitoring

Gas leakage detection equipment should be installed at the place of use of C_4F_7N, especially in a closed environment or gas filling equipment. Although C_4F_7N has a lower environmental impact, leakage may still lead to equipment performance degradation or the generation of hazardous substances. Therefore, the detection equipment should monitor the changes in gas concentration in real time and issue an alarm when the gas concentration exceeds the preset threshold.

SF_6 also needs to install a leak detection system, especially for SF_6 gas, leakage will not only affect the operation of the equipment, but also cause serious environmental pollution. Therefore, leakage monitoring of SF_6 equipment usually requires higher sensitivity and is equipped with an emergency shutdown and recovery system.

2) Ventilation and environmental control

C_4F_7N mixed gas needs to ensure that the use environment has a good ventilation system, especially in closed spaces. Although C_4F_7N mixed gas has low toxicity, it may still react with oxygen in the air to produce harmful gases in the event of leakage.

Therefore, it is very important to ensure that the gas does not accumulate in the working area.

SF_6 leakage is more harmful to the environment, so stricter ventilation management is required. In addition to good ventilation, a professional gas recovery system is also required to prevent SF_6 from leaking into the atmosphere and causing a greenhouse effect.

3) Personal protection

C_4F_7N: Workers should wear appropriate personal protective equipment, such as protective gloves, eye protection, and respirators. Although the toxicity of C_4F_7N mixed gas is low, direct exposure to the gas environment should still be avoided in the event of leakage or high concentration.

4) Equipment maintenance and gas recovery

When maintaining and overhauling C_4F_7N equipment, first confirm whether the gas is completely discharged to avoid exposure of operators to the leaked gas. The recovery measures for the leaked gas can refer to the recovery specifications of SF_6, but since the environmental impact of C_4F_7N is relatively small, its recovery requirements can be appropriately relaxed.

The recovery and disposal requirements of SF_6 gas are very strict, and usually special equipment is required to recover the leaked gas and prevent it from entering the atmosphere. The maintenance of SF_6 equipment should be operated by professionals, and special recovery equipment and monitoring systems should be used.

5) Emergency response and leakage disposal

When C_4F_7N gas leaks, the leak point should be quickly identified and appropriate measures should be taken to control it. The leaked gas can be diluted by methods such as increasing ventilation or using gas absorption devices. During the emergency response process, the treatment measures need to be adjusted according to the gas concentration, environmental conditions, and the scale of the leak.

The emergency response to SF_6 leaks must be strictly carried out in accordance with national or industry specifications. After a leak occurs, the leak source must be quickly blocked and the gas recovery device must be started. Emergency personnel need to be equipped with efficient respiratory protection equipment and isolate the leak area.

In general, C_4F_7N and SF_6 are similar in many protection measures, especially in terms of leak detection, ventilation, personal protection, and equipment maintenance. However, due to the greater greenhouse effect potential and more serious environmental impact of SF_6, its protection recommendations are generally more stringent. In contrast, although C_4F_7N is still a strong greenhouse gas, its impact on the environment is relatively small, so the protection requirements can be appropriately relaxed, especially in terms of leak disposal and gas recovery.

9.4 Eco-Friendly Insulating Gas State Evolution Mechanism

9.4.1 Effect of Temperature on Thermal Decomposition of Eco-Friendly Insulating Gas

In order to study the influence of temperature on the decomposition characteristics of C_4F_7N-CO_2 mixed gas, this section carried out superheat decomposition experiments under three temperature conditions of 350, 450, and 550°C. The experimental period was 10 hours, the gas pressure was 0.15 MPa, and the C_4F_7N content in the fixed mixed gas was 5%.

From the meteorological chromatogram after the thermal decomposition of the mixed gas, it can be seen that the decomposition of the C_4F_7N-CO_2 mixed gas produces CF_4, C_2F_6, C_3F_6, C_2N_2, C_6F_{14}, CF_2CFCN, and other components. Among them, since the characteristic peak time of CO_2 gas is within 5–5.5 minutes and its content is relatively high (95%), there is cross-interference with decomposition components such as CF_3CN, resulting in some decomposition components cannot be detected. The following is an analysis of the changes in the content of characteristic decomposition components of mixed gas at different temperatures.

In order to study the influence of temperature on the decomposition characteristics of C_4F_7N-CO_2 mixed gas, this project carried out superheat decomposition experiments under three temperature conditions of 350, 450, and 550 °C. The experimental period was 10 hours, the gas pressure was 0.15 MPa, and the content of C_4F_7N in the fixed mixed gas was 5%.

From the meteorological chromatogram after the thermal decomposition of the mixed gas, it can be seen that the decomposition of the C_4F_7N-CO_2 mixed gas produces CF_4, C_2F_6, C_3F_6, C_2N_2, C_6F_{14}, CF_2CFCN, and other components. Among them, since the characteristic peak time of CO_2 gas is within 5–5.5 minutes and its content is relatively high (95%), there is cross-interference with decomposition components such as CF_3CN, resulting in some decomposition components cannot be detected. The following is an analysis of the changes in the content of characteristic decomposition components of mixed gas at different temperatures.

This project uses the external standard method to quantify the characteristic decomposition components of the mixed gas. The standard curve is made using the standard gas purchased for the characteristic decomposition components, and then the experimental sample gas is collected every 2 hours for testing. According to the test results, the content of each characteristic decomposition component of the mixed gas shows an increasing trend with the simulation time. Under the temperature conditions of 350 and 450°C, C_3F_6 has the highest content among the four types of characteristic decomposition components, followed by C_3F_8, and the contents of C_2F_6 and CF_4 are the lowest. When the temperature reaches 550 °C, the content of C_3F_8 in the mixed gas is the highest, followed by C_3F_6. This may be related to the poor thermal stability of C_3F_6 and the re-decomposition of C_3F_6.

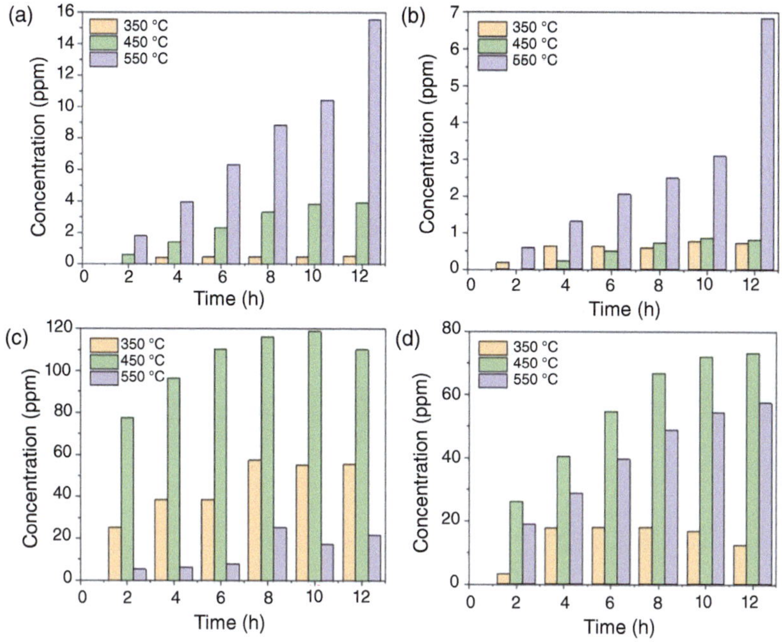

Figure 9.8 Effect of temperature on the yield of characteristic decomposition components. (a) CF_4; (b) C_2F_6; (c) C_3F_6; (d) C_3F_8. *Source:* Adapted from [34, 35].

Figure 9.8 shows the effect of temperature on the characteristic decomposition components of the mixed gas. It can be seen that with the increase in temperature, the contents of the two decomposition components CF_4 and C_2F_6 both show an increasing trend. Under the temperature conditions of 350 and 450°C, the contents of both products are below 4 ppm; when the temperature rises to 550°C, the contents of CF_4 and C_2F_6 increase significantly. Since the above two substances are mainly formed by the combination of CF_3 and F, the increase in their contents indicates that a large number of free radicals with smaller mass numbers such as CF_3 and F are generated under high temperature. The content of C_3F_6 reaches a peak value at 450°C and decreases significantly to a minimum value at 550°C. This shows that when the mixed gas undergoes thermal decomposition, the substance C_3F_6 produced at a lower temperature undergoes re-decomposition as the temperature increases, so the content decreases. With the increase in temperature, the content of C_3F_8 also increases, but its concentration also decreases under high temperature conditions.

In general, among the various products of thermal decomposition of C_4F_7N-CO_2 mixed gas at temperatures below 450°C, the contents of C_3F_6 and C_3F_8 are relatively high, while the contents of C_2F_6 and CF_4 are relatively low; when the temperature reaches 550°C, the contents of C_3F_6 and C_3F_8 begin to decrease, while the contents of C_2F_6 and CF_4 increase rapidly.

9.4.2 Effect of Mixing Ratio on Thermal Decomposition of Eco-Friendly Insulating Gas

In order to explore the influence of the mixing ratio on the thermal decomposition characteristics of C_4F_7N-CO_2 mixed gas, the temperature was fixed at 400 °C and the gas pressure was 0.15 MPa. The mixed gas with C_4F_7N content of 5%, 12.5% and 20% was subjected to a 10-hour thermal decomposition test.

Figure 9.9a–c shows the changes in the content of characteristic decomposition products of C_4F_7N-CO_2 mixed gas under different mixing ratios. It can be seen that

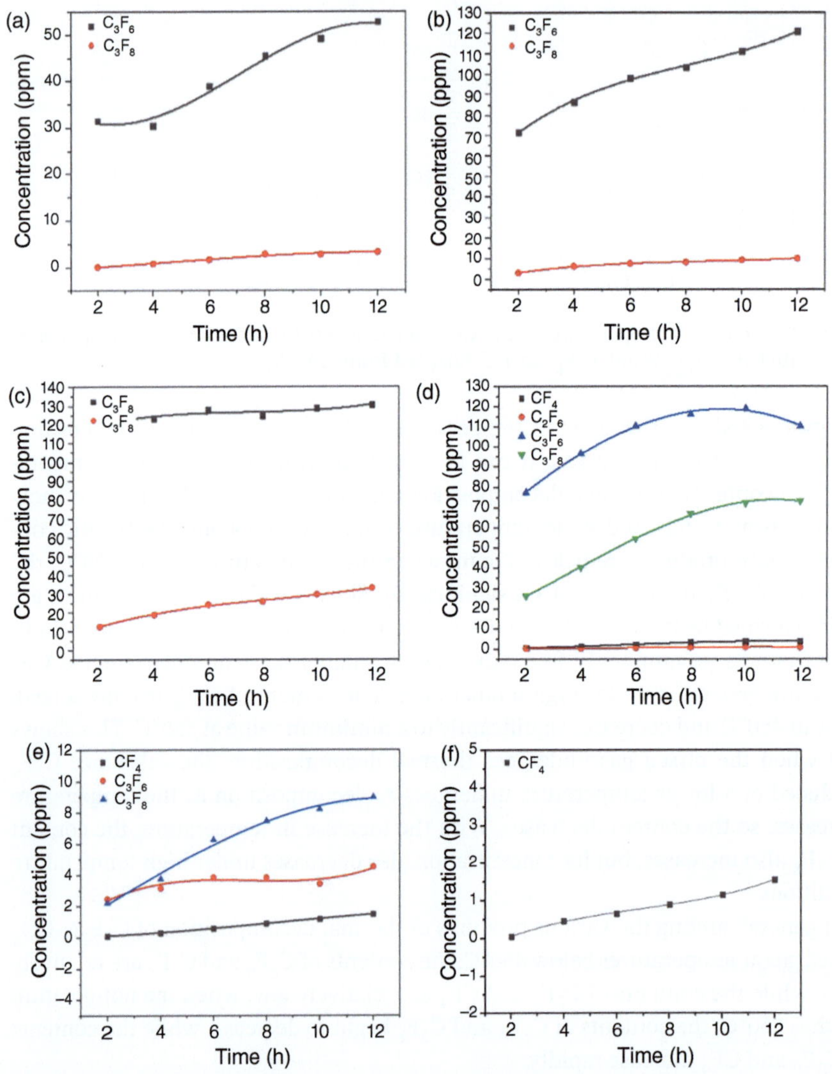

Figure 9.9 Changes of characteristic decomposition components of mixed gas at different pressures over test time. (a) 5%; (b) 12.5%; (c) 20%; *Source:* Adapted from [36].

the thermal decomposition of gases with different mixing ratios mainly produces two types of decomposition components, C_3F_6 and C_3F_8, and CF_4 and C_2F_6 are not detected. As the reaction time continues, the content of both decomposition products, C_3F_6 and C_3F_8, shows an increasing trend, and the content of C_3F_6 is always higher than that of C_3F_8, which is the main product of the superheated decomposition of the mixed gas.

According to the content changes of the two types of characteristic decomposition products under different mixing ratios, it can be seen that with the increase of the mixing ratio, the content of C_3F_6 and C_3F_8 tends to increase, that is, the decomposition of the mixed gas is intensified under high mixing ratio conditions. In fact, with the increase of the mixing ratio, the content of C_4F_7N in the system increases, and the number of particles participating in the reaction increases, so the decomposition of the C_4F_7N mixed gas is intensified under the same conditions.

9.4.3 Effect of Air Pressure on Thermal Decomposition of Eco-Friendly Insulating Gas

Three pressure conditions of 0.15, 0.3, and 0.45 MPa were selected for testing. The experimental control heat source temperature was 450°C and the C_4F_7N content was 5%. According to the test results, it can be seen that the decomposition components of the mixed gas under different pressure conditions show an increasing trend with the simulation time. The contents of C_3F_6 and C_3F_8 in the thermal decomposition products of the C_4F_7N mixed gas under low pressure are higher than other characteristic decomposition components. However, with the increase of gas pressure, the product C_2F_6 was not found during the 0.3 MPa experiment, while only the product CF_4 was detected under 0.45 MPa. On the whole, with the increase of gas pressure, the thermal decomposition of the mixed gas slows down, that is, the thermal decomposition characteristics of the mixed gas under high pressure are better than those under low pressure.

9.4.4 Effect of Discharge Energy on Partial Discharge Decomposition of Eco-Friendly Insulating Gas

In order to explore the discharge decomposition characteristics of C_4F_7N mixed gas and the influence of various factors, this project conducted experimental tests on C_4F_7N mixed gas under various conditions. Stable partial discharge was generated by the constructed needle-plate electrode model, and the experiment lasted for 96 hours. Gas samples in the gas chamber were collected every 12 hours for detection to obtain the composition and content information of the decomposition products.

For the influence of discharge energy on the discharge decomposition characteristics of mixed gas, this project selected two discharge conditions. First, the PDIV of the mixed gas was measured, and then 1.6 times the PDIV value was applied as one group of experiments, and 2 times the PDIV value was applied in the other group.

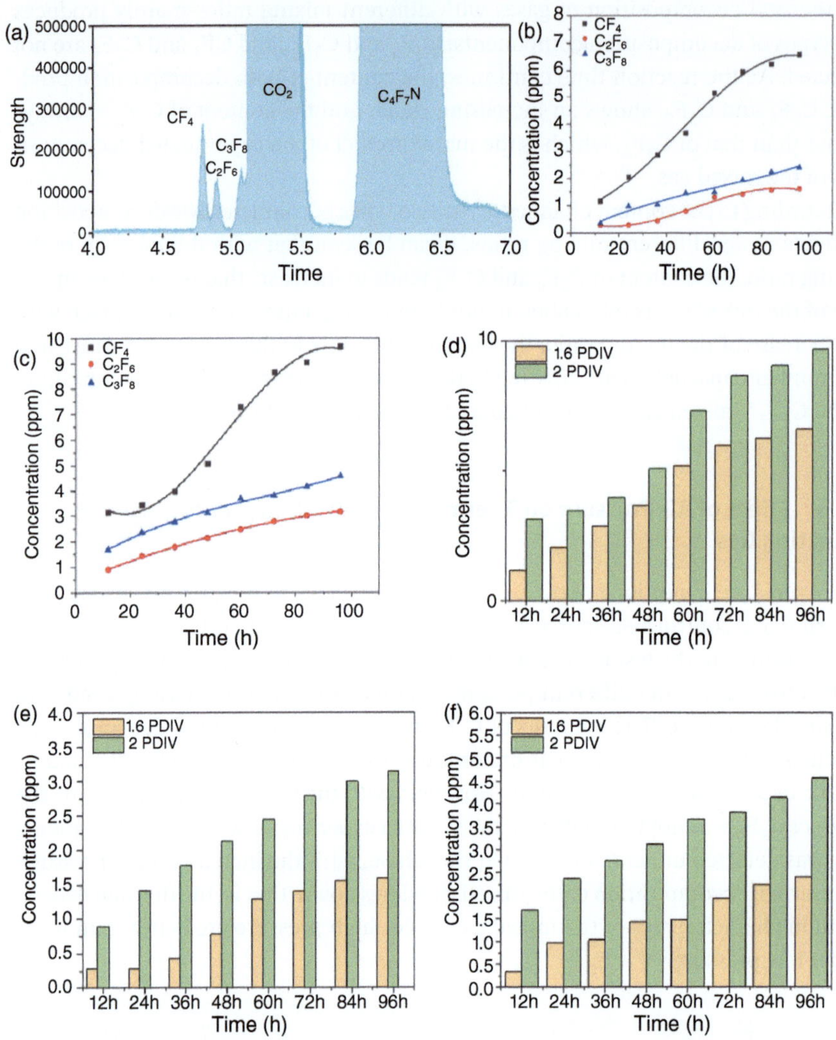

Figure 9.10 Effect of discharge volume on the yield of characteristic and changes of characteristic decomposition components. (a) Gas chromatogram of mixed gas discharge decomposition; (b) 1.6PDIV; (c) 2 PDIV; (d) CF_4; (e) C_2F_6; (f) C_3F_8. *Source:* Adapted from [37].

The distance between the needle-plate electrodes was 10 mm, the content of C_4F_7N in the mixed gas was 5%, and the gas pressure was 0.15 MPa.

Figure 9.10a shows the gas chromatogram after discharge decomposition of C_4F_7N mixed gas. It can be seen that the main decomposition products of C_4F_7N-CO_2 mixed gas under partial discharge conditions are CF_4, C_2F_6, and C_3F_8. In addition, due to the cross-interference of CO_2 gas, some products cannot be detected. Unlike overheating decomposition, the local discharge decomposition of C_4F_7N mixed gas did not detect the product C_3F_6.

Combined with the changes in the characteristic decomposition components of the mixed gas under different applied voltages, with the test time given in Figure 9.10b–c, it can be seen that the content of CF_4 is the highest among all decomposition products, and the output of C_2F_6 and C_3F_8 is low and similar. In addition, it can be seen that the growth rate of CF_4 content is high before 72 hours, and the growth rate slows down and tends to saturation after 72 hours. The output of C_2F_6 and C_3F_8 shows a linear growth trend throughout the test period.

Figure 9.10d–f shows the effect of discharge energy on the content of characteristic decomposition products of mixed gas. It can be seen that the content of various characteristic decomposition products of C_4F_7N-CO_2 mixed gas increases with the increase of applied voltage. In fact, the increase of applied voltage essentially changes the intensity of partial discharge. For mixed gas under the same conditions, it is equivalent to increasing the ionization energy. Therefore, the content of each characteristic decomposition product increases with the increase of applied voltage.

9.4.5 Effect of Mixing Ratio on Partial Discharge Decomposition of Eco-Friendly Insulating Gas

In order to explore the influence of the mixing ratio on the discharge decomposition characteristics of the C_4F_7N-CO_2 mixed gas, this project carried out tests under three conditions: 6%, 10%, and 14%. The applied voltage was 1.6 times the PDIV of each mixed gas, and the test pressure was fixed at 0.4 MPa.

Figure 9.11a–c shows the discharge decomposition of the C_4F_7N-CO_2 mixed gas under different mixing ratio conditions. It can be seen that under different mixing ratio test conditions, the CF_4 content produced by the mixed gas discharge decomposition is the highest, followed by C_2F_6, and the C_3F_8 content is the lowest. (At 6% content, the contents of C_2F_6 and C_3F_8 are both around 1 ppm, and there is a certain detection error.) Since the production of CF_4 and C_2F_6 requires the participation of CF_3 groups and F particles, the decomposition of C_4F_7N produces a large number of CF_3 particles.

According to the comparison of characteristic decomposition components under different mixing ratios given in Figure 9.11d–f, it can be seen that the content of each characteristic decomposition product of C_4F_7N mixed gas increases with the increase of mixing ratio. In fact, with the increase of mixing ratio, the content of C_4F_7N increases under the same conditions, and the probability of molecular collision ionization increases, so the content of characteristic decomposition products increases with the increase of mixing ratio.

9.4.6 Effect of Gas Pressure on Partial Discharge Decomposition of Eco-Friendly Insulating Gas

In order to explore the influence of gas pressure on the discharge decomposition characteristics of C_4F_7N mixed gas, this project carried out partial discharge decomposition tests under different gas pressure conditions. During the test, the C_4F_7N

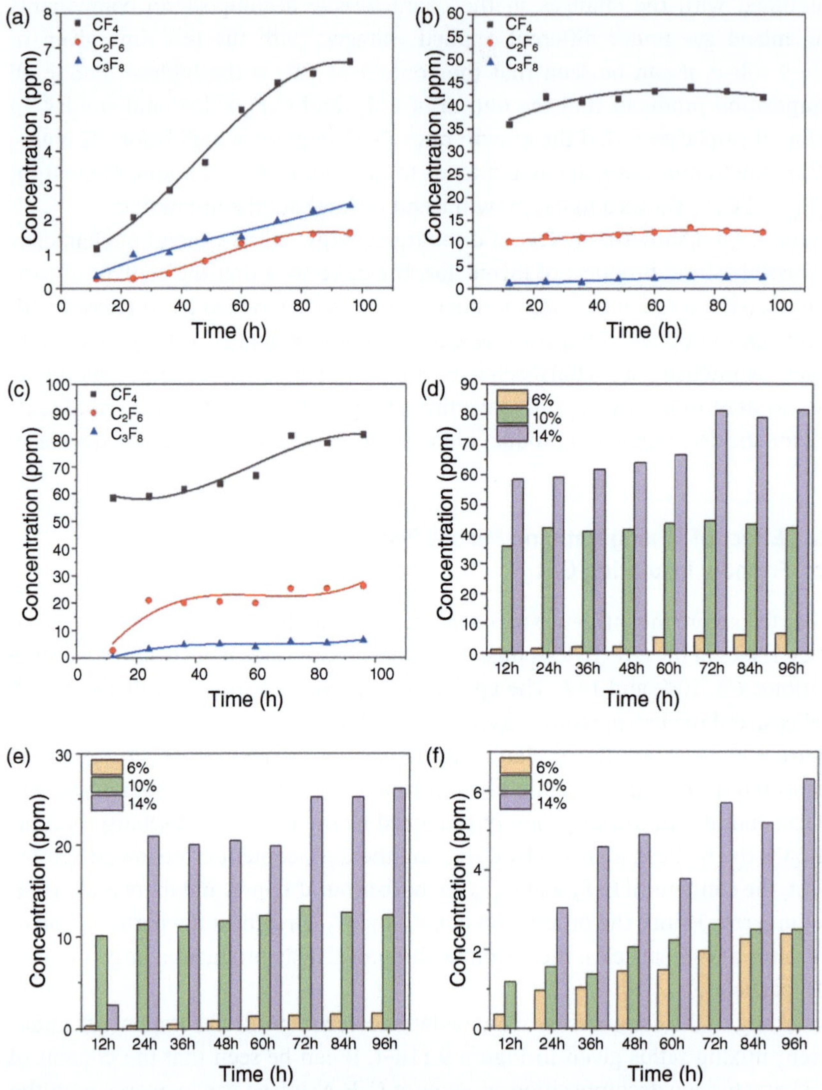

Figure 9.11 Effect of mixing ratio on the yield and variation of characteristic decomposition components. (a) 6%; (b) 10%; (c) 14%; (d) CF_4; (e) C_2F_6; (f) C_3F_8. *Source:* [33] / with permission of John Wiley & Sons.

content was fixed at 6%, and the applied voltage was 1.6 times the partial discharge starting voltage.

According to the quantitative detection results given in Figure 9.12, it can be seen that the decomposition of the mixed gas intensifies with the increase of gas pressure. Among them, under 0.2 MPa, only two types of characteristic decomposition products, CF_4 and C_2F_6, can be detected in the 96-hour test cycle, and C_3F_8 is also produced under 0.4 MPa. With the increase of gas pressure, the content of C_4F_7N increases under the same conditions, and the probability of molecular

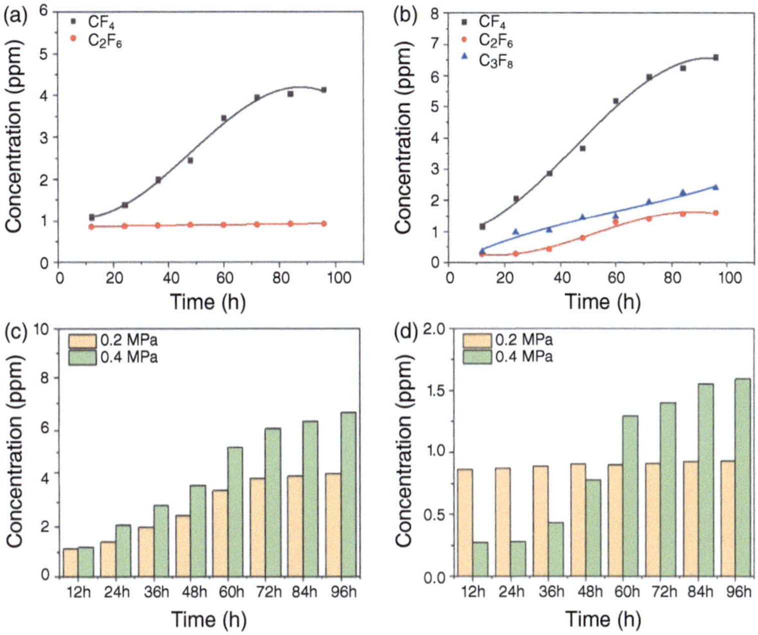

Figure 9.12 Effect of air pressure on the yield and variation of characteristic decomposition components. (a) 0.2 MPa; (b) 0.4 MPa; (c) CF_4; (d) C_2F_6. *Source:* [33] / with permission of John Wiley & Sons.

decomposition caused by collision ionization increases. Therefore, it is manifested as an increase in the content of characteristic decomposition products such as CF_4 on a macro scale.

9.4.7 Effects of Trace Water and Trace Oxygen on Discharge Decomposition of Eco-Friendly Insulating Gas

Since C_4F_7N belongs to fluorocarbon gas and contains carbon in its molecular structure, carbon may be precipitated when used as an arc extinguishing medium, and the precipitation of solid particles will pose a certain threat to the safe operation of the equipment. At present, General Electric has chosen to add a small amount of oxygen to prevent the precipitation of solid carbon when applying C_4F_7N in actual engineering. In order to study the influence of oxygen content on the discharge decomposition characteristics of C_4F_7N mixed gas, this report carried out the discharge decomposition characteristics test of C_4F_7N mixed gas under different oxygen contents.

It should be pointed out that since the addition of CO_2 will submerge the characteristic peaks of oxygen-containing characteristic decomposition products such as COF_2, and in order to better evaluate whether the addition of O_2 will hinder the precipitation of carbon, this project used C_4F_7N-N_2 mixed gas when testing the influence of oxygen on the discharge decomposition characteristics of C_4F_7N. In addition, the test was carried out by 150 power frequency breakdown methods, with an interval of 3 minutes between each two breakdowns, and a group of gas samples was collected every 50 times for quantitative analysis. The test fixed the C_4F_7N content at 6% and the gas pressure at 0.15 MPa.

Combined with the trend curve of the yield of the characteristic decomposition components of the mixed gas under different oxygen contents with the number of breakdowns given in Figure 9.13a–c, it can be seen that with the increase in the

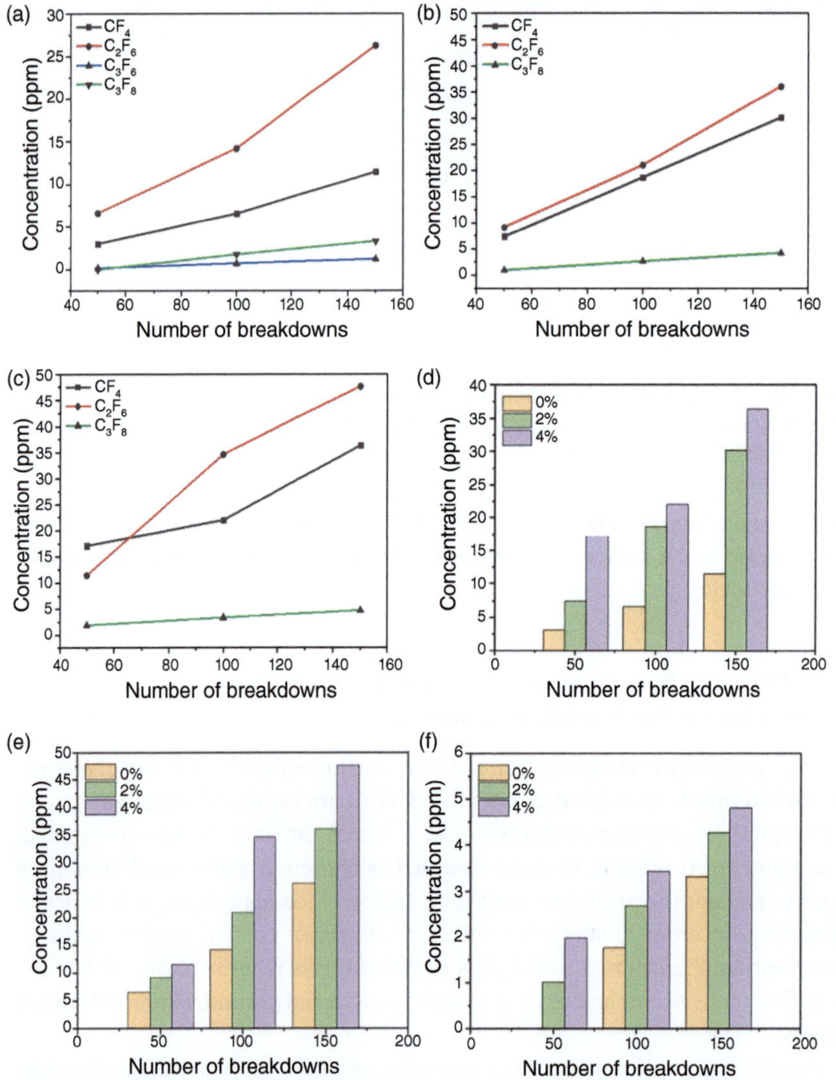

Figure 9.13 Changes of characteristic decomposition components of mixed gas with test time at different oxygen contents: (a) 0%; (b) 2%; (c) 4%. Effect of oxygen content on\nobreak the yield of characteristic decomposition components: (d) CF_4; (e) C_2F_6; (f) C_3F_8. *Source:* Adapted from [38].

number of breakdowns, the contents of CF_4, C_2F_6 and C_3F_8 all increase, and the contents of CF_4 and C_2F_6 are relatively high, while the content of C_3F_8 is below 5 ppm.

In addition, as shown in Figure 9.13d–f, with the increase of oxygen content, the characteristic decomposition components produced by the discharge decomposition of C_4F_7N mixed gas all show an increasing trend, indicating that oxygen has a certain promoting effect on the decomposition of C_4F_7N mixed gas. The addition of oxygen can oxidize the carbon in the mixed gas to generate CO_2, and it is also possible to oxidize the characteristic decomposition product C_3F_6.

In actual engineering applications, the mixed gas will contain a certain amount of water. In order to explore the effect of water content on the discharge decomposition characteristics of C_4F_7N mixed gas, this project carried out a discharge decomposition characteristic test of C_4F_7N mixed gas at 1000 and 2000 ppm micro-water content. The test fixed the C_4F_7N content at 6% and the gas pressure at 0.15 MPa.

The addition of trace water does not have a significant effect on the content of decomposition components of the C_4F_7N mixed gas. The content of the main decomposition products of the mixed gas, CF_4, C_2F_6, and C_3F_8, did not change significantly. In addition, the test found that the breakdown voltage of the mixed gas did not decrease significantly. Overall, trace moisture will not have a significant impact on the decomposition characteristics of the C_4F_7N mixed gas, but in actual engineering, it is still necessary to strictly limit the moisture content in the mixed gas to avoid possible hazards.

In order to explore the effect of oxygen on the thermal decomposition characteristics of the C_4F_7N mixed gas, this project carried out mixed gas thermal decomposition tests under two oxygen content conditions of 2% and 4%. The test fixed the C_4F_7N content at 5% and the gas pressure at 0.15 MPa.

Figure 9.14a–c shows the changes in the characteristic decomposition components of the C_4F_7N mixed gas under different oxygen contents. It can be seen that the addition of oxygen does not change the relative content of each decomposition component of the mixed gas. Among them, the content of C_3F_6 is the highest, and the contents of C_2F_6 and CF_4 are relatively low. Combined with the content comparison of the characteristic decomposition components under different oxygen contents given in the figure, it can be seen that with the increase of oxygen content, the contents of the four types of characteristic decomposition products all show an increasing trend.

Therefore, the addition of oxygen promotes the thermal decomposition of C_4F_7N mixed gas to a certain extent, which is related to the strong oxidizing property of oxygen. Therefore, in later practical applications, the oxygen content should be limited to a certain level to avoid the negative impact of the addition of oxygen on the thermal stability of the mixed gas.

In order to explore the influence of trace water on the thermal decomposition characteristics of C_4F_7N mixed gas, this project carried out the thermal decomposition characteristics test of C_4F_7N mixed gas under 1000 and 2000 ppm trace water content. The test fixed the C_4F_7N content at 6% and the gas pressure at 0.15 MPa.

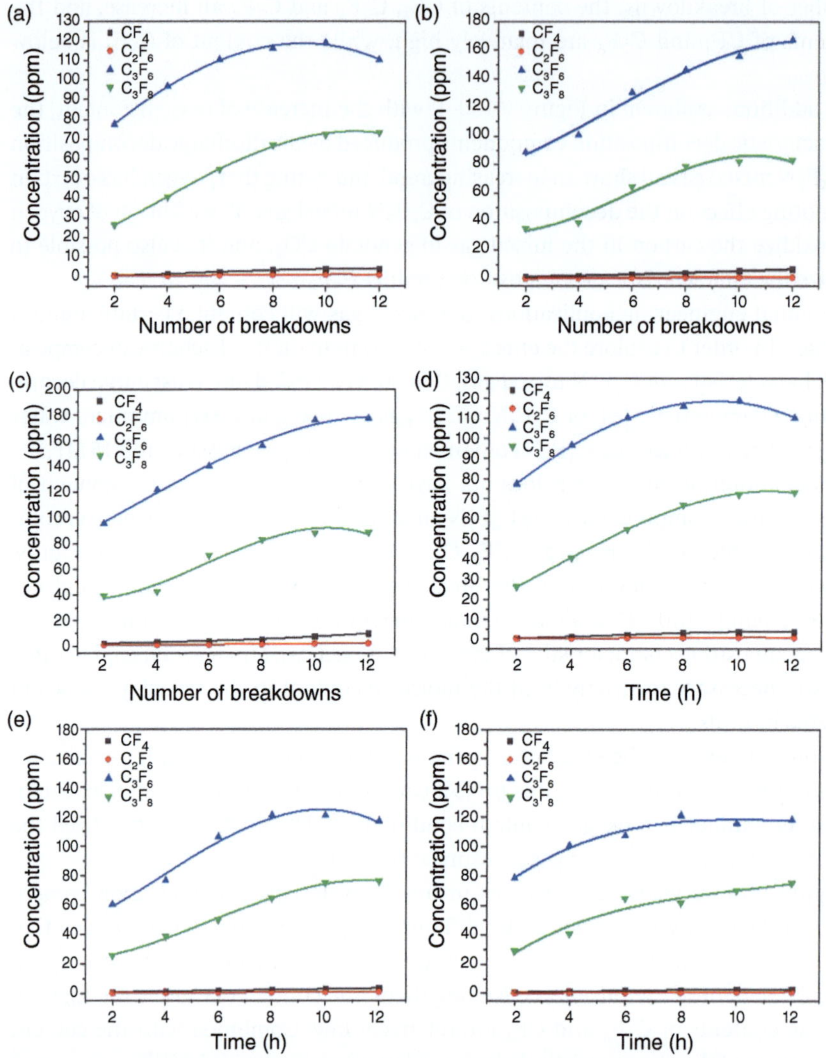

Figure 9.14 Variation of characteristic decomposition components of mixed gas with test time at different oxygen contents. (a) 0%; (b) 2%; (c) 4%; variation of characteristic decomposition components of mixed gas under trace water conditions with test time; (d) 0%; (e) 1000 ppm; (f) 2000 ppm. *Source:* Adapted from [38, 39].

According to the test results given in Figure 9.14d–f, it can be seen that the addition of trace water does not have a significant effect on the decomposition of the C_4F_7N mixed gas. Under each test condition, the contents of C_3F_6 and C_3F_8 are higher than those of CF_4 and C_2F_6, indicating that the thermal decomposition of C_4F_7N produces a certain amount of C_3F_7 particles, and the yield of CF_3 particles is relatively low. Overall, trace water does not have a serious impact on the thermal stability of the C_4F_7N mixed gas, which may be related to the high content of CO_2 in the mixed gas.

That is, although the presence of water may have a negative impact on C_4F_7N, this impact is not enough to change the component type and content of the entire mixed gas.

References

1 Robinson, J.B. and Herbert, D. (2001). Integrating climate change and sustainable development. *International Journal of Global Environmental Issues* 1 (2): 130–149.

2 Li, X., Zhao, H., and Murphy, A.B. (2018). SF_6-alternative gases for application in gas-insulated switchgear. *Journal of Physics D: Applied Physics* 51 (15): 153001.

3 Reilly, J., Prinn, R., Harnisch, J. et al. (1999). Multi-gas assessment of the Kyoto Protocol. *Nature* 401 (6753): 549–555.

4 European Commission Joint Research Centre (JRC)/Netherlands Environmental Assessment Agency (PBL) (2010). Emission Database for Global Atmospheric Research (EDGAR), release version 4.2. 2010; http://edgar.jrc.ec.europa.eu (accessed 13 October 2025).

5 Global Monitoring Laboratory (2023). Global monitoring laboratory earth system research laboratories, Combined dataset. https://www.esrl.noaa.gov/gmd/hats/combined/SF6.html (accessed 13 October 2025).

6 Rabie, M. and Franck, C.M. (2018). Assessment of eco-friendly gases for electrical insulation to replace the most potent industrial greenhouse gas SF_6. *Environmental Science & Technology* 52 (2): 369–380.

7 Chen, L., Zhang, B., Li, X., and Yang, T. (2020). Decomposition pathway of C_4F_7N gas considering the participation of ions. *Journal of Applied Physics* 128 (14): 143303.

8 Kieffel, Y., Biquez, F., Vigouroux, D. et al. (2017). Characteristics of g^3–an alternative to SF_6. *CIRED 24* 2017 (1): 54–57.

9 Kieffel, Y., Irwin, T., Ponchon, P., and Owens, J. (2016). Green gas to replace SF_6 in electrical grids. *IEEE Power and Energy Magazine* 14 (2): 32–39.

10 Xiao, S., Shi, S., Li, Y. et al. (2021). Review of decomposition characteristics of eco-friendly gas insulating medium for high-voltage gas-insulated equipment. *Journal of Physics D: Applied Physics* 54 (37): 373002.

11 Xing, H., Lu, S., Yang, H., and Zhang, H. (2022). Review on research progress of $C_6F_{12}O$ as a fire extinguishing agent. *Fire* 5 (2): 50.

12 Zhang, Y., Zhang, X., Li, Y. et al. (2019). AC breakdown and decomposition characteristics of environmental friendly gas $C_5F_{10}O$/air and $C_5F_{10}O/N_2$. *IEEE Access* 7: 73954–73960.

13 Xiao, S., Li, Y., Zhang, X. et al. (2017). Formation mechanism of CF_3I discharge components and effect of oxygen on decomposition. *Journal of Physics D: Applied Physics* 50 (15): 155601.

14 Widger, P. and Haddad, A. (2017). Solid by-products of a CF_3I–CO_2 insulating gas mixtures on electrodes after lightning impulse breakdown. *Journal of Physics Communications* 1 (2): 025010.

15 Tian, S., Zhang, X., Xiao, S. et al. (2019). Experimental research on insulation properties of $C_6F_{12}O/N_2$ and $C_6F_{12}O/CO_2$ gas mixtures. *IET Generation Transmission and Distribution* 13 (3): 417–422.

16 Yamamoto, O., Takuma, T., Hamada, S. et al. (2002). Applying a gas mixture containing c-C_4F_8 as an insulation medium. *IEEE Transactions on Dielectrics and Electrical Insulation* 8 (6): 1075–1081.

17 Wu, B.T., Xiao, D.M., Liu, Z.S. et al. (2006). Analysis of insulation characteristics of c-C_4F_8 and N_2 gas mixtures by the Monte Carlo method. *Journal of Physics D: Applied Physics* 39 (19): 4204.

18 Yang, Y., Gao, K., Ding, L. et al. (2021). Review of the decomposition characteristics of eco-friendly insulation gas. *High Voltage* 6 (5): 733–749.

19 Pan, B., Wang, G., Shi, H. et al. (2020). Green gas for grid as an eco-friendly alternative insulation gas to SF_6: a review. *Applied Sciences* 10 (7): 2526.

20 Tian, S., Zhang, X., Xiao, S. et al. (2019). Application of $C_6F_{12}O/CO_2$ mixture in 10 kV medium-voltage switchgear. *IET Science, Measurement & Technology* 13 (9): 1225–1230.

21 Li, Y., Zhang, X., Tian, S. et al. (2019). Insight into the decomposition mechanism of $C_6F_{12}O$-CO_2 gas mixture. *Chemical Engineering Journal* 360: 929–940.

22 Rao, X., Li, D., Xia, X. et al. (2021). Study on discharge decomposition characteristics of environmentally friendly gas $C_6F_{12}O/CO_2$. *Vacuum* 186: 110004.

23 Lin, X., Cui, Z., Geng, Z. et al. (2019). Calculation of thermodynamic physical parameters of $C_5F_{10}O/CO_2$ and $C_6F_{12}O/CO_2$ mixtures. *2019 5th International Conference on Electric Power Equipment-Switching Technology (ICEPE-ST)*, 55–59. IEEE.

24 Ahmed, R., Abd Rahman, R., Kamarudin, M.S. et al. (2022). Feasibility of fluoronitrile (C_4F_7N) as a substitute to sulphur hexafluoride (SF_6) in gas insulated application: a review. *2022 IEEE International Conference on Power and Energy (PECon)*, 391–396. IEEE.

25 Cui, G., Wang, C., Yang, Y. et al. (2024). Discharge fault type identification of $C_4F_7N/CO_2/O_2$ mixed insulating gas for engineering based on product components. *Journal of Physics D: Applied Physics* 57 (47): 475503.

26 Han, P., He, Y., Chen, C. et al. (2016). Study on synergistic mechanism of inhibitor mixture based on electron transfer behavior. *Scientific Reports* 6 (1): 33252.

27 Zhang, X., Chen, Q., Zhang, J. et al. (2019). Experimental study on power frequency breakdown characteristics of C_4F_7N/CO_2 gas mixture under quasi-homogeneous electric field. *IEEE Access* 7: 19100–19108.

28 Li, Y., Zhang, X., Chen, Q. et al. (2018). Study on the dielectric properties of C_4F_7N/N_2 mixture under highly non-uniform electric field. *IEEE Access* 6: 42868–42876.

29 Siddagangappa, M.C., Lakshminarasimha, C.S., and Naidu, M.S. (1982). Electron attachment in binary mixtures of electronegative and buffer gases. *Journal of Physics D: Applied Physics* 15 (8): L83.

30 Long, Y., Guo, L., Chen, C. et al. (2020). Measurement of ionization and attachment coefficients in C_4F_7N/CO_2 gas mixture as substitute gas to SF_6. *IEEE Access* 8: 76790–76795.

31 Li, Y., Zhang, X., Zhang, J. et al. (2019). Experimental study on the partial discharge and AC breakdown properties of C_4F_7N/CO_2 mixture. *High Voltage* 4 (1): 12–17.

32 Zhang, B., Xiong, J., Hao, M. et al. (2022). Pulsed townsend measurement of electron swarm parameters in C_4F_7N–CO_2 and C_4F_7N–N_2 mixtures as eco-friendly insulation gas. *Journal of Applied Physics* 131 (3).

33 Zhou, W., Zhang, T., and Wang, L. (2020). Breakdown characteristics of C_4F_7N/CO_2 mixtures under steep-wavefront impulse voltages. *IEEE Access* 8: 29291–29298.

34 Li, Y., Zhang, X., Li, Y. et al. (2020). Interaction mechanism between the C_4F_7N–CO_2 gas mixture and the EPDM seal ring. *ACS Omega* 5 (11): 5911–5920.

35 Zhang, X., Zhang, Y., Huang, Y. et al. (2020). Detection of decomposition products of C_4F_7N-CO_2 gas mixture based on infrared spectroscopy. *Vibrational Spectroscopy* 110: 103114.

36 Li, Y., Zhang, X., Zhang, J. et al. (2020). Study on the thermal decomposition characteristics of C_4F_7N–CO_2 mixture as eco-friendly gas-insulating medium. *High Voltage* 5 (1): 46–52.

37 Li, Y., Zhang, X., Xiao, S. et al. (2018). Decomposition properties of C_4F_7N/N_2 gas mixture: an environmentally friendly gas to replace SF_6. *Industrial & Engineering Chemistry Research* 57 (14): 5173–5182.

38 Li, Y., Zhang, X., Chen, Q. et al. (2019). Influence of oxygen on dielectric and decomposition properties of C_4F_7N N_2-O_2 mixture. *IEEE Transactions on Dielectrics and Electrical Insulation* 26 (4): 1279–1286.

39 Ye, F., Chu, Y., Brault, P. et al. (2024). Mechanism of the effect of micro-water on the thermal decomposition characteristics of $C_4F_7N/CO_2/O_2$ gas mixture. *IEEE Transactions on Dielectrics and Electrical Insulation* 32 (4): 2033–2042.

40 Li, Y., Zhang, X., Zhang, J., Xiao, S., Xie, B., Chen, D., Gao, Y., & Tang, J. (2019). Assessment on the toxicity and application risk of C4F7N: A new SF6 alternative gas. *Journal of Hazardous Materials*, 368, 653–660.

41 Zhang, X., Ye, F., Li, Y., Tian, S., Xie, B., Gao, Y., & Xiao, S. (2020). Acute toxicity and health effect of perfluoroisobutyronitrile on mice: a promising substitute gas-insulating medium to SF6. *Journal of Environmental Science and Health, Part A*, 55 (14), 1646–1658.

10

Synergistic Improvement of the Performance of Green Insulating Materials in Power Equipment

Kangning Wu, Ge Zhao, and Jianying Li

Xi'an Jiaotong University, State Key Laboratory of Electrical Insulation and Power Equipment, No. 28, Xianning West Road, Taiyi Lu Street, Xi'an City, Shaanxi Province, China

10.1 Preface

Insulating materials play a crucial role in power equipment, and their performance directly determines the reliability, efficiency, and service life of the equipment. As a key component of the power system, insulating materials not only need to provide excellent electrical insulation performance to prevent current leakage or breakdown but also must have sufficient mechanical strength to support the conductor structure, withstand thermal loads to maintain stability in high-temperature environments, and be able to resist the erosion of external factors such as humidity, chemical corrosion, and ultraviolet rays.

All kinds of insulating materials complement each other in different application scenarios of power equipment, jointly supporting the safe and stable operation of modern power systems. Polyethylene (PE) has become the preferred material for insulation in medium and low voltage cables due to its low dielectric constant and good flexibility. Especially cross-linked polyethylene (XLPE) has significantly improved heat resistance and corona resistance through the cross-linking process, making it occupy an important position in high voltage direct current cables. Polypropylene (PP) has become an ideal choice for the film material of power capacitors due to its high heat distortion temperature, low dielectric loss, and excellent breakdown field strength. Epoxy resin, with its extremely high mechanical strength, excellent dielectric properties, and outstanding chemical stability, is widely used in solid-sealing components of transformer cast insulation and gas-insulated switchgear (GIS).

In power equipment, most insulating materials need to be modified to meet the strict operational requirements. Common modification methods include chemical modification (such as cross-linking and copolymerization), physical blending (such as adding nano-fillers and elastomers), surface treatment, and the introduction of functional additives (such as antioxidants and voltage stabilizers). For instance, PE can significantly enhance its heat resistance and mechanical

Next-Generation Electrical Insulating Materials: Low-Carbon, Eco-Friendly, and Sustainable Solutions, First Edition. Edited by Jun-Wei Zha.

strength through chemical cross-linking, transforming it from ordinary PE to XLPE, thus making it suitable for medium and high voltage cable insulation. Adding nano-silica (SiO_2) or magnesium oxide (MgO) can suppress the accumulation of space charges and improve its direct current insulation performance [1]. PP is often modified by copolymerization to improve its low-temperature brittleness, or by blending to enhance its low-temperature impact resistance while maintaining good heat resistance. In addition, the introduction of nanoparticles such as MgO can well balance the thermal, mechanical, and electrical properties of the material [2, 3]. Epoxy resins are usually toughened and modified to reduce brittleness, or combined with micro/nano ceramic fillers to simultaneously improve thermal conductivity and dielectric strength, enabling them to meet the strict requirements of high-voltage transformers and GIS insulators.

Some modification methods can only specifically enhance a certain aspect of the material's characteristics. However, the reliable long-term operation of insulating materials under complex working conditions requires the coordinated optimization of electrical, mechanical, and thermal properties (such as temperature resistance grade and thermal conductivity). For instance, merely enhancing the dielectric strength of PP might sacrifice its flexibility, while over-toughening epoxy resin could reduce its insulation performance. Therefore, the modification of modern insulating materials increasingly focuses on the collaborative design of multiple properties. This concept of multiobjective collaborative modification is precisely the key to ensuring that insulating materials maintain long-term stability under the coupled action of electric fields, thermal fields, and mechanical stresses.

10.2 Modification Methods of Polypropylene

Since the 21st century, PP-based thermoplastic materials have attracted extensive attention due to their high temperature stability and good recyclability. Compared with XLPE, PP material not only has a higher working temperature but also offers advantages such as convenient manufacturing and processing, lower cost, high production efficiency, and no limit on cable length. It has high application value in the power cable industry. Therefore, as a material with great development prospects, PP has begun to attract key attention from both the academic and industrial communities. However, there are still some problems in the application of PP materials:

1) PP has high hardness and poor flexibility.
2) PP is highly sensitive to oxygen, and chain growth and breakage are prone to occur on its molecular chains, resulting in poor aging resistance.
3) PP has poor thermal conductivity, which can easily cause excessive temperature differences between the inside and outside of the insulation layer, leading to electric field distortion, inducing partial discharge and branch discharge in the insulation layer.

Therefore, to achieve the safe, stable, and reliable operation of PP insulated power equipment, it is necessary to effectively improve the mechanical properties, thermal conductivity, and dielectric properties of PP insulating materials. The complex multiscale structure of PP insulating materials is a key factor determining their macroscopic electrical and mechanical properties. Therefore, scholars at home and abroad have regulated the complex multiscale structure of PP through three types of methods: chemical synthesis modification, physical doping modification, and physical field-induced molding modification, to optimize and improve performance in certain aspects. Meanwhile, the application of bio-based and other methods will further improve the environmental pollution problem during the preparation process of PP.

10.2.1 Chemical Synthesis Modification

Chemical synthesis modification generally refers to the modification method of optimizing the molecular chain structure of PP through chemical means to enhance its macroscopic properties. At present, two synthetic modification methods, namely chemical grafting modification and olefin copolymerization modification, are relatively common in the research related to PP cable insulation.

10.2.1.1 Chemical Grafting

Chemical grafting modification mainly achieves the regulation of trap distribution by grafting deep trap-type side branches or functional groups onto the main chain of PP molecules, thereby promoting the improvement of electrical resistance. Taking grafted styrene as an example, the general principle of chemical grafting modification of PP is introduced, as shown in Figure 10.1. Chemical grafting generally involves two main reaction steps. The first is the activation of the PP molecular chain, that is, by means of initiators or high-energy ray irradiation, etc., the more active hydrogen atoms (usually the hydrogen on the tertiary carbon atoms) on the PP molecular chain are seized to form free radicals that can be used for the grafting reaction. The second is the reaction between the activated PP molecular chain and the graft monomer. The graft monomer usually contains unsaturated chemical bonds. Under certain temperature conditions, the unsaturated bonds will open and chemically react with the free radicals formed on the PP molecular chain to form covalent bonds, thereby enabling the monomer to be grafted onto the PP molecular chain.

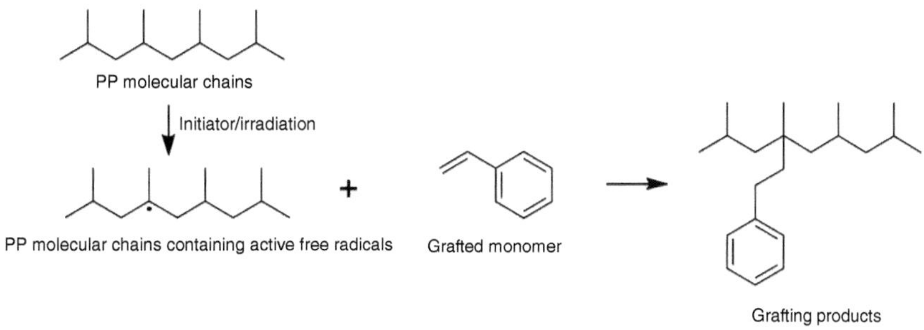

Figure 10.1 Schematic diagram of the principle of chemical grafting modification of polypropylene.

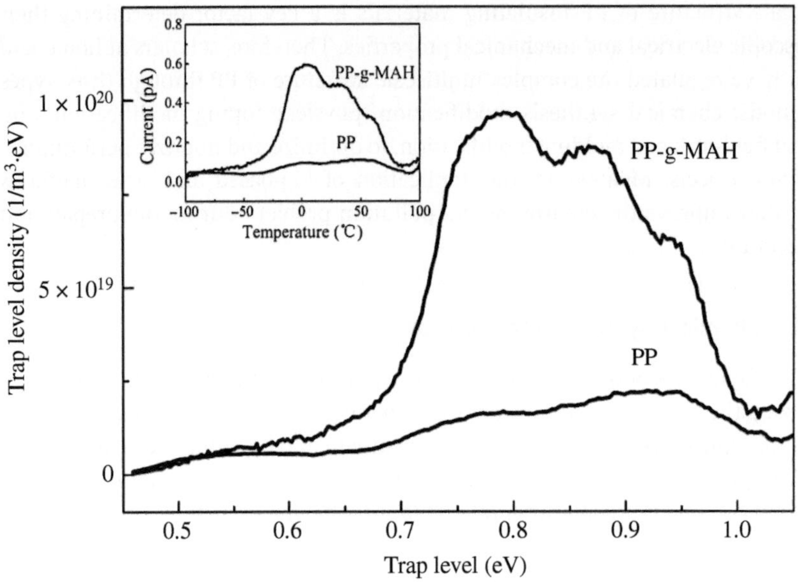

Figure 10.2 Trap level distribution derived from TSDC measurements in PP and PP-g-MAH. The inset is the TSDC spectra of PP and PP-g-MAH. *Source:* Reproduced [4] / with permission of IOP Publishing.

Depending on the generation of free radicals and the reaction conditions, common chemical grafting methods include the melt method, radiation method, solid-phase method, solution method, and aqueous suspension method.

The commonly used grafting groups for PP grafting modification include several types. The first type is monomers with polar groups, such as maleic anhydride and silane derivatives [4, 5]. The second type is monomers with conjugated structures like benzene rings, such as styrene and 4-acetyloxyethylene [6, 7]. The third type is some monomers containing antioxidant functional groups, such as β-(3,5-di-tert-butyl-4-hydroxyphenyl) n-octadecyl propionate (antioxidant 1076), and aromatic amine antioxidants [8, 9].

Grafting PP with maleic anhydride can enhance the interfacial interaction between PP and other inorganic fillers. This strategy is also used to improve the compatibility of PP with other polar polymers. This is because maleic anhydride is a polar molecule that can regulate the local electric field distribution in PP. As shown in Figure 10.2, the introduction of maleic anhydride induces a charge trap, which in turn will change the electrical properties of PP. For example, maleic anhydride can introduce shallow traps and thereby promote the dissipation of space charges, thereby enhancing their performance. It should be noted that the grafting of maleic anhydride to PP is usually achieved through a melt reaction in the presence of peroxides. Peroxides can cause chain breakage in PP, while unreacted maleic anhydride may affect the long-term performance of PP, such as increasing the dielectric constant and loss of PP and reducing the volume resistivity at room temperature [4].

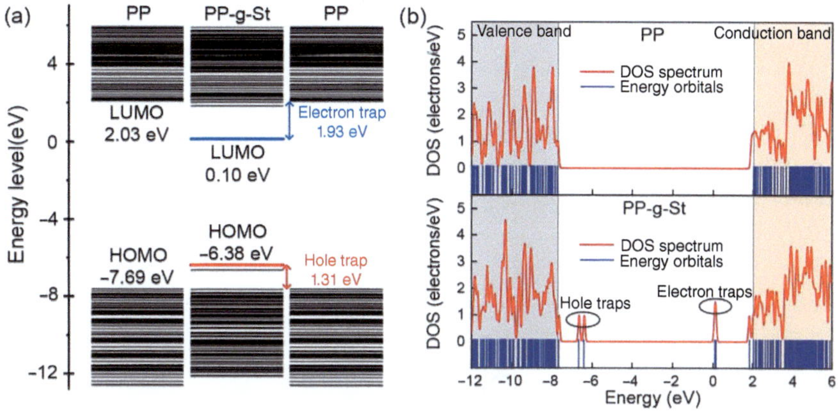

Figure 10.3 Electronic band structure of PP and PP-g-St simulated by DFT. (a) Energy level distribution of PP and PP-g-St, (b) density of states in PP and PP-g-St. *Source:* Reproduced [6] / with permission of IOP Publishing.

Similar to grafting maleic anhydride, grafting monomers with conjugated structures such as benzene rings onto PP can also inhibit the accumulation of space charges and enhance the breakdown strength. This may also be related to the deep traps introduced by the grafting groups. The simulation results in Figure 10.3 show that there are differences in the potential distribution and band structure of the molecular chains after PP and PP grafted with styrene (PP-g-St), that is, grafted styrene can change the band structure of PP. This is regarded as the reason for grafting into a deep trap. Figure 10.3a shows the electronic band structure of PP and PP-g-St. Due to the differences in electronic band structure, the electron trap depth and hole trap depth between PP and PP-g-St molecular chains are calculated to be 1.93 and 1.31 eV, respectively. The electron and hole traps introduced by styrene grafting can also be obtained from the density of states spectra of PP and PP-g-St in Figure 10.3b. So the grafted functional group offers strong localized deep charge traps which help to improve the electrical performance of the grafted polymers [6].

Due to the instability of protons on the tertiary carbon atoms in the main chain of PP, they will separate from these atoms under the action of light and heating, leading to the degradation of PP and endowing it with relatively low thermal oxidation stability. Introducing antioxidants is an effective method to delay the thermal oxidation aging of PP. However, low-molecular-weight antioxidants will migrate from the PP body to the surface, causing the antiaging effect to decrease over time. As shown in Figure 10.4, the researchers grafted the antioxidant onto the main chain of PP using the Steflich esterification reaction, significantly enhancing the thermal oxidation stability of PP. It should be noted that the grafting of hindered phenolic groups will slightly increase the dielectric constant of PP. Further studies have shown that the methylene units in the hindered phenolic groups can cross-link with each other, thereby enhancing the antioxidant property and mechanical strength of PP. Since the main chain of PP can accept the highly grafted hindered phenolic groups, introducing hindered phenolic groups grafted PP can also improve the thermal oxidation stability of PP.

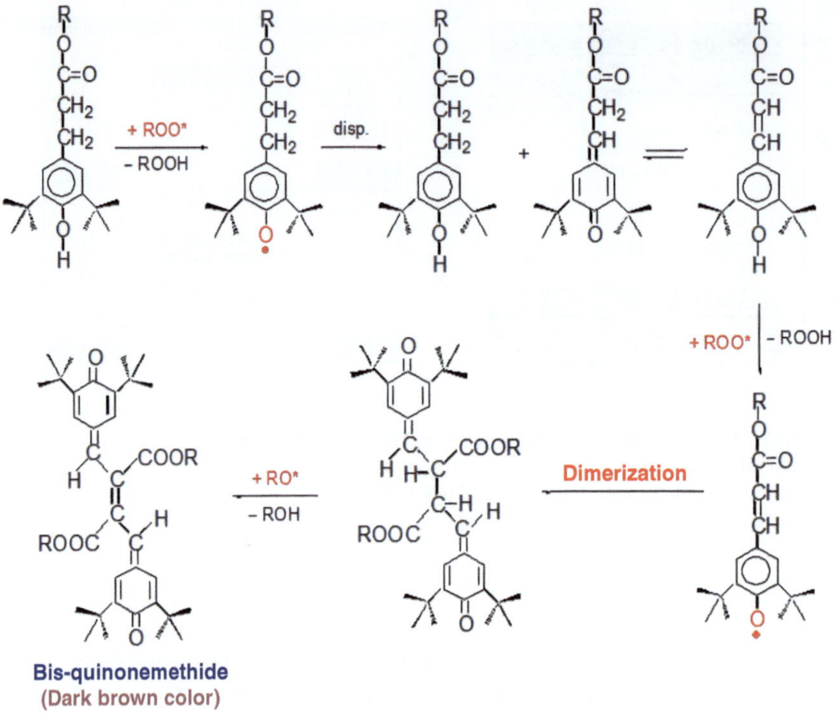

Figure 10.4 Hindered phenol antioxidant and in situ coupling reaction mechanism. *Source:* Reproduced from [10] / with permission of American Chemical Society.

It can be seen that chemical grafting of specific functional groups can play a significant regulatory role in the electrical properties of PP, while its mechanism of action has not been fully revealed at present, and the improvement of mechanical properties is relatively limited. The chemical grafting method has certain application potential in synergistically improving the overall performance of PP. However, during chemical grafting, peroxide initiators are generally used for auxiliary reactions. The influence of residual reaction byproducts and their removal methods still need to be further explored in future studies.

10.2.1.2 Olefin Copolymerization

Olefin copolymerization modification refers to the modification method of introducing other olefin monomers for copolymerization during the synthesis process of PP. For copolymer PP, its copolymerization form, monomer content, isotactic degree, and other chain structure parameters will all have an impact on the overall performance.

Due to the randomness of the polymerization reaction, copolymer PP is generally a mixture composed of copolymer molecular chains with multiple different components [11, 12]. Take the copolymerization of propylene and ethylene as an example to introduce the conventional molecular chain structure in copolymerized PP. The molecular chain structures that may be formed during the copolymerization

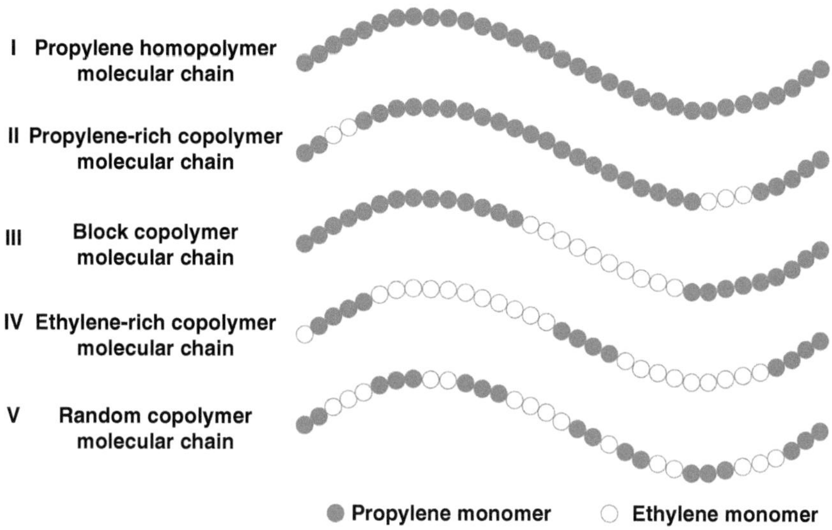

Figure 10.5 Schematic diagram of the molecular chain structure of copolymer PP.

reaction of propylene and ethylene can be roughly classified into five categories, as shown in Figure 10.5. The I type is the most common one, which is the homopolymer chain segment of propylene. This part of the molecular chain mainly contains only the propylene monomer, with good regularity and strong crystallization ability. The II type is the copolymer molecular chain rich in propylene. This type of molecular chain is still mainly composed of propylene monomers, with only a small amount of copolymer olefin monomers inserted, causing limited damage to the regularity and crystallization ability of the molecular chain. The III type is block copolymer molecular chains. In this type of molecular chain, both the propylene monomer and the copolymer olefin monomer form relatively long chain segments. The ethylene and propylene parts on the same molecular chain can both form crystals. The precise synthesis of this type of molecular chain is rather difficult, and generally, only a small amount can be formed during the copolymerization reaction process. The IV type is ethylene-rich molecular chains. This type of molecular chain has a similar structure to the second category. PE segments can form crystals under certain conditions. The V type is random copolymer segments. In this type of molecular chain, propylene and ethylene monomers are completely disordered in arrangement, the crystallization ability of the molecular chain is lost, and it exists in a rubber-like state, which can play a role in improving the toughness of PP.

Based on the differences in the types of molecular chains contained, copolymer PP can generally be classified into three categories: random copolymer PP, impact copolymer PP (IPC), and block copolymer PP. Random copolymerization of PP is generally synthesized by adding a small amount of ethylene, octene, etc. during the polymerization of propylene (about 3–5%), mainly including the I, II, and V types of molecular chains in Figure 10.5, among which the II type accounts for the highest proportion. Therefore, the crystallinity of random copolymer PP can generally still be maintained at a relatively high level. However, due to the introduction of copolymer monomers, the isotropy of random copolymer PP has decreased, and it is not easy to form thick flake crystals. The melting temperature is about 10–30 °C lower than that of isotactic PP. The synthesis of PP IPC is generally carried out in two steps. Taking propylene and ethylene as examples, the propylene monomer is added first to synthesize propylene homopolymer or a slightly modified copolymer. Then ethylene monomer is added and the copolymerization reaction is carried out under the action of the same catalyst system. The obtained products generally include the above five types of molecular chains, among which the first and second types account for the largest proportion. Generally, it exceeds 70%, with the proportion of type V molecular chains being approximately 20%, and the rest being type III and type IV molecular chains. Impact-resistant copolymer PP, due to the presence of a large number of elastomers formed by Class V molecular chains, has better toughness. Moreover, Class III and IV molecular chains can act as compatibilizers, enhancing the compatibility between the elastomers and the PP matrix and further optimizing mechanical properties. Strictly speaking, block copolymer PP should refer to copolymer PP mainly composed of the above-mentioned type III molecular chains. However, the synthesis of such molecular chains requires a specific catalyst system and process control, and the synthesis is rather difficult. At present, a large-scale production and application environment has not yet been formed in China. In actual production and research, impact-resistant copolymer PP is sometimes referred to as block copolymer PP.

By comparison, random copolymers exhibit better flexibility, higher breakdown strength, and superior space charge suppression characteristics. In addition, the mechanical properties of PP copolymers have a strong correlation with the processing methods. Among them, the tensile strength and elongation at break of random copolymers and block copolymers prepared by extrusion molding are both higher than those prepared by compression molding.

Studies have shown that in situ copolymerizing long-chain branched structures into copolymer PP to form "H-shaped" branched structures can achieve a synergistic improvement in both electrical and mechanical properties. This is because PP-H-EPC enhances the compatibility of the matrix-elastomer interface, resulting in a significant reduction in the low-temperature modulus. PP-H-PP and EPC-H-EPC also enhance the entanglement of the chain, endowing the material with excellent thermal, mechanical, and electrical strength, and significantly improving the temperature stability. Moreover, the "H-shaped" branching can be obtained through catalytic polymerization. The density and distribution of long-chain branching are controllable. In the multiphase structure of copolymer PP, multiple long-chain branching structures as shown in Figure 10.6, can be formed, which have great potential in enhancing the electrical strength of the elastomer itself and its interaction with the PP matrix. At present, three key mutually restrictive issues of thermoplastic high-voltage insulation have been initially achieved: the relatively low mechanical modulus at low temperatures, the improvement of mechanical strength at extremely high temperatures, and the enhancement of electrical properties within a wide operating temperature range [13–15].

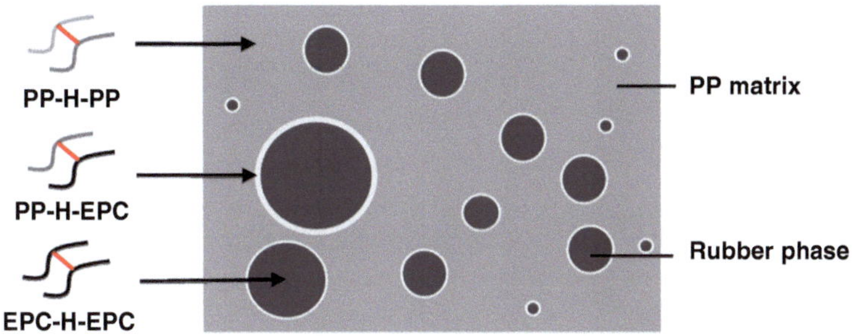

Figure 10.6 Copolymer polypropylene containing multiple long-chain branched structures.

Olefin copolymerization modification has a significant effect on the regulation of the mechanical and electrical properties of PP. However, limited by the complexity of precise control in the chemical synthesis process, most of the existing research on olefin copolymerization modification of PP is based on commercial materials. Information such as synthesis process parameters and monomer content is unknown, making it difficult to ensure the complete consistency of other parameters except for the research variables. Based on laboratory synthetic materials, conducting quantitative research on the relationship between the insulation structure and performance of olefin copolymer PP should be an important direction in the future. Long-chain branched copolymer PP will have broad application prospects based on its comprehensively optimized performance.

10.2.2 Physical Doping Modification

Physical doping modification refers to the method of adding functional additives to the PP matrix through mechanical blending to enhance the material's performance. The most common ones include elastomer blending, nano-blending, and nucleating agent blending.

10.2.2.1 Elastomer Blending

Due to its high hardness and poor toughness, PP often needs to be toughened and modified when used as cable insulation. Mechanical blending is the most convenient method to operate and has been widely applied in both industrial production and experimental research. The blending objects are generally elastomers with good mechanical toughness, such as ethylene propylene rubber (EPR), ethylene propylene diene monomer (EPDM), ethylene-octene copolymer (EOC), allyl elastomer (PBE), and polyolefin elastomer (POE). The blending of single elastomers can significantly optimize the high modulus and low toughness of PP, enabling its mechanical properties to meet the insulation requirements of cables. At the same time, the introduction of elastomers will lead to the deterioration of electrical performance, including the reduction of breakdown field strength and the increase of space charge accumulation, threatening the insulation performance of the material.

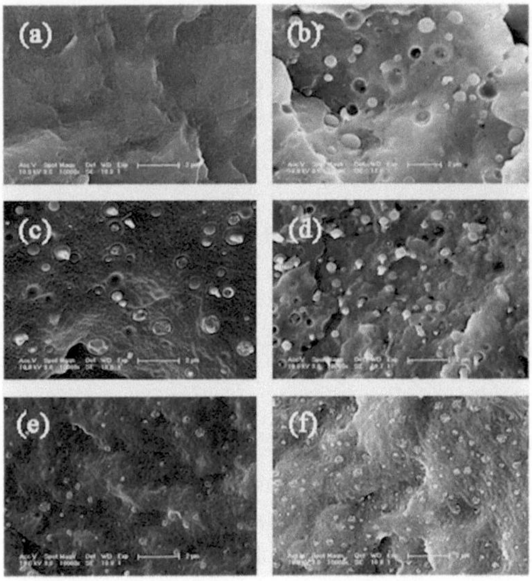

Figure 10.7 SEM images of fracture surface of PP and PP/PS blends: (a) PP, (b) PP-PS, (c–f) the content of PP-g-PS gradually increases. *Source:* Reproduced with permission from [16] / Elsevier.

Considering that the comprehensive performance of multiphase materials is related to interphase compatibility, some scholars have attempted to introduce compatibilizers to construct a three-phase blend system to further optimize the performance of PP. The ternary blend composed of iPP, EOC, and EPR has significantly stronger impact resistance than iPP at −40 °C, and the mechanical modulus does not show a significant decrease at 90 °C, presenting better temperature stability of mechanical properties. When the total content of the two elastomers is guaranteed to be constant, the volume resistivity and breakdown strength of the three-phase blend are significantly higher than those of the iPP/EOC and iPP/EPR two-phase blends, indicating the effectiveness of the three-phase blend in improving the comprehensive insulation performance of PP. For another example, PP-g-PS graft copolymer is added to PP and polystyrene (PS) as a compatibilizer to increase the compatibility of the blends. Figure 10.7 shows the microstructure of the above-mentioned blend observed by scanning electron microscopy. When the mass fraction of PP-g-PS is within 24%, the size of the PS particles keeps decreasing as the amount of compatibilizer increases. The presence of PP-g-PS graft copolymer reduces the interfacial tension, enhances the interfacial adhesion, and makes the PS phase disperse uniformly, which may have a significant impact on the electrical properties [16].

In addition to elastomers, some crystalline polymers with better mechanical toughness than PP are also often used to blend and modify PP, such as random copolymer PP, branched PP, and PE. Some scholars have conducted research on the blending of isotactic PP with long-chain branched PP and found that long-chain branched PP has a certain nucleation effect. After blending, the spherulite size in the material is significantly reduced. Moreover, long-chain branched PP can enhance the entanglement between molecular chains, improve the structural stability at high temperatures, and thereby enhance the high-temperature dielectric properties of PP [17].

It can be seen that elastomer blending can significantly optimize the mechanical properties of PP. The multiphase blending system, in particular, shows a better state at the material level and has the potential for practical application. However, it should be noted that the elastomer phase with low modulus and high toughness generally has strong molecular chain activity. Under the long-term combined action of electricity and heat, the phase structure stability of the blend system still needs to be further evaluated, and the methods to improve the long-term structural stability of the multiphase system also need to be further studied.

10.2.2.2 Nano Blending

The traditional toughening of PP is generally achieved by directly blending with elastomers. This simple melt blending will form micron-scale phase separation in the PP matrix, significantly damaging the electrical properties of the material. Therefore, to balance the mechanical and electrical properties of the material, researchers will add nanoparticles to the blend to prevent the accumulation of space charges, thereby enhancing its electrical properties.

Common nanoparticles include MgO, SiO_2, Al_2O_3, and TiO_2. The influences of different nanoparticles on parameters such as dielectric loss and volume resistivity of PP vary. Selecting the appropriate nanoparticles and their addition amounts to synergistically enhance the electrical and other properties of PP is a huge challenge. Breakdown strength is the most widely studied electrical parameter in the development of insulating materials. Studies show that the appropriate introduction of nanoparticles can enhance the AC breakdown strength or DC breakdown strength of polymers. The influence of nanoparticles on the aggregated state structure and the induced charge traps are regarded as the main factors [18, 19].

Inorganic nanoparticles are usually incompatible with nonpolar PP, making it difficult for nanoparticles to be uniformly dispersed. The modification of the surface of nanoparticles is not only for achieving uniform dispersion but also for adjusting the interface region between the nanoparticles and the PP matrix. Studies show that trimethylsilane-modified silica nanoparticles increase the AC and DC breakdown strengths of the composite materials by 19.9% and 52.3% respectively. The octamethylpolysesquisiloxane with a mass fraction of 3% increased the AC breakdown strength of the composite material by 22% [20, 21]. There are also studies on improving the thermal conductivity of PP by adding BN, AlN, and Al_2O_3 [22].

The modification method of simultaneously doping elastomers with nanoparticles can reduce the mechanical modulus of PP, and increase the elongation at break, while maintaining or even enhancing the electrical properties of PP, enabling it to meet the insulation requirements of power equipment at the material level. However, in the actual production process, the dispersion control of nanoparticles, elastomers, etc., during processing remains a difficult problem, and the influencing factors of the extrusion stability of their multiphase structures still need to be further studied.

10.2.2.3 Nucleating Agent Blending

PP is a typical semicrystalline polymer, and its crystallinity is closely related to the regularity of chain segments. During the polymerization process of propylene, due to the asymmetry of propylene monomers, there may be different monomer connection methods in the PP molecular chain. The connection mode of monomers will affect the stability of the helical conformation of the PP molecular chain. PP chains with different helical conformations interact with each other during the crystal stacking process and may eventually form different crystal structure forms. The most common ones include monoclinic (α) crystal form, hexagonal (β) crystal form, and triclinic (γ) crystal form. Among them, the α crystal form is the most common. Under conventional molding conditions, the crystals formed by the vast majority of isotactic PP are mainly in the α crystal form. α-crystalline PP has excellent insulation performance, but it has high hardness and poor brittleness at low temperatures. The stability of the β crystal form under conventional crystallization conditions is relatively poor, and in most cases, only a small amount of β crystal form can be formed, coexisting with other crystal forms. β-crystalline PP has excellent impact resistance. γ crystals are generally regarded as the result of simple shearing of monoclinic α unit cells along their A-axis, and can also be seen as defective α crystals. The thermodynamic stability of γ crystals is poor and they are difficult to obtain. Therefore, they have not received significant attention in the related research on modified PP.

Inducing the generation of β crystals as much as possible in PP may play a positive regulatory role in mechanical properties, making it adapt to the relevant requirements of insulation. At present, the most efficient method for inducing β-crystal formation is to add β-crystal nucleating agents. B-nucleating agents can be classified into inorganic nucleating agents and organic nucleating agents: Inorganic β-nucleating agents include some inorganic salts and metal oxides. Organic β-nucleating agents mainly include polycyclic compounds, amide compounds, dicarboxylate salts, rare earths, and polymers [23]. In practical applications, two to three of them may also be mixed and applied to pursue a better nucleation effect.

1) Inorganic β-nucleating agents: These nucleating agents have a wide range of sources. For example, $CaCO_3$, ZnO, and TiO_2 can all induce the formation of β-PP [24, 25]. The raw materials of inorganic nucleating agents are readily available and inexpensive. However, they generally have poor compatibility with PP and require a large amount of addition when inducing the formation of β-PP, which greatly limits their own applications.
2) Polycyclic compound β-nucleating agents: In the 1960s, Leugering et al. first reported a polycyclic organic nucleating agent, γ-quinacridone (E3B), which can induce the formation of β-PP [26]. Subsequently, researchers successively discovered that a variety of polycyclic compounds, such as thiodiphenylamine (PT), 2-mercaptobenzimidazole (MBIM), triphenyldithiazide (TPDT), lycitin golden yellow (IGK), and lycitin gray (IBL), also have the ability to induce the formation of β-PP [27]. Their chemical structures are shown in Figure 10.8. However, polycyclic β-nucleating agents are prone to staining PP products and have a relatively low nucleation efficiency, which is why they have not been widely used.
3) Amide compound β-nucleating agents: Amide β-nucleating agents can be classified into monoamide β-nucleating agents, bisamide β-nucleating agents, and

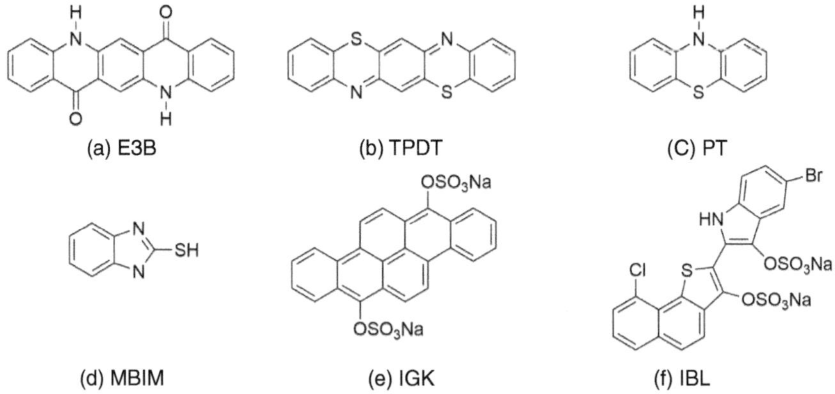

Figure 10.8 Chemical structure of fused ring *β*-nucleating agents. (a) E3B; (b) TPCT; (c) PT; (d) MBIM; (e) IGK; (f) IBL.

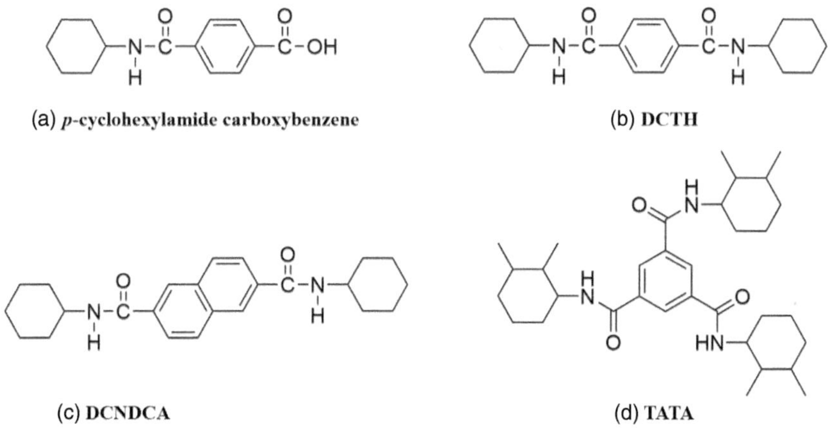

Figure 10.9 Chemical structure of amide *β*-nucleating agents. (a) *p*-cyclohexylamide carboxybenzene; (b) DCHT; (c) DCHN.

polyamide *β*-nucleating agents based on the number of amide groups. The chemical structural formulas of common amide compound *β*-nucleating agents are shown in Figure 10.9. Most amide nucleating agents have highly efficient *β*-phase crystallization induction capabilities. A small amount of addition can significantly enhance the toughness of PP products. However, their preparation processes are rather complex [28, 29].

4) Dicarboxylate *β*-nucleating agents: Dicarboxylate *β*-nucleating agents are mainly divided into aliphatic dicarboxylate and aromatic dicarboxylate. The addition schemes are mainly divided into two types: The first involves preparing the nucleating agent in advance by acid-base neutralization or the double decomposition method, and then add the nucleating agent to the PP matrix during the melt blending process to induce the formation of the *β* phase. The second option is to adopt the in situ reaction method. The nucleating agent raw material is

added to the PP matrix, so that the nucleating agent raw material synthesizes the β-nucleating agent during the melt blending process and directly acts on the PP matrix to form the β-phase. Compared with the first scheme, the in situ reaction method eliminates the storage process of the nucleating agent, eliminating concerns about it becoming ineffective due to improper storage. Moreover, existing studies have also proved that the nucleating agent prepared by the in situ reaction method has better dispersion in the PP matrix and is more conducive to improving the efficiency of crystallization induced by the nucleating agent. In practical applications, both the addition method of nucleating agents and the source of metal cations affect the efficiency of crystallization induced by such nucleating agents.

5) Rare earth β-nucleating agents: Rare earth β-nucleating agents are rare earth organic complexes formed by the combination of rare earth elements and organic compounds, and they have a good ability to induce the formation of β-PP. Rare earth elements can be lanthanide light rare earth elements, such as lanthanum, cerium, praseodymium, and neodymium. Organic compounds include polycyclic compounds, organic carboxylic acids and their derivatives, or other organic heterocyclic compounds. Rare earth type β-nucleating agents have the characteristics of high nucleation efficiency and an obvious toughening effect [30]. However, this type of nucleating agent contains rare earth elements, the raw material price is relatively high, and the preparation process is complex.
6) Polymer-based β-nucleating agents: Unlike the relatively small-molecular-weight organic or inorganic nucleating agents mentioned above, some macromolecular polymers containing benzene ring structures can also be used as β-nucleating agents. For instance, main-chain liquid crystal polyester (PBDPS) is an efficient β-nucleating agent, and its chemical structure is shown in Figure 10.10 [31]. The nucleation efficiency of polymer-based β-nucleating agents is generally lower than that of small-molecular-weight organic and inorganic nucleating agents, which may be related to the lower dispersibility of polymer-based β-nucleating agents in PP. The mechanism of crystallization induced by polymer-based β-nucleating agents is still unclear at present. However, existing studies generally suggest that the ordered stacking of benzene rings may be a key factor leading to the β-complementary nucleation ability of polymers.

The nucleation efficiency of different types of nucleating agents varies. The presence of β crystals can indeed have a positive impact on insulation properties such as volume resistivity, space charge, and breakdown strength, while also optimizing the

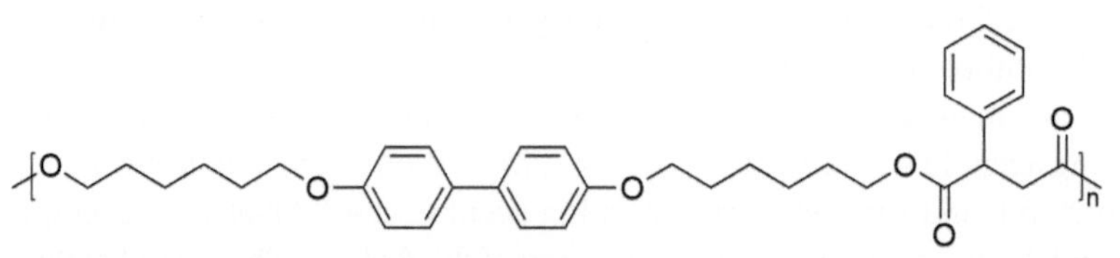

Figure 10.10 Chemical structure of PBDPS.

mechanical toughness of PP. Considering that most β-crystal nucleating agents are small-molecule substances with certain polarity, they may undergo molecular chain rotation or even migration under the influence of high temperature or electric field, affecting the insulation performance of the material at high temperatures. On the other hand, the β crystal will transform into the α crystal form under certain temperature conditions. The stability of its crystal form and performance under the long-term operating temperature of the cable still needs to be further evaluated. The development of nucleating agents with good dispersibility, high stability, and economic applicability is of great significance for optimizing the overall performance of PP preparation.

10.2.3 Physical Field-Induced Molding Modification

The crystal structure of PP and the distribution state of the added phase after the introduction of elastomers or nanoparticles are both related to the physical fields during processing and molding. Starting from the extrusion processing conditions of power cables, the main physical fields are the shear force field and the temperature field.

10.2.3.1 Shear Force Field Regulation Molding

Typical applications of PP in power equipment include insulation of power cables. The processing technology of the cable determines that it will continuously be subjected to shear forces parallel to the extension direction of the cable during the extrusion process. Therefore, it is necessary to pay attention to the structural and performance evolution process of PP insulation under the action of shear stress.

Benjamin S. Hsiao et al. studied the crystallization behavior of isotropic PP under shear stress at 140 °C. They found that under the action of shear stress, a large number of crystal nuclei were oriented along the shear direction and closely arranged, promoting the growth of flake crystals in the direction perpendicular to the shear force. The schematic diagram of crystal growth is shown in Figure 10.11. The shear rate is closely related to the crystal form of PP. After the application of shear force, the α crystal form immediately appears in the material, followed by a certain amount of β crystal form. When the shear rate is relatively high, the relative content of β crystal can reach 65–70%, indicating that an appropriate shear rate has a very significant inducing effect on the crystal form [32]. In addition, the crystallization process of nanocomposites is more sensitive to shear induction. The higher the content of nanoparticles, the shorter the crystallization induction time.

It can be seen that the aggregation state structure of PP is closely related to the shear force field it is subjected to during the molding process, and it is one of the decisive factors for its macroscopic properties. Therefore, the crystallization behavior of PP composite insulated cables during the extrusion process and its influence on the performance deserve attention.

10.2.3.2 Temperature Field Regulation Molding

The processing of thermoplastics generally includes four stages: heating and melting, holding, extrusion or compression molding, and cooling and molding.

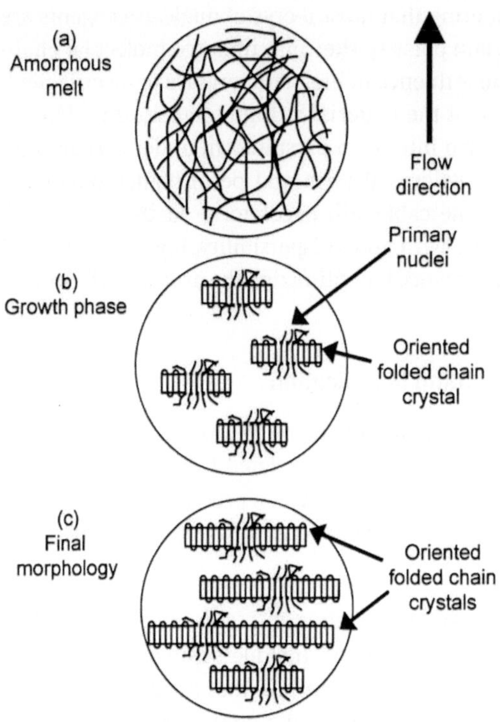

Figure 10.11 Schematics of the nature of shear-induced primary nuclei and subsequent growth of oriented crystals at the surface undergoing maximum orientation in the step-shear experiment: (a) before shear (amorphous melt, no orientation), (b) at short time after the step shear (primary nuclei and initial growth of the oriented crystals), and (c) final morphology after the step shear (stacks of oriented crystals, kebabs). *Source:* Adapted from [32].

Generally speaking, since the material needs to be maintained at the melting processing temperature for a period of time before extrusion or compression molding to fully plasticize and eliminate the residual stress of the granular material, the temperature conditions in the heating stage have little impact on the final performance of the material. The insulation and temperature conditions during the cooling and forming stages have become two more important influencing factors.

When studying the influence of insulation conditions on structures, variables are generally controlled by changing the temperature and time of melt heat treatment during the molding process. Heat treatment in the melt state can affect the multiphase structure of the material after molding, thereby influencing the properties such as elongation at break and impact strength of PP [13, 14].

The rate and form of molding cooling will also have a significant impact on the structural properties of the polymer. Relevant studies show that as the pressure during the crystallization process increases, the intermediate phase crystal forms within the polymer decrease, while the stable crystal forms increase. Moreover, the optimal cooling rates corresponding to different crystal forms of PP vary greatly. A slow cooling rate is conducive to improving the unit cell structure in the material, enhancing crystallinity, restricting the injection and transport processes of carriers, and thereby improving the material's electrical properties [33]. In addition, the effect of the nucleating agent is also significantly correlated with the cooling rate during molding. The β-nucleating agent can only achieve a good crystal form induction effect at a relatively low cooling rate. When the cooling rate is too high, only the α crystal form can be formed in PP [34].

It can be seen that the thermal fields at each stage of the polymer material molding process significantly regulate the aggregated state structure, and will further affect the macroscopic properties of the material. Most importantly, the method of enhancing material performance through thermal field regulation does not involve the use of additives. It can be integrated into the material processing and forming process, avoiding postmodification of insulating materials and the risk of introducing impurities through multiple processing steps.

10.2.4 Bio-Based Polypropylene

According to the latest life cycle assessment of PP products, over 90% of environmental impacts are attributed to the manufacturing stage [35]. Therefore, people are constantly striving to produce industrial-level bio-based PP through eco-friendly processes. At present, bio-PP is still in the pilot production stage and has not been fully commercialized yet.

Propylene is a major component of PP. Traditionally, it has been a byproduct of the naphtha and gasoline vapor cracking process for ethylene production, or a byproduct of the fluid catalytic cracking process in oil refineries. Both of these processes emit large amounts of greenhouse gases. People are researching sustainable methods for obtaining propylene from renewable biological resources to reduce the impact on the environment and the consumption of fossil fuels. Propylene can be produced from various bio-based sources in multiple ways, among which the most popular raw material is bioethanol produced by the fermentation of sugarcane. As shown in Figure 10.12, bioethanol is first converted to ethylene through dehydration and then to butene through dimerization. Finally, propylene monomer is obtained through the metathesis reaction of ethylene and butene. Another method is to vaporize different types of carbon-rich biomass raw materials (such as corn, grass, and agricultural waste) through thermochemical processes to produce biosyngas. However, the gasification method is more costly than the fermentation route. In addition, the hydrogenation treatment of vegetable oil or used edible oil can also be used to produce green propylene.

Once bio-based propylene monomers are obtained, current polymerization techniques can produce high-molecular-weight PP with the same characteristics as fossil-based PP. However, unmodified bio-based PP may be slightly inferior to petroleum-based PP in terms of high-temperature resistance and antiaging performance. In addition, the main challenge faced by bio-based PP producers is to develop process technologies that are cost-competitive with PP produced from fossil fuels.

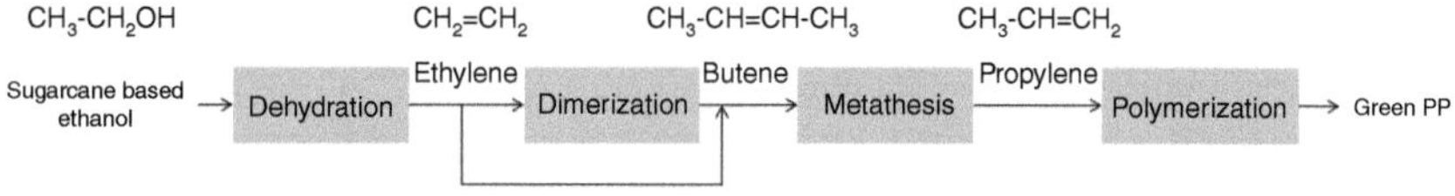

Figure 10.12 Production route of bio-based PP from bio-ethanol. *Source:* [36] / with permission of John Wiley & Sons.

10.2.5 Summary

PP is an important basic material in the field of power insulation materials. The development process of its modification technology fully reflects the deep integration of materials science and the demands of power engineering. From the perspective of molecular structure, the modification of PP mainly unfolds along three dimensions: In the dimension of molecular chain design, through copolymerization modification (such as introducing ethylene monomers to form random copolymers or block copolymers), not only is the low-temperature brittleness of the material significantly improved but also the dielectric properties are optimized by regulating the crystallinity. In the supramolecular structure dimension, physical field-induced molding enables molecular chains to be orderly arranged in a specific direction, and at the same time, mechanical strength is further improved by controlling the crystalline morphology (such as β-crystal form induction). In the dimension of composite materials, nano-modification technology has demonstrated unique advantages. The environmental pollution problem of PP during the manufacturing stage is also being addressed through methods such as bio-based solutions.

It is worth noting that modern modification technology has evolved from a single performance improvement stage to a multiparameter collaborative optimization stage. For instance, the long-chain branching technology has initially achieved three key mutually restrictive issues for thermoplastic high-voltage insulation: a relatively low mechanical modulus at low temperatures, an increase in mechanical strength at extremely high temperatures, and an improvement in electrical properties over a wide operating temperature range. With the development of artificial intelligence-assisted molecular design and in situ polymerization technology, PP insulating materials will make a leap from "empirical modification" to "precise customization" in the future, providing more superior insulating solutions for key equipment such as ultra-high voltage direct current cables and new power capacitors. The advancements in these modification technologies not only expand the application boundaries of PP in the power industry but also provide an important paradigm for the development of new multifunctional insulating materials.

10.3 Modification Methods of Epoxy

Generally speaking, compounds with two or more epoxy groups in their molecular structure can all be called epoxy resins. Epoxy resin was first synthesized in 1891 and was first industrialized in the United States in 1947. Epoxy resins can usually be synthesized in two ways: (1) Oxidizing olefins with peroxides to obtain epoxy groups. (2) Epoxy resin is obtained by the reaction of phenolic hydroxyl groups with epichlorohydrin. Epoxy resin alone generally has no application value and needs to be used in combination with a curing agent. Curing agents can directly affect the quality of epoxy resin after curing, such as viscosity, color, and transparency. The types of curing agents are much more complex than those of epoxy resins and can be roughly classified into several major categories, such as aromatic amines, fatty amines, and anhydrides.

In the field of insulation for power equipment, epoxy resin has long been regarded as an indispensable key material due to its outstanding mechanical properties, excellent electrical insulation, and stable chemical characteristics. However, as a typical thermosetting polymer, the permanent cross-linked network structure of traditional epoxy resin makes it impossible to reshape or recycle the material after curing, which not only causes resource waste but also brings about increasingly serious waste disposal problems. Given the urgent need for global carbon neutrality and sustainable development, developing a new generation of green epoxy insulating materials with both high performance and environmental protection features has become a focus of common attention in both academic and industrial circles.

In recent years, revolutionary progress has been made in the green modification of epoxy resins, mainly reflected in two major directions: dynamic bond recyclable epoxy resins and bio-based epoxy resins. By introducing dynamic covalent bonds (such as imine bonds, disulfide bonds, and borate ester bonds) and supramolecular interactions (such as hydrogen bonds and metal coordination bonds), researchers have successfully endowed traditional thermosetting epoxy resins with reversible cross-linking properties. This type of material can achieve depolymerization and reprocessing under heat, light, or chemical stimulation while maintaining excellent insulation performance, providing the possibility for the circular economy model of power equipment. Meanwhile, the epoxy resin system synthesized from bio-based raw materials such as plant-derived tannic acid, cashew phenol, and lignin has achieved insulation performance and thermal stability comparable to traditional petroleum-based materials through molecular structure design.

This section will systematically elaborate on the design strategies, preparation methods, and performance optimization paths of two types of green insulating materials, namely dynamic recyclable epoxy and bio-based epoxy.

10.3.1 Epoxy Based on Dynamic Covalent Bonds

In 2011, Leibler et al. first proposed the "vitrimer" concept, which was later translated as a glass-like polymer. They developed the vitrimer for the first time by curing epoxy resin and fatty acid/anhydride as raw materials with zinc salt catalyst and heating [37]. The ester bond and the hydroxyl group in the cross-linking structure undergo a dynamic transesterification (TE) reaction at high temperatures (>100 °C) and the promotion of the catalyst, causing the cross-linking network to rearrange and the material to show solid-state plasticity. The types of vitrimer dynamic covalent bonds have expanded in recent years, from the original hydroxyl ester to disulfide bonds, Schiff bases, urethanes, silyl ethers, disulfides, and others.

10.3.1.1 Diels–Alder Reaction

As shown in Figure 10.13a, the Diels–Alder (DA) reaction is a simple [4+2] cycloaddition reaction involving conjugated dienes and modified alkene molecules, generating stable cyclohexene admixtures. It is one of the most widely studied and promising processes in organic chemistry. The cross-linked thermoreversible polymer based on the DA cycloaddition mixture of furan and maleimide was

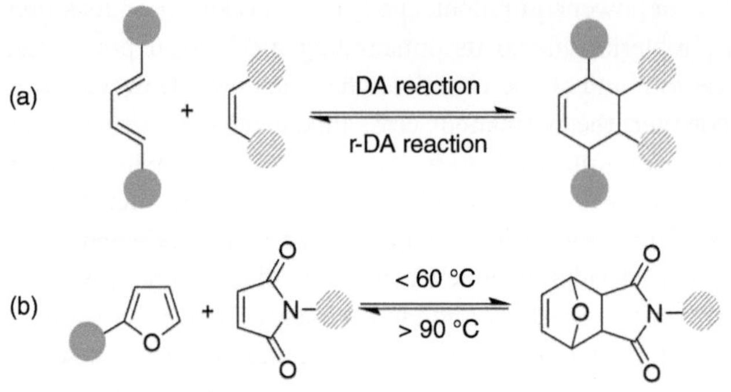

Figure 10.13 (a) Thermotriggerable Diels–Alder reaction. (b) Retro-Diels–Alder reaction involving primarily used furan and maleimide groups. *Source:* Reproduced from [38] / with permission of Elsevier.

first published in DuPont's patent in 1969 (Figure 10.13b). At low temperatures, conjugated dienes and substituted alkenes will undergo cross-linking to form a dense network structure. At high temperatures, a reversible DA reaction occurs, and the cross-network structure dissociates. This temperature-regulated reversible reaction without the need for a catalyst provides ideas for the reprocessing of epoxy resin materials.

At present, there are mainly two ideas for preparing recyclable epoxy resins containing DA structure: One is to modify the curing agent and introduce the DA structure into the curing agent. The modified curing agent in this way can not only undergo ring-opening reactions with epoxy groups to generate cross-linked network structures but also the DA structure can control the dissociation and regeneration of the network structure. For example, a diamine cross-linking agent with a DA structure was synthesized by using diaminooctane, maleic anhydride, and furfuramine, and then reacted with bisphenol A diglycidyl ether to prepare epoxy resin containing a DA structure [39].

The second approach is to first introduce conjugated dienes into the epoxy resin reaction monomers to prepare the corresponding oligomers, and then react them with the corresponding derivatives containing substituted alkenes to prepare the epoxy resin network structure. At this time, the conjugated dienes and substituted alkenes in the epoxy resin form a dynamic DA structure, and the dissociation and regeneration of the network structure are completed at different temperatures. For instance, FAN et al. reacted bisphenol A epoxy resin containing oxyethylene groups with furfuramine to synthesize polyhydroxyamino ether prepolymers bearing furan functional groups. These prepolymers were then cross-linked via D–A reation with 1,5-bis(maleimide)-2-methylpentane. This material can be reshaped at 135 °C and 10 MPa and can be dissolved for recycling at 120 °C [40].

Although the DA addition reaction has the advantages of simple synthesis and mild conditions, the bond energy of the generated product of the DA reaction is relatively low, which reduces the thermal stability of this type of epoxy resin. In the case of local high temperature, it may cause the DA structure to fracture, resulting in the decline of the mechanical and electrical properties of the material.

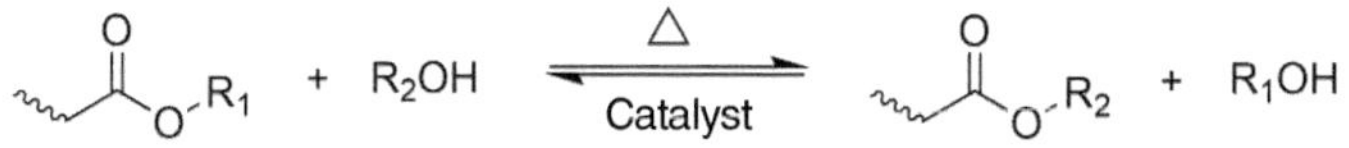

Figure 10.14 Transesterification of ester bonds and hydroxyl groups.

10.3.1.2 Ester Bond

The typical chemical reaction used in commercial and laboratory environments is TE. Unless equilibrium is reached, a dynamic TE reaction will occur between the ester and the alcohol to form new esters and alcohols, which usually requires high temperature and a strong catalyst, as shown in Figure 10.14.

In 2011, Leibler and colleagues opened a window for glass polymer chemistry by using zinc acetate–catalyzed TE in epoxy thermosetting resins through the reaction of traditional DGEBA and anhydride curing agents. In addition, these epoxy materials can be reused, repaired, and recycled without affecting the integrity of the network. This is because TE undergoes association exchange reactions at high temperatures and maintains similar physical properties after multiple recycling [37].

Although epoxy thermosetting resins based on ester exchange have good processability and toughness, to achieve the best processing efficiency, a relatively high catalyst content and operating temperature are required. The catalyst used in this reaction will reduce the consistency of each component and accelerate their degradation [41]. Due to the hydrolysis of the ester bond, the stability of the material in a humid and hot environment will also be affected. Moreover, the epoxy resin based on ester exchange may cause local network reorganization under high stress cycles, resulting in a slightly lower fatigue life than that of traditional epoxy resins.

10.3.1.3 Disulfide Bond

The disulfide bond is a relatively weak covalent bond. As shown in Figure 10.15, under the action of ultraviolet light or heat, disulfide exchange usually proceeds along anion or radical routes. The decomposition and reconstruction of disulfide bonds can occur at relatively low temperatures, and degradation can take place at room temperature in solvents such as mercaptans. This enables it to be widely applied in the field of degradable epoxy resins. In 1990, Tesor et al. first synthesized disulfide bonds containing curing agents and cured traditional epoxy monomers to form epoxy networks containing disulfide bonds. This represents that fully cured networks can be destroyed by reduction and restored by oxidation under mild conditions [42]. Inspired by this pioneering research and rapid exchange reactions, scientists have focused extensively on integrating disulfide bonds into epoxy molecular structures to provide curability, reprocessability, and recyclability under mild conditions.

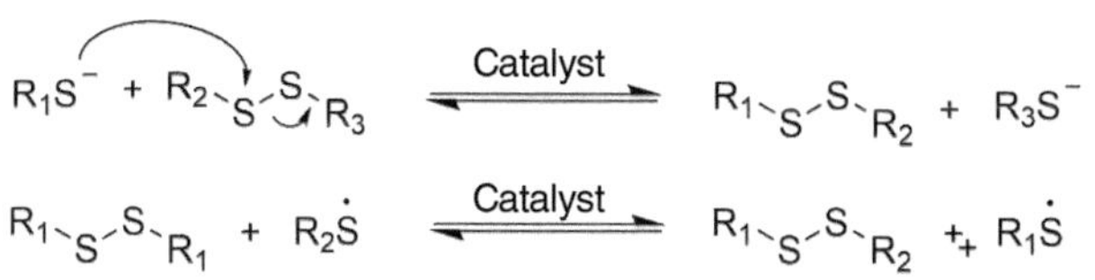

Figure 10.15 Exchange mechanism of dynamic disulfide bonds.

Among the several dynamic covalent bonds in the design of recyclable epoxy resins, the degradation behavior of the disulfide bond is very excellent. However, high temperatures will lead to the thermal degradation and oxidation of the polymer network, resulting in poor resistance to humidity and heat aging. Therefore, in practical applications, the thermal stability of the disulfide bond needs to be considered. In addition, compared with traditional epoxy resins, the volume resistivity and tensile strength of this type of epoxy resin are slightly lower.

10.3.1.4 Imine Bond

The condensation of primary amines and aldehydes to form imine bonds, sometimes also known as the Schiff base reaction, was developed by Huc and Lehn in 1997 [43]. As shown in Figure 10.16a, the imine bond can be re-hydrolyzed into an aldehyde group and an amino group under the action of water, which is a reversible reaction. In the absence of water, as shown in Figure 10.16b, c, the imine group can undergo reversible exchange reactions with the amino group, or reversible exchange (metathesis) reactions can occur between two imine groups. The two reactions under anhydrous conditions are regarded as the main mechanisms by which imine-based thermosetting epoxy resins can be thermally reshaped and chemically degraded.

Taynton et al. reported for the first time a degradable thermosetting resin based on Schiff base. Without additional catalysts, the polymer cross-linked network was decomposed into small fragments through an isoalanine exchange reaction between the active small molecules, and the polymer network could be regenerated by removing the active small molecules [44]. To further enhance the thermal performance and sustainability of Schiff base thermosetting resins, Wang et al. prepared a biomass Schiff base diphenol compound and introduced it into an epoxy resin network. Due to the π-conjugated Schiff base structure and the high content of related hydrogen bonds, the glass transition temperature of Schiff base epoxy resin is 172 °C, the tensile strength is 81 MPa, the tensile modulus is 2112 MPa,

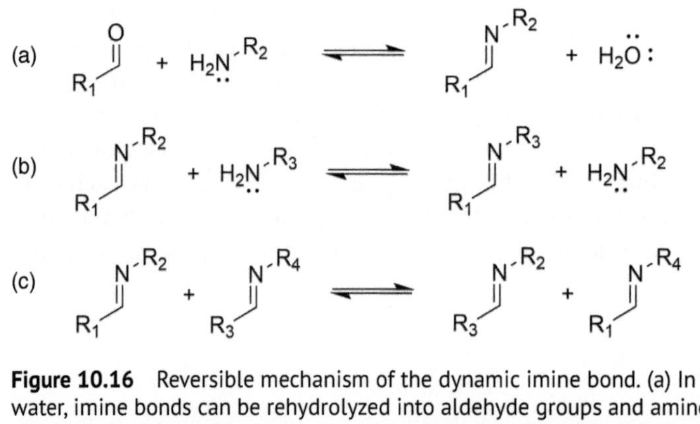

Figure 10.16 Reversible mechanism of the dynamic imine bond. (a) In the presence of water, imine bonds can be rehydrolyzed into aldehyde groups and amino groups. (b) In the absence of water, another primary amine can be introduced into the transimine reaction through covalent exchange. (c) In the presence of a catalyst, an imine exchange reaction can occur between two imines through amine exchange, thereby generating a new imine. *Source:* Reproduced from [38] / with permission of Elsevier.

the 5% thermostatic weight loss temperature is 323 °C, and the elongation at break is higher than or similar to that of Dow Chemical's (DER331) high-performance bisphenol A epoxy resin [45].

As reversible covalent bonds, Schiff bases can endow thermosetting resins with degradability, ductility, and posttreatment recyclability, thereby expanding their application range. In addition, the preparation process of Schiff base thermosetting resin is simple, with excellent performance and good stability, which is conducive to commercial application.

As a reversible covalent bond, the imine bond can endow thermosetting resins with degradability, ductility, and posttreatment recyclability, while ensuring their insulation properties and mechanical properties. Moreover, the preparation process of Schiff base thermosetting resins is simple, with excellent performance, good stability, and is conducive to commercial applications. However, it should be noted that the exchange temperature of the imine bond is relatively low, which facilitates low-temperature reshaping but may limit high-temperature applications.

10.3.1.5 Acetal Bond

As shown in Figure 10.17, the acetal structure is reversible under acidic conditions and has sufficient stability in nonacidic environments. Due to this characteristic, acetal structural bonds are introduced into the cross-linked network of thermosetting resins to achieve degradability.

In order to develop degradable thermosetting resins, Buchwalter and Kosber added acetal bonds to epoxy resins in 1996. This epoxy retains the positive properties of thermosetting epoxy to expand its application, while eliminating its difficulty in degradation [46]. After this, many acetal epoxy thermosetting resins are made by adding acetal bonds to the curing agent.

The degradation and reaction conditions of degradable epoxy resins containing acetal bonds are relatively mild. When the structure is reasonable, its heat resistance and mechanical properties are also superior to those of the traditional bisphenol A epoxy resin. However, it is prone to double decomposition reactions in weakly acidic aqueous solvents. To enhance its water resistance, hydrophobic groups can be grafted to improve the hydrophobic performance of the epoxy resin network structure itself, thereby increasing its water resistance to a certain extent. However, acidic environments are inevitable during the operation of electrical equipment. This uncertain operating environment and the extremely high reliability requirements of electrical equipment will restrict the use of epoxy resins containing acetal bonds.

10.3.1.6 Boronic-Ester Bond

As shown in Figure 10.18, the condensation reaction of boric acid with 1,2-diol or 1,3-diol can generate reversible boron ester bonds under environmental conditions.

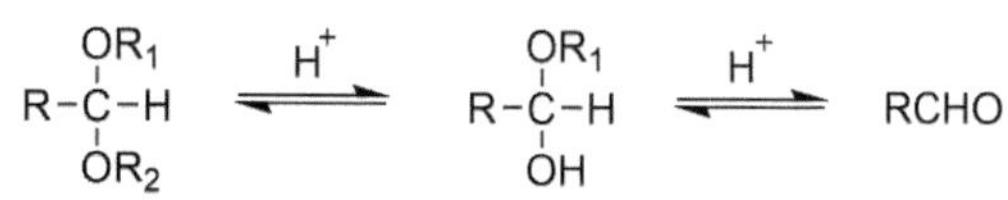

Figure 10.17 The decomposition of acetals under acidic conditions.

Figure 10.18 Reversible mechanism of the dynamic boronic-ester bond.

When the pH value is higher than the pKa of boric acid, the equilibrium in the aqueous solution is conducive to the formation of borate esters. When it is lower than this pKa, the equilibrium tends to form boric acid and diols. Boron-ester exchange occurs rapidly and is affected by external factors, including moisture, pH, and the content of diols in the system.

This reaction has been used for the full recovery of thermosetting materials and their composites. Cromwell et al. prepared a degradable thermosetting resin based on borate esters for the first time through the borate-ester TE reaction, but its mechanical properties were poor [47]. Zeng et al. synthesized a curing agent containing dynamic boron ester bonds and cured conventional epoxy resins to form reversible epoxy thermosetting resins. This material exhibits excellent mechanical properties and thermal stability. Due to the TE rearrangement of the borate ester network, the crushed and cut samples of this material can be restored by hot pressing [48].

Thermosetting epoxy resins containing dynamic boron ester bonds have complete recyclability, good mechanical and thermal properties, and are easy to process. However, in alcohol, the rapid bond exchange at room temperature causes the material to creep under similar mild conditions, affecting stability and limiting application. New methods are needed to balance repairability/recyclability and creep resistance.

10.3.1.7 Silicon-Oxygen Bond

Organosilicon has excellent heat resistance and oxidation resistance. Among them, the Si—O bond has a relatively high dissociation energy and can be introduced into the polymer through free hydroxyl and alkoxy groups. As shown in Figure 10.19, the decomposition and reconstruction process of Si—O—Si bonds in the network structure under specific conditions is also an effective way for polymer materials to achieve self-healing, recycling, and reuse.

Based on the dynamic equilibrium of silicon-oxygen bonds, Liu et al. introduced the silyl ether bond and used KOH as the catalyst to prepare a new type of thermosetting epoxy resin containing dynamic silyl ether bonds, demonstrating its excellent recyclability and reprocessing performance. However, this material is difficult to ensure the uniform dispersion of free KOH catalysts, thereby affecting the thermal and mechanical properties of the material [49]. To optimize this issue, Wu et al. prepared a high-performance recyclable epoxy resin based on siloxane equilibrium using oligosiloxane containing potassium silolate end groups and aminopropyl side groups as the catalyst and curing agent. Since potassium

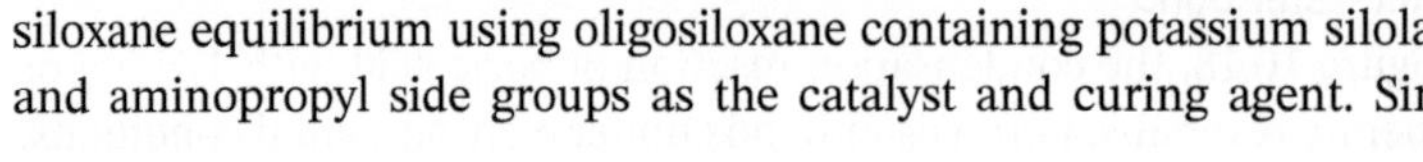

Figure 10.19 Reversible mechanism of the dynamic silicon oxygen bond.

silolate, which plays a catalytic role, is bonded to the long chains of polysiloxane, it can be uniformly distributed in the matrix to promote the decomposition and reconstruction of silicon-oxygen bonds. This material exhibits excellent mechanical properties and recyclability [50].

Epoxy based on silicon-oxygen bonds has unique advantages in electrical, mechanical, and thermal properties, and is particularly suitable for applications in harsh environments such as high temperatures and high insulation. The decomposition and reconstruction of silicon-oxygen bonds in the currently reported recyclable epoxy resin systems based on silicon-oxygen bonds all rely on the promoting effect of catalysts. Although multiple reversible repairs and reuses can be achieved, and excellent mechanical properties can still be maintained after reuse, due to the poor compatibility of inorganic salt catalysts such as potassium salts with epoxy resins. Whether the catalyst is free or bonded in the system, it will affect the relevant properties of the material, especially dielectric properties, electrical properties, etc. This directly affects the application of recyclable epoxy resin in the fields of electrical equipment and insulating materials. Therefore, developing a catalyst-free siloxane epoxy resin system is one of the ideas for achieving the synergistic improvement of multiple properties.

10.3.1.8 Other Dynamic Covalent Bonds

In addition to the above-mentioned several widely used dynamic covalent bonds, the related research also involves other types of dynamic chemical bonds, which have also made significant contributions to the development of recyclable epoxy resins.

Hindered urea bonds (HUBs) can be dynamically dissociated into related amines and isocyanates without a catalyst. Chen et al. attempted to integrate HUBs into the epoxy cross-linked network and successfully obtained a dynamic covalent epoxy resin based on HUBs-EP. Compared with traditional epoxy resins, the greatest advantage of HUBs EP is that it can have a fast covalent bond exchange rate and reprocessing recyclability without any catalyst, while maintaining good thermal stability [51].

The acylhydrazone bond is a dynamic covalent bond that is highly sensitive to pH. It is very stable under slightly alkaline conditions, but can undergo decomposition and exchange reactions under acidic conditions. Its dynamic reversible property under specific pH conditions also gives it certain advantages in the preparation of recyclable epoxy resins. However, it should be noted that its volume resistivity is slightly lower than that of traditional epoxy resin, and its dielectric loss is relatively high.

The di-n-benzylaniline bond can be prepared by oxidizing imines under mild conditions and can be hydrolyzed and degraded. Lo et al. reported the formation of di-n-benzylaniline from the imine intermediate generated by the heat of intermolecular alkylation of phenoxazine, and the copolymerization with commercial dibenzazine and acetyl epoxy monomers to prepare thermosetting epoxy resins with excellent thermal and mechanical properties [52].

Hexahydrotriazine is formed by the condensation of amine and formaldehyde and can be hydrolyzed under strongly acidic conditions. You et al. synthesized thermosetting epoxy resin using hexahydrotriazine. Due to the rigid structure of hexahydrotriazine, the resin exhibited high thermal stability, good mechanical properties, and degradability [53].

10.3.2 Epoxy Based on Supramolecular Interaction

Supramolecular chemistry is characterized by noncovalent bonds, represented by hydrogen bonds, metal coordination, π–π, ions, host–guest, and van der Waals interactions. Although these interactions are relatively weak compared with covalent bonds, on the whole, they form mechanically strong and very dynamic systems. The field of supramolecular chemistry has existed for several decades. Due to its reversibility, directionality, and sensitivity, it is attractive for the development of self-healing materials.

10.3.2.1 Hydrogen Bond

Hydrogen bonds are usually one of the strongest noncovalent interactions. Their directionality and high concentration per unit volume endow the material with acceptable mechanical strength, and their dynamic response to external stimuli endows the material with restorable properties. For example, Liu et al. used 2-urea-4 [1H]-pyrimidinone (UPy) as the quadruple hydrogen bond unit and grafted it to the main chain of the epoxy matrix through an amine-terminated oligomer (propylene glycol) linker. Without any intervention, the epoxy resin matrix exerts its self-repairing function within 5 minutes [54].

Epoxy resins based on hydrogen bonds usually have a slightly lower volume resistivity than traditional epoxy, and have higher dielectric loss and lower tensile strength. Meanwhile, the thermal stability of hydrogen bonds themselves is relatively low, so they usually need to act in conjunction with covalent networks.

10.3.2.2 Metal–Ligand Interactions

When metal ions combine with appropriate ligands, they form a coordination complex that links the polymer chain together. The advantage of metal–ligand complexes lies in their ability to coordinate with different metal ions and ligand substitutes, thereby forming different binding strengths. When mechanical force is applied, these complexes dissociate, and their recombination leads to self-repair. Another attractive feature is that there are minimal or no side effects during repeated damage repair cycles.

Coordination bonds are usually applied in combination with dynamic covalent bonds. Smulders reported that the creep resistance of polyimine covalent adaptive network materials can be significantly improved by forming metal coordination bonds with dynamic covalent imine groups. The cross-linking formed by metal coordination bonds has a significant influence on the imine exchange reaction. Furthermore, this kind of metal coordination material can be completely recovered through postthermal treatment, and it has no significant impact on the performance of the recovered material [55].

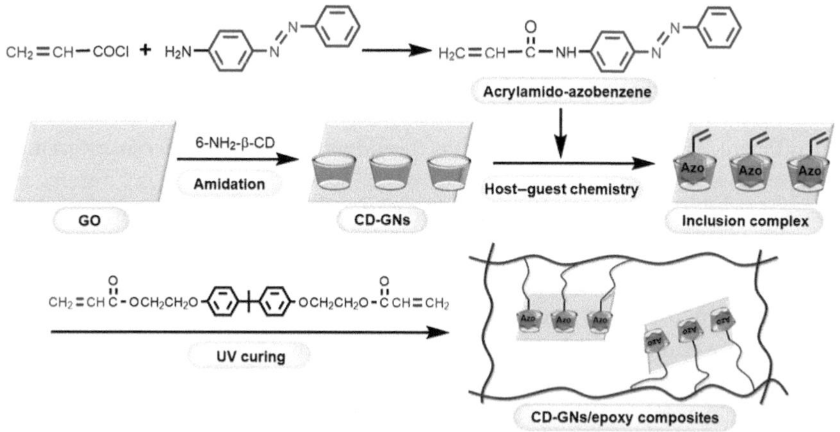

Figure 10.20 Schematic diagram of epoxy resin synthesis based on host–guest action of acrylamide azobenzene/β-CD. *Source:* Reproduced from [56] / with permission of American Chemical Society.

10.3.2.3 Host–Guest Interactions

Host–guest interaction is a process in which the host molecules and guest molecules selectively combine through noncovalent interactions under conditions such as shape and structure complementarity and energy matching. Under the stimulation of heat, moisture, light, etc., the guest molecules incorporate into the internal cavities of the host to form inclusion complexes. This endows the material with regenerative ability and is therefore often used in self-healing materials. For example, Hu et al. prepared epoxy resins based on reversible host–guest networks by introducing acrylamide-azophenyl groups as the guest group and the amideated β-CD/ graphene group as the subject through copolymerization reaction into the epoxy resin system, as shown in Figure 10.20 [56].

Although the method of introducing the host–guest structure into the epoxy resin system can endow the material with properties such as recyclability and self-healing, it also reduces the mechanical and thermal properties of thermosetting materials. Therefore, judging from the current research progress of epoxy resins based on host–guest structure, this type of epoxy resin is still relatively difficult to apply in the field of epoxy resins for electrical use. How to enhance the mechanical and thermal properties of materials and find the appropriate balance point will become one of the research directions of epoxy resins based on host–guest interaction.

10.3.3 Epoxy Based on Hybrid Bonds

The research results in recent years show that the synergistic effect of multiple dynamic bonds is one of the effective methods. The synergy of multiple dynamic bonds refers to the construction of two or more dynamic bonds (generally a combination of strong dynamic bonds and weak dynamic bonds) in the cross-linked network structure of materials, jointly endowing the mechanical and reprocessing properties of the materials. This is because the strong dynamic bonds in the cross-linked network structure can ensure the mechanical properties of the material, such as significantly increasing tensile strength, modulus. However, if only strong dynamic bonds exist, the toughness of the cross-linked network structure may be insufficient. Weak dynamic bonds, on the other hand, can undergo reversible bond-breaking – reconstruction or transformation when the material is subjected to stress, eliminating stress and enhancing the material's toughness. Meanwhile, the dissociation and reconstruction conditions of weak dynamic bonds are relatively mild, which endows the material with relatively mild recycling performance and ductility. Therefore, by combining strong dynamic bonds and weak dynamic bonds, while ensuring mechanical properties, it can also endow the material with recyclability and reshaping properties. If a dual dynamic network structure polymer with hydroxyl esters and disulfide bonds is formed by curing tung oil-based triglycerides (TOTGE) and 4-aminophenyl disulfide, its tensile strength can reach 59.6 ± 6.1 MPa. In addition, it also shows excellent self-repairing and recyclable properties [57]. Epoxy resins that combine imine bonds with disulfide bonds also have excellent reprocessability, degradability, and self-healing ability [58].

In addition to the combined action of double dynamic covalent bonds, the joint preparation of self-healing materials using dynamic covalent bonds and supramolecular interactions has also been a research hotspot in recent years. The introduction of dynamic noncovalent bonds can promote the dynamic exchange performance of materials. However, due to the rapid occurrence of dynamic covalent bond exchange reactions in polymer networks, they are often not very stable. The use of dynamic covalent bond synergistic reactions can make up for this defect. As shown in Figure 10.21, the synthetic route of introducing the imidazole-zinc ion coordination bond and the dynamic disulfide bond together into the epoxy network is presented. Due to the cross-linking effect of the zinc ion coordination bond and stress dissipation, the elastic modulus, mechanical strength, and toughness of the material have been enhanced. Meanwhile, the co-cross-linking network of zinc ion coordination bonds and dynamic disulfide bonds also endows the material with the ability to be reprocessed through a hot press [59].

It is worth noting that although dynamic structures have valuable thermal reprocessability and self-healing ability, their incorporation may damage dielectric strength, mechanical properties, or thermal stability. Accurate molecular design in the early stage is indispensable to mitigate these effects. The selection of dynamic bond types must take into account their bond dissociation energy, reversibility conditions, and compatibility with the epoxy resin network. For example, disulfide exchange provides room-temperature repairability, but it may lower the glass transition temperature. The selection of reactants affects the cross-linking density and the content of polar groups, directly influencing the electrical insulation performance [60]. The optimal formula requires a systematic assessment of the dynamic bond concentration, network topology, and their impact on key parameters such as breakdown strength, volume resistivity, and flexural modulus. It is necessary to reveal the structure-performance relationship with the aid of advanced

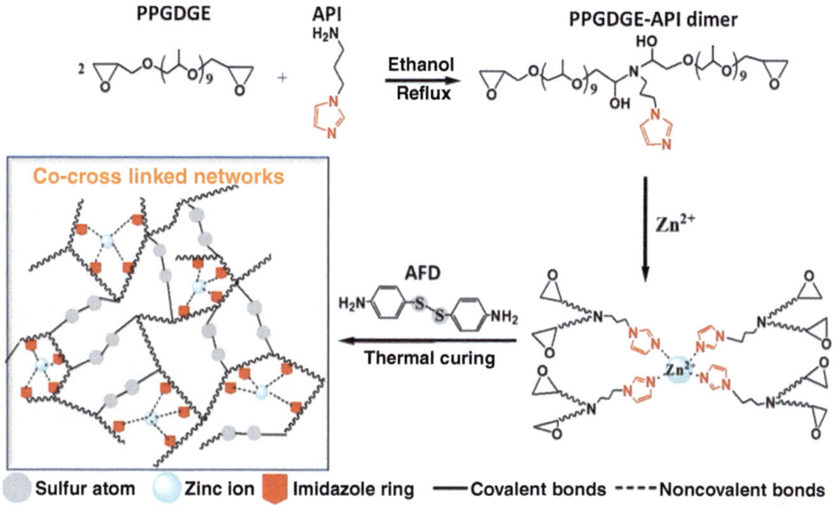

Figure 10.21 Schematic diagram of epoxy resin synthesis containing dual dynamic bonds of imidazole-zinc ion coordination bonds and disulfide bonds. *Source:* Reproduced from [59] / with permission of Elsevier.

characterization techniques to guide the precision engineering of this recyclable high-performance insulating material.

10.3.4 Bio-Based Epoxy

The preparation of the vast majority of epoxy resins relies on petroleum resources. The demand for epoxy resins is increasing year by year. To alleviate the pressure of resource shortage, people have paid attention to renewable resources. Bio-based, using plant-derived renewable resources as raw materials, has successfully achieved electrical strength, mechanical properties, and thermal stability comparable to traditional petroleum-based epoxy resins through molecular structure design and synthetic process innovation. The core advantages of bio-based epoxy resins are not only reflected in the renewability of raw materials but also in the designability of their molecular structure, which provides a broad space for performance optimization.

10.3.4.1 Synthesis of Bio-Based Epoxy Monomers

The synthetic route of bio-based inherent recyclable epoxy monomers is very close to that of traditional epoxy monomers, and there are mainly two ways.

The first is a hydroxyl with epoxy chloropropane (ECH) condensation reaction under the catalysis of sodium hydroxide, generating epoxy monomers, as shown in Figure 10.22a. In this direction, before reacting with ECH, the key is to produce phenols with recyclable dynamic covalent bonds, such as imines, esters, and disulfides. This is the most widely used technology for creating bio-based epoxy monomers.

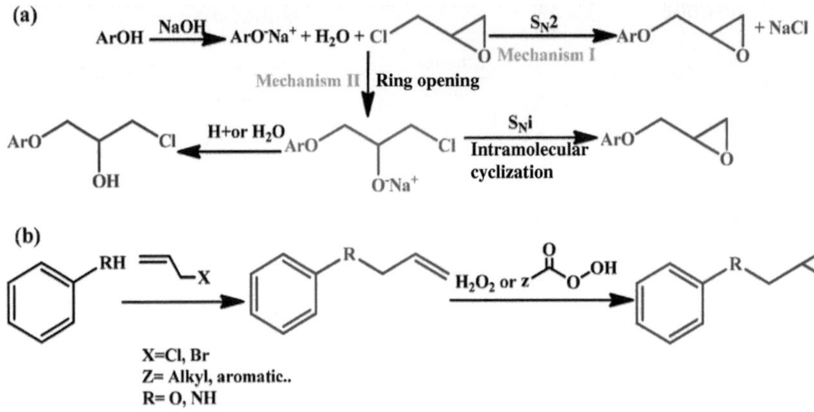

Figure 10.22 Synthesis route of epoxy resins from (a) hydroxyl group-containing aromatic via glycidylation reaction and (b) double bond oxidation. *Source:* [61] / Taylor & Francis Group.

The peroxidation of carbon–carbon double bonds (C=C) is another technology for the production of bio-based inherently recyclable epoxy monomers (Figure 10.22b). This process is typically used to obtain intrinsically recyclable epoxy monomers from biomass with carbon-carbon double bonds.

As shown in Figure 10.23, various raw materials for the preparation of renewable epoxy resins have been reported, such as vanillin, eugenaldehyde, vegetable oil, rosin, and lignin.

10.3.4.2 Partially Bio-Based Epoxy

The epoxy thermosetting resin is made by the chemical reaction between epoxy monomers and curing agents to form a three-dimensional cross-linked network. Partially bio-based epoxy refers to epoxy resins made from biomass epoxy monomers and fossil fuel-based curing agents or fossil fuel-based epoxy monomers and biomass curing agents.

The bio-based recyclable epoxy monomers in Figure 10.23 can be used for the synthesis of epoxy resins. For instance, eugenol has an active aromatic hydroxyl group and an allyl group, which can be converted into epoxy resin through glycidyl acylation reaction with ECH or the double bond oxidation reaction. Liu et al. prepared bioderived glass polymers by epoxidation of eugenol and succinic anhydride in the presence of zinc catalyst at different molar ratios of epoxidation anhydride. Due to the TE reaction carried out at high temperatures, these glass polymers exhibit excellent crack rectification, shape exchange, and shape memory, as well as physical and chemical posttreatment properties [62].

Commonly used cross-linking agents or curing agents for epoxy resins include amines, acids, amides, anhydrides, phenols, and polyphenols. At present, there are still relatively few studies on bio-based recyclable curing agents. As shown in Figure 10.24, different researchers have developed three main types of vanillin-based imine-containing bond curing agents [61]. Due to the presence of dynamic imine

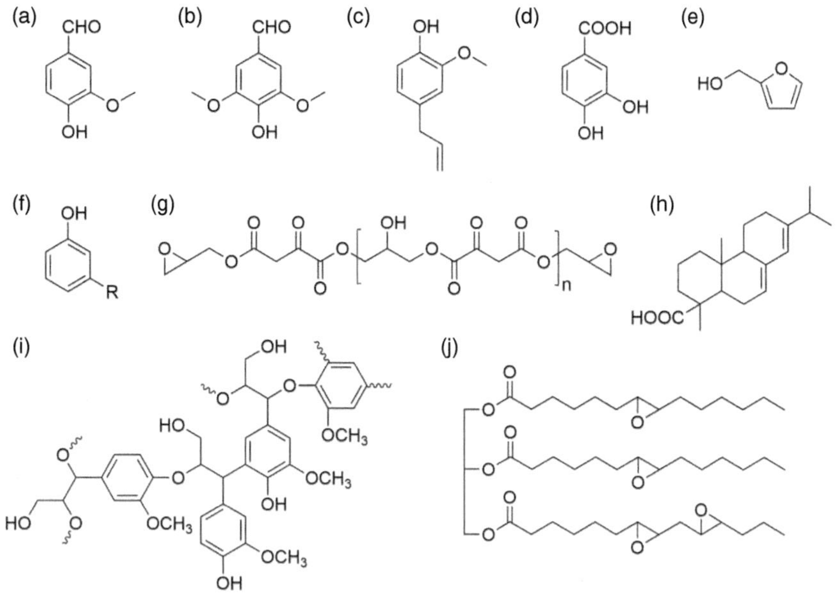

Figure 10.23 Common biological-based monomers and epoxy resin: (a) vanillin, (b) syringaldehyde, (c) eugenol, (d) protocatechuic acid, (e) furan, (f) cardanol, (g) epoxidized itaconic acid, (h) rosin, (i) lignin, (j) epoxy-modified soybean oil (ESO).

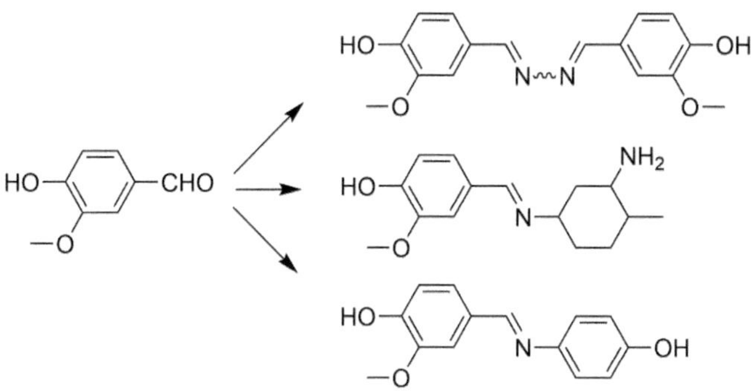

Figure 10.24 Possible formation of recyclable curing agents from bio-based vanillin. *Source:* [61] / Taylor & Francis Group.

bonds, epoxy resins treated with this type of curing agent are remanufacturable and chemically degradable. In addition, there are curing agents based on lignin, vegetable oil, bioderived tannic acid, biomass itacanhydride, etc.

It should be noted that some bio-based components may introduce polar groups, resulting in a slight increase in dielectric loss. The flexible long chains of bio-based monomers can enhance the elongation at break, but they also lead to an increase in the thermal expansion coefficient. The core of its performance lies in the selection of the chemical structure of the biomass components and the coordinated ratio with the petroleum-based components.

10.3.4.3 Completely Bio-Based Epoxy

Due to the increasing concern for environmental sustainability, the biomass content of the product should be as high as possible. Therefore, recyclable epoxy thermosetting materials based entirely on bio-based materials are very attractive. The synthesis of this type of epoxy uses bio-based epoxy monomers and bio-based curing agents, and generally introduces dynamic structures to make the recycling process easier and more efficient. For instance, epoxy soybean oil and vanillin modified curing agent, epoxy soybean oil and glycyrrhizic acid, and palm oil epoxy monomer and citric acid. They have excellent reprocessability, repairability, and degradability, but the glass transition temperature, heat resistance, and mechanical properties of these thermosetting materials are lower than those of traditional thermosetting materials. In addition, the mechanical and thermal properties of these materials after posttreatment are also relatively low.

10.4 Summary

Epoxy resin, as an indispensable insulating material in power equipment, its green modification has become a current research hotspot. By introducing dynamic covalent bonds and supramolecular interactions, researchers have successfully developed a new type of epoxy material with reversible cross-linking properties. Accurate selection of reactants and design of reaction processes in the early stage can appropriately enable such materials to maintain excellent electrical, mechanical, and thermal properties while achieving functions such as recyclability, reconfigurability, and self-healing, providing an innovative solution to the problem of difficult recycling of traditional thermosetting epoxy resins. Meanwhile, the development of bio-based epoxy resins has opened up new avenues for the sustainability of insulating materials. Epoxy systems synthesized from renewable resources such as lignin and vegetable oil not only reduce carbon emissions but also achieve comparable performance to traditional materials through molecular structure design.

The combination of dynamic modification and biological base key strategy represents the epoxy insulating material to the new trend in the development of high performance and environmental protection. In the future, by optimizing the reversible response conditions of dynamic bonds, enhancing the performance stability of bio-based monomers, and developing large-scale production processes, this type of green epoxy insulating material is expected to be widely applied in power equipment such as transformers and GIS.

10.5 Conclusions

In the field of power equipment, the research and development of green insulating materials has become an important direction for achieving sustainable development. Through the environmentally friendly modification of traditional insulating materials, researchers have successfully developed a new generation of material systems that combine excellent electrical performance with environmentally friendly characteristics.

In the field of thermoplastic materials, PP has achieved the coordinated optimization of electrical properties, mechanical strength, and thermal stability through means such as copolymerization modification, nanocomposites, and physical processing. Its recyclability also provides the possibility for the circular economy. In the field of thermosetting epoxy resins, the introduction of dynamic bonds endows the material with reversible cross-linking capabilities, making it recyclable and self-healing. Meanwhile, the rise of bio-based insulating materials has further reduced the carbon footprint of the material. By replacing traditional petroleum-based monomers with renewable raw materials, a breakthrough has been achieved in reducing environmental impact from the source.

These modification technologies not only enhance the comprehensive performance of the materials but also promote the transformation of the insulation system of power equipment toward green and low carbon. In the future, through multistrategy collaboration (such as "bio-based + dynamic bond" composite modification), intelligent design (such as AI-assisted molecular optimization), and the development of large-scale preparation processes, green insulating materials will play a more significant role in new energy power equipment, smart grids, and other fields, providing key material support for the sustainable development of global energy.

References

1. Fu, Y.W., Sun, W.F., and Wang, X. (2020). UV-initiated crosslinking reaction mechanism and electrical breakdown performance of crosslinked polyethylene. *Polymers* 12 (2): 420.
2. Hosier, I.L., Cozzarini, L., Vaughan, A.S., and Swingler, S.G. (2009). Propylene based systems for high voltage cable insulation applications. *Journal of Physics Conference Series* 183 (1): 012015.
3. Yang, K., Liu, Y., Yan, Z. et al. (2020). Enhanced morphology-dependent tensile property and breakdown strength of impact polypropylene copolymer for cable insulation. *Materials* 13 (18): 3935.
4. Zhou, Y., Hu, J., Dang, B., and He, J. (2016). Mechanism of highly improved electrical properties in polypropylene by chemical modification of grafting maleic anhydride. *Journal of Physics D: Applied Physics* 49 (41): 415301.
5. Zha, J.W., Wu, Y.H., Wang, S.J. et al. (2016). Improvement of space charge suppression of polypropylene for potential application in HVDC cables. *IEEE Transactions on Dielectrics and Electrical Insulation* 23 (4): 2337–2343.
6. Yuan, H., Zhou, Y., Zhu, Y. et al. (2020). Origins and effects of deep traps in functional group grafted polymeric dielectric materials. *Journal of Physics D: Applied Physics* 53 (47): 475301.

7 Zhang, Y., Shi, K., Zang, C. et al. (2022). Improved insulation properties of polypropylenes in HVDC cables using aqueous suspension grafting. *Materials* 15 (18): 6298.

8 Zhang, G., Nam, C., Petersson, L. et al. (2018). Increasing polypropylene high temperature stability by blending polypropylene-bonded hindered phenol antioxidant. *Macromolecules* 51 (5): 1927–1936.

9 Zhang, C., Wang, T., Sun, W. et al. (2021). Grafting of antioxidant onto polyethylene to improve DC dielectric and thermal aging properties. *IEEE Transactions on Dielectrics and Electrical Insulation* 28 (2): 541–549.

10 Zhang, G., Nam, C., Chung, T.C.M. et al. (2017). Polypropylene copolymer containing cross-linkable antioxidant moieties with long-term stability under elevated temperature conditions. *Macromolecules* 50 (18): 7041–7051.

11 Li, R., Xing, Q., Zhao, Y. et al. (2015). Correlation between chain microstructure and mechanical properties of two polypropylene/poly(ethylene-co-propylene) in-reactor alloys. *Colloid and Polymer Science* 293 (4): 1011–1021.

12 Wang, J., Niu, H., Dong, J. et al. (2012). Morphology and mechanical properties of polypropylene/poly(propylene-1-octene) in-reactor alloys prepared by Metallocene/Ziegler–Natta hybrid catalyst. *Polymer* 53 (7): 1507–1516.

13 Tian, Y., Song, S., Feng, J., and Yi, J. (2012). Phase morphology evolution upon melt annealing treatment and corresponding mechanical performance of impact-resistant polypropylene copolymer. *Materials Chemistry and Physics* 133 (2): 893–900.

14 Luo, J., Liang, Y., Yang, J. et al. (2012). Effects of liquid–liquid phase separation on crystallization kinetics and morphology of isotactic polypropylene/poly(ethylene-co-octene) in-reactor alloy. *Polymer* 53 (12): 2465–2475.

15 Sui, H., Wu, K., Zhao, G. et al. (2024). Greatly enhanced temperature stability of eco-friendly polypropylene for cable insulation by multifold long-chain branched structures. *Chemical Engineering Journal* 485: 149811.

16 Guo, Q., Chen, Y., Zhang, J., and Yao, Z. (2017). Large impact in electrical properties of polypropylene by improving the filler-matrix interface effect in PP/PS blends. *Polymer Testing* 1 (63): 587–595.

17 Ran, Z., Du, B., Xiao, M., and Liu, H. (2022). Breakdown performance of polypropylene and its long-chain branched blending for metalized film capacitors. *IEEE Transactions on Dielectrics and Electrical Insulation* 29 (1): 145–152.

18 Li, S., Min, D., Wang, W., and Chen, G. (2016). Linking traps to dielectric breakdown through charge dynamics for polymer nanocomposites. *IEEE Transactions on Dielectrics and Electrical Insulation* 23 (5): 2777–2785.

19 Min, D., Wang, W., and Li, S. (2015). Numerical analysis of space charge accumulation and conduction properties in LDPE nanodielectrics. *IEEE Transactions on Dielectrics and Electrical Insulation* 22 (3): 1483–1491.

20 Takala, M., Karttunen, M., Salovaara, P. et al. (2008). Dielectric properties of nanostructured polypropylene-polyhedral oligomeric silsesquioxane compounds. *IEEE Transactions on Dielectrics and Electrical Insulation* 15 (1): 40–51.

21 Takala, M., Ranta, H., Nevalainen, P. et al. (2010). Dielectric properties and partial discharge endurance of polypropylene-silica nanocomposite. *IEEE Transactions on Dielectrics and Electrical Insulation* 17 (4): 1259–1267.

22 Huang, X., Jiang, P., and Tanaka, T. (2011). A review of dielectric polymer composites with high thermal conductivity. *IEEE Electrical Insulation Magazine* 27 (4): 8–16.

23 Guo, J.Q., Li, Q., Luo, Z. et al. (2022). Research progress of polypropylene β nucleating agents polymer bulletin. *Polymer Bulletin* 7: 11–21. (in Chinese).

24 Guan, Z., Lin, Z., and Mai, K. (2013). *Monetaria moneta* as a novel β-nucleating agent for isotactic polypropylene. *Composites Science and Technology* 18 (87): 58–63.

25 Liu, J. and Liang, H. (2021). Heterogeneous nucleation and self-nucleation of isotactic polypropylene with addition of nano-ZnO. *Journal of Thermal Analysis and Calorimetry* 146 (5): 2115–26.

26 Leugering, V.H.J. (1967). Einfluß der kristallstruktur und der überstruktur auf einige eigenschaften von polypropylen. *Makromolekulare Chemie* 109 (1): 204–216.

27 Garbarczyk, J. and Paukszta, D. (1981). Influence of additives on the structure and properties of polymers: 2. Polymorphic transitions of isotactic polypropylene caused by aminosulphur compounds. *Polymer* 22 (4): 562–564.

28 Wu, Y.H., Zha, J.W., Li, W.K. et al. (2015). A remarkable suppression on space charge in isotatic polypropylene by inducing the β-crystal formation. *Applied Physics Letters* 107 (11): 112901.

29 Kawai, T., Iijima, R., Yamamoto, Y., and Kimura, T. (2002). Crystal orientation of β-phase isotactic polypropylene induced by magnetic orientation of *N,N′*-dicyclohexyl-2,6-naphthalenedicarboxamide. *Polymer* 43 (26): 7301–7306.

30 Xu, W., Yu, M., Shi, S. et al. (2018). Simultaneous durability and strength enhancement of β-polypropylene through montmorillonite and melt-soluble β-nucleating agent addition. *Polymer Testing* 1 (65): 150–155.

31 Yang, R., Ding, L., Chen, W. et al. (2018). Molecular-weight dependence of nucleation effect of a liquid crystalline polyester β-nucleating agent for isotactic polypropylene. *Industrial and Engineering Chemistry Research* 57 (19): 6734–6740.

32 Somani, R.H., Hsiao, B.S., Nogales, A. et al. (2000). Structure development during shear flow-induced crystallization of i-PP: in-situ small-angle X-ray scattering study. *Macromolecules* 33 (25): 9385–9394.

33 van Drongelen, M., van Erp, T.B., and Peters, G.W.M. (2012). Quantification of non-isothermal, multi-phase crystallization of isotactic polypropylene: the influence of cooling rate and pressure. *Polymer* 53 (21): 4758–4769.

34 Mollova, A., Androsch, R., Mileva, D. et al. (2013). Crystallization of isotactic polypropylene containing beta-phase nucleating agent at rapid cooling. *European Polymer Journal* 49 (5): 1057–1065.

35 Mannheim, V. and Simenfalvi, Z. (2020). Total life cycle of polypropylene products: reducing environmental impacts in the manufacturing phase. *Polymers* 12 (9): 1901.

36 Wang, S., Muiruri, J.K., Soo, X.Y.D. et al. (2023). Bio-polypropylene and polypropylene-based biocomposites: solutions for a sustainable future. *Chemistry An Asian Journal* 18 (2): e202200972.

37 Montarnal, D., Capelot, M., Tournilhac, F., and Leibler, L. (2011). Silica-like malleable materials from permanent organic networks. *Science* 334 (6058): 965–968.

38 Rashid, M.A., Islam, M.A., Hasan, M.N. et al. (2024). A critical review of dynamic bonds containing curing agents for epoxy resin: synthesis, challenges, and emerging applications. *Polymer Degradation and Stability* 1 (229): 110980.

39 Bai, N., Saito, K., and P. Simon G. (2013). Synthesis of a diamine cross-linker containing Diels–Alder adducts to produce self-healing thermosetting epoxy polymer from a widely used epoxy monomer. *Polymer Chemistry* 4 (3): 724–730.

40 Fan, M., Liu, J., Li, X. et al. (2014). Recyclable Diels–Alder furan/maleimide polymer networks with shape memory effect. *Industrial and Engineering Chemistry Research* 53 (42): 16156–16163.

41 Capelot, M., Montarnal, D., Tournilhac, F., and Leibler, L. (2012). Metal-catalyzed transesterification for healing and assembling of thermosets. *Journal of the American Chemical Society* 134 (18): 7664–7667.

42 Engelberg, P.I. and Tesoro, G.C. (1990). Mechanical and thermal properties of epoxy resins with reversible crosslinks. *Polymer Engineering and Science* 30 (5): 303–307.

43 Huc, I. and Lehn, J.M. (1997). Virtual combinatorial libraries: dynamic generation of molecular and supramolecular diversity by self-assembly. *Proceedings of the National Academy of Sciences of the United States of America* 94 (6): 2106–2110.

44 Taynton, P., Yu, K., Shoemaker, R.K. et al. (2014). Heat- or water-driven malleability in a highly recyclable covalent network polymer. *Advanced Materials* 26 (23): 3938–3942.

45 Wang, S., Ma, S., Li, Q. et al. (2018). Robust, fire-safe, monomer-recovery, highly malleable thermosets from renewable bioresources. *Macromolecules* 51 (20): 8001–8012.

46 Buchwalter, S.L. and Kosbar, L.L. (1996). Cleavable epoxy resins: design for disassembly of a thermoset. *Journal of Polymer Science Part A: Polymer Chemistry* 34 (2): 249–260.

47 Cromwell, O.R., Chung, J., and Guan, Z. (2015). Malleable and self-healing covalent polymer networks through tunable dynamic boronic ester bonds. *Journal of the American Chemical Society* 137 (20): 6492–6495.

48 Zeng, Y., Liu, S., Xu, X. et al. (2020). Fabrication and curing properties of *o*-cresol formaldehyde epoxy resin with reversible cross-links by dynamic boronic ester bonds. *Polymer* 21 (211): 123116.

49 Liu, Q., Jiang, L., Zhao, Y. et al. (2019). Reprocessable and shape memory thermosetting epoxy resins based on silyl ether equilibration. *Macromolecular Chemistry and Physics* 220 (13): 1900149.

50 Wu, X., Yang, X., Yu, R. et al. (2018). A facile access to stiff epoxy vitrimers with excellent mechanical properties via siloxane equilibration. *Journal of Materials Chemistry A* 6 (22): 10184–10188.

51 Chen, B., Liu, X., Liu, J. et al. (2022). Intrinsically self-healing, reprocessable and recyclable epoxy thermosets based on dynamic reversible urea bonds. *Reactive and Functional Polymers* 1 (172): 105184.

52 Lo, J.N., Nutt, S.R., and Williams, T.J. (2018). Recycling benzoxazine–epoxy composites via catalytic oxidation. *ACS Sustainable Chemistry & Engineering* 6 (6): 7227–7231.

53 You, S., Ma, S., Dai, J. et al. (2017). Hexahydro-s-triazine: a trial for acid-degradable epoxy resins with high performance. *ACS Sustainable Chemistry & Engineering* 5 (6): 4683–4689.

54 Liu, T., Zhao, H., Zhang, D. et al. (2021). Ultrafast and high-efficient self-healing epoxy coatings with active multiple hydrogen bonds for corrosion protection. *Corrosion Science* 15 (187): 109485.

55 Schoustra, S.K. and Smulders, M.M.J. (2023). Metal coordination in polyimine covalent adaptable networks for tunable material properties and enhanced creep resistance. *Macromolecular Rapid Communications* 44 (5): 2200790.

56 Hu, Z., Zhang, D., Lu, F. et al. (2018). Multistimuli-responsive intrinsic self-healing epoxy resin constructed by host–guest interactions. *Macromolecules* 51 (14): 5294–5303.

57 Xu, Y., Ma, H., Zhang, H. et al. (2023). A dual dynamic network self-healing bio-based vitrimer and its application in multiply recyclable carbon fiber reinforced polymers. *Industrial Crops and Products* 1 (198): 116755.

58 Xu, X., Ma, S., Feng, H. et al. (2021). Dissociate transfer exchange of tandem dynamic bonds endows covalent adaptable networks with fast reprocessability and high performance. *Polymer Chemistry* 12 (36): 5217–5228.

59 Liu, J., Liu, Y., Wang, Y. et al. (2017). Disulfide bonds and metal-ligand co-crosslinked network with improved mechanical and self-healing properties. *Materials Today Communications* 1 (13): 282–289.

60 Zhao, G., Wu, K., Zhang, Z. et al. (2024). Novel recyclable epoxy resin with releasable residual stress and synergistically enhanced dielectric properties and healing ability. *Polymer* 9 (311): 127569.

61 Rashid, M.A., Liu, W., Wei, Y., and Jiang, Q. (2022). Review of intrinsically recyclable biobased epoxy thermosets enabled by dynamic chemical bonds. *Polymer-Plastics Technology and Materials* 61 (16): 1740–1782.

62 Liu, T., Hao, C., Wang, L. et al. (2017). Eugenol-derived biobased epoxy: shape memory, repairing, and recyclability. *Macromolecules* 50 (21): 8588–8597.

11

Prospects and Innovation of Green Insulating Materials

Jun-Wei Zha[1] and Baoquan Wan[2]

[1] *North China Electric Power University, State Key Laboratory of Alternate Electrical Power System with Renewable Energy Sources, School of Electrical and Electronic Engineering, No.2 Beinong Road, Huilongguan, Changping District, Beijing, 102206, P. R. China*
[2] *Tsinghua University, State Key Laboratory of Power System Operation and Control, Department of Electrical Engineering, No. 30 Shuangqing Road, Haidian District, Beijing, 100084, P. R. China*

11.1 Introduction

The transition toward sustainable energy systems has intensified the need for green electrical insulating materials that are low-carbon, eco-friendly, and high-performance. Traditional electrical insulators, including solid polymeric dielectrics, liquid transformer oils, and gaseous media like SF_6, have long underpinned power and electronic equipment, but they pose environmental and health challenges [1, 2]. Petrochemical-based solids and mineral oils (MOs) are derived from nonrenewable resources and can produce toxic waste. Growing concerns over climate impact, resource depletion, and waste management have driven researchers and industry to explore sustainable, renewable, biodegradable, and nontoxic insulating materials [3, 4]. Electrical equipment manufacturers are now facing new standards for low-carbon and recyclable materials. Many are replacing traditional insulators with more environmentally friendly alternatives. These alternatives offer comparable or better performance as well as inherent ecological benefits.

Insulation materials are extremely important in electrical and electronic systems, ensuring reliability, efficiency, and safety by preventing unnecessary current flow and dielectric breakdown. Insulation materials are ubiquitous, from high-voltage (HV) transmission equipment (transformers, cables, switchgear) to microelectronic devices (printed circuit boards, capacitors, device packaging) [5, 6]. Over the past century, a wide variety of insulation materials have been developed. For example, cross-linked polyethylene (XLPE) for cables, epoxy resin for transformers, mineral oil, and SF_6 for high-voltage equipment. However, many traditional insulators pose environmental issues. Mineral oil leaks can contaminate soil and water and degrade slowly. Thermosetting polymers like epoxy resin are nonrecyclable and contribute to electronic waste. Although SF_6 emissions are small, they have a significant

Next-Generation Electrical Insulating Materials: Low-Carbon, Eco-Friendly, and Sustainable Solutions, First Edition. Edited by Jun-Wei Zha.

impact on the climate. Therefore, current policies are supporting the development of low-carbon, environmentally friendly insulating materials.

Based on this current situation, the research field is actively developing the next generation of green insulating materials. These studies include solid, liquid, and gas insulating materials. The goal is to replace or improve the unsustainability of traditional insulators. In terms of solid dielectrics, innovations include bio-based polymers, recyclable thermosetting networks, and self-healing materials that extend the service life of equipment [7]. In terms of liquid insulation, natural esters (plant oils) have become viable transformer fluids. Advanced nanofluids are being formulated to enhance dielectric strength while maintaining biodegradability [8]. In the realm of gas insulation, some SF_6-free gas mixtures, as well as vacuum or solid insulation alternatives, are being used to significantly reduce global warming potential [9]. Additionally, cross-disciplinary approaches are accelerating the discovery and optimization of materials. For example, dynamic covalent chemistry enables thermosetting polymers to be reprocessed or repaired, and artificial intelligence (AI) tools are being utilized to design novel dielectric materials with targeted characteristics [10, 11].

This chapter provides a comprehensive overview of the prospects and innovations in green insulating materials for electrical and electronic systems. Section 11.2 discusses solid insulating materials, including traditional polymers and emerging environmentally friendly solutions such as bio-derived insulators, dynamic covalent networks, and smart/self-healing polymers. Section 11.3 focuses on liquid dielectrics, covering the evolution from mineral oils and their drawbacks to natural ester fluids, synthetic alternatives, and high-performance nanofluids. Section 11.4 introduces gas insulation, detailing SF_6 issues and the rise of alternative low-GWP (global warming potential) gases and technologies, including vacuum. In Section 11.5, we delve into specific innovative pathways shaping the future of insulation, including bio-based raw materials, recyclable thermosetting materials, self-healing systems, and AI-assisted material design. Throughout, we highlight technical prospects, future outlooks, and emerging applications, including power transformers and cables, switchgear, capacitors, and electronic packaging. This chapter aims to demonstrate how next-generation insulating materials can make power infrastructure not only high-performance and reliable but also low-carbon and environmentally friendly.

11.2 Solid Insulating Materials: Toward Sustainable Dielectrics

11.2.1 Conventional Solid Insulators and Sustainability Challenges

Solid insulating materials are foundational in electrical systems, providing mechanical support and electrical isolation in components such as bushings, insulators, cable insulation, printed circuit boards (PCBs), and electronic device packaging [12, 13]. Petroleum-based polymers have dominated this category due to their excellent electrical properties and moldability. For example, XLPE is widely used in high-voltage cable insulation and transformer winding insulation for its high dielectric strength and favorable thermal behavior. Epoxy resins, often reinforced with fiberglass or fillers, are used in transformer bushings, instrument transformers, and PCB substrates because of their rigidity, adhesion, and thermal stability. Other common polymers include polyvinyl chloride (PVC) for wire insulation, polyesters and phenolics for laminates, and high-temperature materials like polyimide (Kapton) in aerospace and electronics. While these materials have enabled vast electrification, they come with sustainability issues. They are derived from finite fossil resources, often involve energy-intensive manufacturing, and produce hazardous byproducts if burned or improperly disposed. Most conventional polymers (especially thermosets like XLPE and epoxy) are not recyclable or biodegradable, contributing significantly to electrical equipment waste at end-of-life. For instance, millions of tons of XLPE cable scrap and discarded PCB laminates accumulate globally, since the cross-linking that gives XLPE its desirable properties also makes it chemically inert to recycling processes. Moreover, some legacy materials contain toxic additives (e.g. halogenated flame retardants in older PVC or PCB laminates), which pose environmental and health risks.

XLPE, a thermoset polyethylene, cannot be remelted for recycling. The cable manufacturing scraps and retired XLPE cables often end up in landfills or incinerators. In contrast, polypropylene (PP) thermoplastic polymers can be recycled. However, their high crystallinity and melting point complicate processing and make them susceptible to electrical stress. Another major class of solid insulation is cellulose-based materials, such as paper and pressboard, which have been used in transformers and cables for over a century. These are bio-based and biodegradable, made from wood pulp. Kraft paper insulation, when impregnated with oil, has excellent dielectric strength and has been key in transformer insulation systems. The issue with cellulose is not environmental harm. It's quite eco-friendly but it has performance limitations (e.g. sensitivity to moisture, limited thermal class) and issues with long-term aging. Nonetheless, cellulose exemplifies that greener solid insulators are feasible. The goal today is to extend such eco-friendly concepts to polymeric insulators that can meet modern performance demands.

Conventional solid insulating materials have superb electrical and mechanical properties but often fall short on sustainability, being neither renewable nor recyclable [14, 15]. This has prompted researchers to use bio-based or recyclable raw materials while maintaining or improving the performance of insulating materials. The following will discuss key developments, namely bio-based polymers, dynamic covalent (reprocessable) networks, recyclable thermoset alternatives, and self-healing smart materials.

11.2.2 Bio-Based Polymeric Insulators and Composites

One promising approach to green solid insulation is the use of bio-based raw materials in polymer dielectrics. Bio-based polymers are derived partly or wholly from renewable biological resources (plants, microbes, etc.), reducing reliance on petroleum and improving biodegradability [16]. For example, polylactic acid (PLA) and other polyesters are extracted from corn or sugar raw materials. Polyamides extracted from castor oil. Epoxy resins are synthesized from vegetable oils or lignin derivatives, rather than petrochemical bisphenol A. These materials can significantly lower the carbon footprint and toxicity associated with traditional plastics. For instance, recent studies have formulated bio-epoxy resins using components like epoxidized soybean oil, cardanol (cashew nut shell oil), or vanillin (an aromatic compound obtainable from lignin) [17]. Such bio-epoxies can achieve dielectric properties on par with conventional epoxies while being partially renewable. Yang et al. summarized the bio-based epoxy with intrinsic molecular "charge traps" to improve its breakdown strength and low-loss behavior [18]. By engineering the molecular structure (e.g. incorporating polar groups that act as deep traps for charge carriers), they achieved a high breakdown strength and low dielectric dissipation in a plant-derived epoxy system. This indicates that bio-based insulating polymers can meet demanding electrical requirements through smart chemistry.

As shown in Figure 11.1, natural lignocellulosic fibers (e.g. flax, hemp, jute) and cellulose nanofibers are also being explored as fillers or reinforcements in insulating composites [19]. These fibers, derived from plants, are renewable and have excellent mechanical strength. When embedded in a polymer matrix, they can improve mechanical properties and potentially impart insulation benefits. Raza et al. discussed lignocellulose-based insulation materials, noting that thermoset polymer composites reinforced with natural fibers can yield sustainable high-performance insulation for construction and possibly electrical uses [20]. Bio-composites can replace some synthetic fiber composites. For example, a transformer component like an insulating support could be made from a resin reinforced with flax fibers instead of fiberglass, reducing the environmental impact and weight.

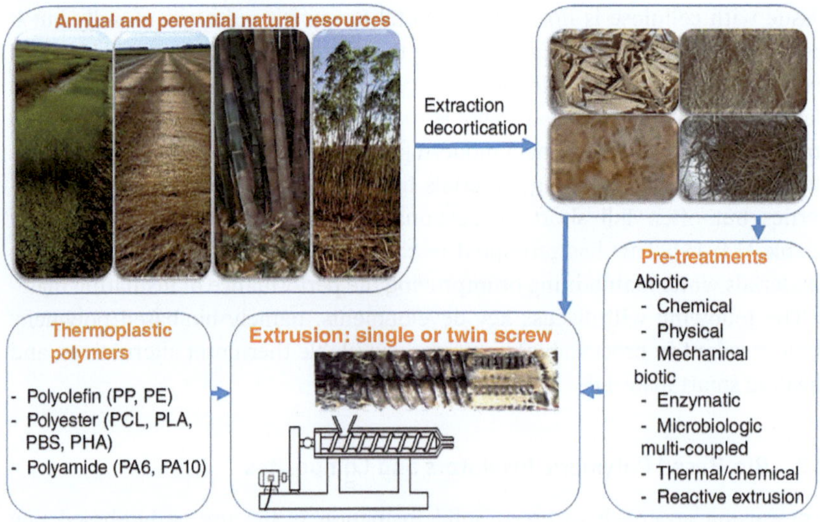

Figure 11.1 Synthetic overview of the preparation and extrusion processes of lignocellulosic (LC) fiber composites. *Source:* [19] / The Royal Society of Chemistry / CC BY-NC 3.0.

Nanocellulose (cellulose nanofibrils extracted from wood pulp) can form thin, strong, and smooth films that serve as dielectric substrates. These films are flexible, have low dielectric loss, and come from renewable sources. Researchers have demonstrated paper-based PCBs and capacitors using cellulose paper or nanocellulose as the insulating substrate, achieving functional devices that are largely biodegradable. Santoso et al. developed a carboxymethyl cellulose (CMC)-based hydrogel dielectric with high permittivity (~78) for low-voltage transistors [21]. The CMC polymer, combined with a natural cross-linker (citric acid) and starch, produced a thin insulating film that is biocompatible, biodegradable, and exhibited excellent capacitance. This demonstrates that even in electronic applications requiring good dielectric performance, bio-based materials like modified cellulose can compete with traditional insulators (like SiO_2 or synthetic polymers) while offering sustainability advantages. As flexible and wearable electronics rise, such green dielectrics that are not only environmentally friendly but also possibly biocompatible have strong appeal.

It is important to note that bio-based does not automatically mean biodegradable or nontoxic. In many cases, it does imply a reduction in ecological impact. For insulating materials, a careful balance must be struck. Biopolymers should meet dielectric strength, permittivity, thermal class, and longevity requirements for the intended application. Using bio-based raw materials in solid insulators addresses the front-end (resource origin) and back-end (end-of-life) sustainability of electrical equipment. These materials draw from renewable feedstocks and are often more amenable to degradation or recycling. Current research is focused on optimizing their electrical properties and stability. As shown in Table 11.1, multiple bio-derived polymers and composites have been proposed to replace conventional insulators, ranging from transformer pressboard (wood-based) to advanced bioplastic dielectrics in electronics [22, 23]. The integration of such materials can significantly reduce the carbon footprint of insulation without compromising performance.

11.2.3 Recyclable Thermosets and Dynamic Covalent Networks

One of the most important innovations in polymer science in recent years has been the development of dynamic covalent networks, commonly referred to as vitrimers. Vitrimers are a class of polymers that behave like traditional thermosetting polymers (permanently cross-linked networks) at operating temperatures, but can undergo bond exchange reactions at high temperatures, allowing the network to reconfigure without losing cross-link density. Essentially, glass polymers can be reprocessed, reshaped, or self-healed after heating, making them recyclable thermosetting materials. This technology directly addresses the recyclability issues of traditional thermosetting insulators such as epoxy resin and XLPE.

Many dynamic covalent chemistry reactions have been explored, such as ester bond exchange, imine exchange (Schiff base bonds), disulfide exchange, urethane exchange, and Diels-Alder reversible links [24]. For example, as shown in

Table 11.1 Solid insulating materials: conventional and green alternatives.

Type	Conventional materials	Green innovations
Thermoset polymers (transformers, bushings, PCBs)	Epoxy resins, polyesters, phenolics (nonrenewable feedstock, cannot be reprocessed or recycled, generate hazardous e-waste)	Bio-epoxies (from plant oils, lignin, e.g. epoxidized soybean oil-based resin); vitrimer epoxies (dynamic bonds enabling recycling); recyclable thermoset composites (using reversible cross-links)
Thermoplastic polymers (cable and wire insulation, device enclosures)	XLPE, polyvinyl chloride (PVC), polyethylene terephthalate (PET), polytetrafluoroethylene (PTFE). (nonrecyclable or degradable, contains chlorine, both from petroleum)	Polypropylene (PP) blends for HV cables; Bio-PET; halogen-free flame-retardant polymers
Natural fiber laminates (structural insulators)	Fiberglass-reinforced plastics (FRP) (nonbiodegradable, limited use)	Cellulose fiber composites; lignocellulosic board; flax or hemp fiber-reinforced resins
Electrical paper and films (capacitors, transformers, electronics)	Kraft paper (cellulose), polyester film (PET), and polypropylene film (petro-based and not biodegradable)	Nanocellulose films; starch-based dielectric films; paper-based printed circuit substrates.
Elastomers and others (cable jackets, sealants)	Petroleum rubber (not eco-friendly to produce)	Natural rubber; bio-based polyurethanes; silicone alternatives

Figure 11.2, Luo et al. designed a novel hardener that combines disulfide bonds and imine bonds through a Schiff base reaction, which was then used to prepare an epoxy vitrimer with dual dynamic covalent bonds [25]. The vitrimer's body maintains high strength and modulus, and its performance recovery rate is high after recycling. This indicates that vitrimers-based insulators can be used in equipment such as casings or insulating gaskets, then melted down and reshaped into new components, thereby greatly reducing waste. The dynamic covalent bond mechanism can be illustrated with the common transesterification chemistry in polyester or epoxy-acid networks. The transesterification exchange occurs when an –OH group on one polymer segment can swap with an –OR group on another, effectively exchanging network linkages [26]. At room temperature, these reactions are negligible, so the material is a stable cross-linked solid. But when heated (>150 °C) in the presence of a catalyst or by inherent bond lability, the exchanges occur, and the network can flow like a very viscous fluid without depolymerizing. This unique behavior grants malleability and healing. Vitrimers thus bridge the

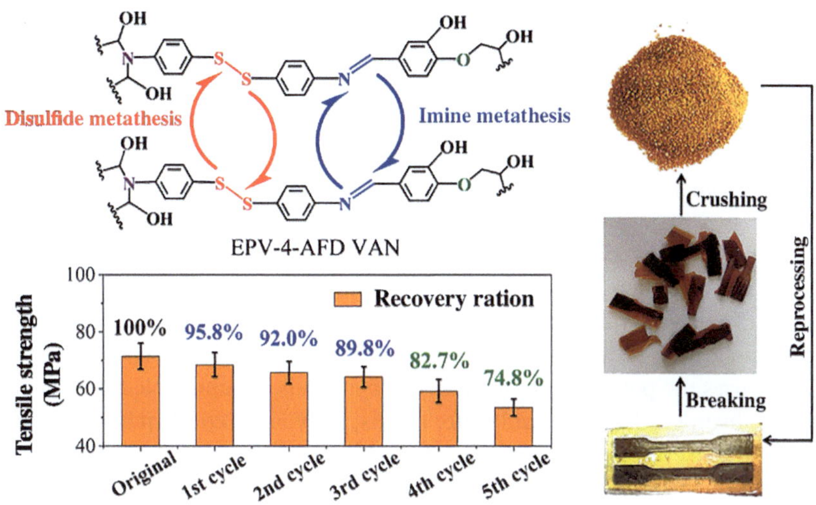

Figure 11.2 Schematic diagram of molecular structure and recovery process of epoxy vitrimer with dual dynamic covalent bonds. *Source:* [25] / with permission of American Chemical Society.

gap between thermosets and thermoplastics, offering the mechanical strength and solvent resistance of thermosets but the reprocessability of thermoplastics.

Dynamic covalent networks in electrical insulation offer multiple advantages:

(1) Recyclability: scrap material (from manufacturing or end-of-life) can be remolded into new shapes or recycled as feedstock, aligning with circular economy goals. (2) Self-healing: if a micro-crack or electrical tree forms, the material can potentially be healed by applying heat, reforming bonds across the crack (with or without external healing agents). (3) Stress relaxation: under high electrical or thermal stress, the network can relieve built-up stress via bond exchange, potentially reducing crack initiation. (4) Longer service life: minor damages need not accumulate irreversibly; maintenance protocols could include in-situ heating to rejuvenate insulation.

Dynamic covalent networks (vitrimers) represent a cutting-edge innovation pathway for green solid insulation. They directly tackle the recyclability challenge of thermoset polymers, enabling a new generation of insulators that are thermoset when in service, but thermoplastic on demand. When combined with bio-based chemistry, the sustainability profile improves even further. We can expect that the application of vitreous base materials will become increasingly widespread in the coming years. Starting with lower voltage or less critical applications, confidence in their long-term stability will eventually lead them to find their way into high-voltage applications.

11.2.4 Smart and Self-Healing Polymeric Insulators

Except for recyclability, researchers aim to imbue insulating materials with **smart functions**. The ability to sense damage, respond to stress, or self-heal after electrical or mechanical injury. Self-healing dielectric polymers are an especially exciting class of materials that can recover their insulating properties after events like electrical treeing or partial discharge, which are common precursors to failure in high-voltage insulation. The goal is to enhance reliability and safety. A self-healing insulator could prevent a minor insulation defect from growing into a catastrophic breakdown.

There are two main approaches to self-healing in polymers, namely extrinsic healing and intrinsic healing. Extrinsic self-healing involves embedding a healing agent (like a liquid resin or monomer) in microcapsules or hollow fibers within the polymer [27]. When a crack or electrical tree forms, it ruptures the capsules, releasing the healing resin, which then polymerizes and fills the crack. Intrinsic self-healing uses the polymer's inherent reversible bonds or supramolecular interactions (like hydrogen bonding, ionic interactions, or dynamic covalent bonds as in vitrimers) to close cracks when stimulated [28].

For electrical insulation, both approaches face unique challenges. Electrical treeing damage typically starts as sub-micron channels created by high-field stress and partial discharges. Traditional microcapsule healing may not readily work for such tiny defects because the capsules are much larger than the initial tree channels. Indeed, early attempts to heal electrical trees in epoxy using microencapsulated monomer and Grubbs' catalyst showed limited success. The tree channels were too fine and didn't reliably rupture capsules until the damage was already extensive. Moreover, adding many microcapsules can degrade the dielectric strength of the material, since they introduce interfaces and possibly conductive impurities. Despite these issues, researchers have creatively adapted extrinsic healing to electrical contexts. Gao et al. reported self-healing polymers composed of microcapsules [29]. As shown in Figure 11.3, these microcapsules can fully restore their dielectric properties, thereby directly responding to the degradation caused by electrostatic trees. The microcapsules they designed have a higher dielectric constant than the polymer matrix, guiding the propagation of electrostatic trees. In situ electroluminescence triggers the cross-linking of healing agents released from the capsules during the electrostatic treeing process of the polymer, thereby repairing electrically damaged areas under environmental conditions. Notably, the fracture strength of the repaired region is significantly enhanced during healing. Integrating self-healing functionality into dielectric materials opens new avenues for smart polymers in electronic and energy applications.

Intrinsic self-healing via dynamic bonds is closely related to the vitrimer concept discussed earlier. A polymer with reversible bonds can heal a cut or crack if given mild heat or pressure to mobilize the chains. For electrical damage, intrinsic healing has shown promise, especially in softer dielectric polymers where chain mobility is higher. For instance, self-healing silicone rubber has been studied for outdoor HV insulators. Silicone insulators suffer from surface discharges and aging. A self-healing silicone formulation with reversible bonds can heal surface degradation, extending service life. Xiang et al. designed a novel silicone rubber containing disulfide bonds and hydrogen bonds by introducing dynamic reversible hydrogen bonds to construct a synergistic dynamic cross-linked

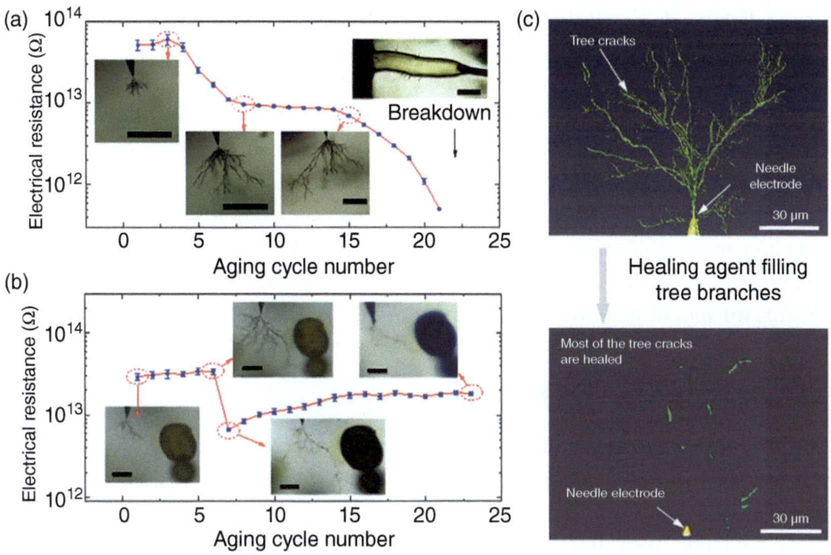

Figure 11.3 (a) A typical degradation curve of the pristine polymer. (b) A typical degradation and healing curve of the self-healing polymer. (c) 3D images of the electrical tree channels and healed tree channels obtained from the 3D CT tests. *Source:* [29] / with permission of Elsevier.

network [30]. The interactions and reversible cleavage-recombination between dynamic hydrogen bonds and disulfide bonds confer the elastomer with excellent mechanical and self-healing properties. The tensile strength of the optimal elastomer reached 1.82 MPa, surpassing that of the original silicone rubber. The self-healing efficiencies for mechanical damage and electrical damage of the elastomer reached 91% and 67%, respectively. This provides a feasible pathway for the commercialization of self-healing silicone rubber, facilitating its application in advanced electrical devices.

The benefits of using self-healing insulation materials in electrical equipment are extended service life and improved safety. If the insulation between transformers and cables heals itself, they can potentially tolerate more partial discharge events, thereby delaying the accumulation of damage that leads to failure. However, achieving self-healing in high-voltage equipment remains an active area of research and development and has not yet been widely commercialized. Challenges include ensuring that the healing mechanism does not compromise the initial insulation quality and verifying that healing can occur in actual service environments. Self-healing insulating materials represent an innovation pathway aimed at reliability and longevity. Whether via embedded microcapsules that release repair agents or via intrinsic bond reformation, these materials can autonomously mitigate damage from electrical stress. This not only enhances equipment lifespan but also reduces maintenance and downtime. The combination of self-healing with the previously discussed recyclable/bio-based approaches could yield insulators that are both sustainable and extremely durable, aligning perfectly with the needs of next-generation electrical grids.

11.3 Liquid Insulating Materials: Green Dielectric Fluids

Liquid dielectrics are critical in many electrical systems because they fulfill dual roles, providing insulation and acting as a coolant. The archetypal example is transformer oil, which insulates high-voltage components inside power transformers while also removing heat from the core and windings. Other uses include oils in high-voltage capacitors, circuit breakers, and as impregnates for paper insulation. This section reviews conventional liquid insulators and the innovations leading to eco-friendly alternatives, focusing on natural esters, synthetic fluids, and advanced nanofluids for sustainable high-voltage insulation.

11.3.1 Mineral Oil and its Limitations

For over a century, mineral oil, a refined petroleum product, has been the dominant insulating liquid in transformers and related apparatus. Mineral oil (MO) possesses a favorable combination of high dielectric strength, good thermal conductivity, and low cost. It can withstand electric fields on the order of tens of kV/mm when properly purified and has served reliably in millions of transformers worldwide. However, mineral oil comes with significant environmental drawbacks that motivate the search for alternatives. As discussed earlier, MO is toxic and nonbiodegradable. Spills can persist in soil and water for long periods, and even a small leak can contaminate vast quantities of water. Additionally, mineral oil has a relatively low fire point, which poses fire hazards. If a transformer ruptures or overheats, MO can ignite and its combustion produces toxic smoke. This is a safety concern, especially for transformers located indoors or in sensitive areas. From a sustainability perspective, MO is a fossil resource. There is concern that petroleum reserves are finite and that relying on them for specialty products like transformer oil is not sustainable in the long term. Some analysts suggest that high-quality crude stocks for such oils may diminish in a matter of decades. Moreover, the refining of mineral oil has a carbon footprint, and the disposal of used oil at the end of its life involves energy and chemicals.

Several alternatives emerged, including silicone oils (polydimethylsiloxane fluids) introduced in the 1970s for specialized applications. They are fire-resistant but much more expensive, and they too are not biodegradable. Chlorinated hydrocarbons (like the infamous PCBs, polychlorinated biphenyls) were used mid-20th century for their excellent stability and nonflammability, but were banned in the 1970s due to extreme toxicity and persistence. By the late 20th century, attention returned to natural products. Specifically, vegetable oils as promising insulating fluids.

11.3.2 Natural Ester Oils (Vegetable-Based Insulating Fluids)

Vegetable oils (also called natural esters (NE)) have emerged as the leading eco-friendly alternative to mineral oil in transformers [31, 32]. These oils are derived from plant seeds (e.g. soybeans, rapeseed/canola, sunflower, coconut, palm kernel) and consist primarily of triglycerides (esters of glycerol and fatty acids).

Natural ester fluids offer several environmental and safety advantages. They are highly biodegradable, typically more than 95% biodegradable in standard tests. They are made from renewable crops and have very high flash and fire points, making them far less prone to igniting. Moreover, they are generally nontoxic to flora and fauna.

From a performance standpoint, natural ester oils have comparable dielectric strength to mineral oil and actually can absorb more moisture while still maintaining good insulation. Natural esters also have higher viscosity than mineral oil, which can slightly affect heat transfer and filling in cold climates, but formulations and proper design mitigate these issues. The main technical challenge for vegetable oils has been oxidation stability. Unsaturated fatty acid chains in vegetable oils can oxidize to form acids and sludge, especially at elevated temperatures and in the presence of oxygen. Mineral oil, being fully saturated hydrocarbons, is inherently more oxidation stable. To address this, natural ester fluids use high-oleic oils, additives, and often hermetically sealed transformer designs. Industry standards (like IEEE C57.147 and IEC 62770) govern the use of natural esters and specify tests for oxidation, and modern natural ester fluids have demonstrated adequate stability for typical transformer operating temperatures. Nonetheless, oxidation stability remains a limiting factor for using natural esters in high-temperature or very large transformers without additional care. Research continues to improve this via partial hydrogenation of oils, blending with synthetic esters (SE), or even adding nanofillers as oxidation inhibitors.

Given these advantages, the use of natural ester insulating oils is forecast to grow, especially as the industry seeks to reduce the environmental footprint of the electric grid. They are not a panacea, large power transformers (500 kV class) still mostly use mineral oil due to the stringent requirements and lack of long-term track record of esters in such units. But even there, a few high-voltage transformers (up to 245 kV) filled with natural ester have been successfully put in service in recent years as demonstration units, indicating confidence is increasing.

11.3.3 Synthetic Esters and Other Alternatives

Alongside natural esters, synthetic ester fluids form the other major category of alternative insulating liquids. Synthetic esters (SE) are man-made molecules, typically derived from polyol alcohols (like pentaerythritol or trimethylolpropane) fully esterified with fatty acids. An example is pentaerythritol tetrahexanoate, a synthetic ester fluid used under brand names like Midel® 7131. Synthetic esters share many of the biodegradability and fire-safety advantages of natural esters. They are often more than 90% biodegradable and have high fire points. Because they are synthesized with specific fatty acids and tailored structures, synthetic esters generally have better oxidation stability and low-temperature properties than unmodified natural esters. Midel 7131, for instance, remains fluid at lower temperatures than some vegetable oils and can be used in colder climates without pour point depressants.

Apart from esters, other alternative liquids include silicone oil, hydrocarbon blends, and novel fluorinated. Silicone oil is excellent in high temperature stability and was historically used in furnace transformers and aircraft transformers due to its inertness and fire resistance. It is also biodegradable to some extent, but its high cost and tendency to dissolve more gases are drawbacks. Halogenated hydrocarbons are mostly phased out or avoided due to toxicity and environmental persistence, so they are not considered "green." However, one interesting category is perfluorocarbon fluids, which are extremely inert and were sometimes used in specialized capacitors or electronic cooling. These are not environmentally friendly either, because they have very high GWPs and persist indefinitely. A newer concept in development is the use of blends or nano-enhanced fluids. For example, mixing a natural ester with some synthetic ester can yield a fluid that combines the best attributes of both. These blends can be tuned to meet various standards. Some patents in recent years cover such mixtures.

Nanofluids represent a technical innovation aimed at improving dielectric and thermal performance. Adding nanoparticles like TiO_2, Al_2O_3, BN, or SiO_2 to oils in small concentrations (≤1%) has been shown to increase breakdown voltage and suppress streamer propagation. The particles act as electron scavengers or field distortion agents, raising the inception voltage for discharges. A green variant of this is using environmentally friendly base fluids (e.g. vegetable oil) with nanoparticles made from benign materials. As shown in Figure 11.4, Siddique et al. formulated a vegetable oil-based nanofluid with SiO_2 nanoparticles that significantly outperformed plain mineral oil in dielectric tests [33]. By combining castor, flaxseed, and blackseed oils as a hybrid base and adding SiO_2 particles, they achieved higher relative permittivity and lower dissipation factor than mineral oil. This underscores that beyond simply replacing oil, we can enhance natural fluids to meet or exceed

(a)

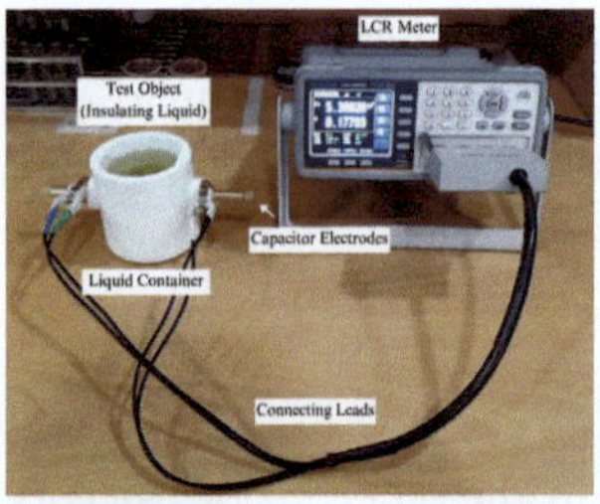

Nanofluids	Relative permittivity (ε_r)
NF 1	2.1219
NF 2	2.2110
NF 3	2.3092
NF 4 (OS)	2.5277
NF 4 (SG)	2.6994
NF 4 (GS)	2.8844
NF 5	1.9949
MO	1.7920

(b)

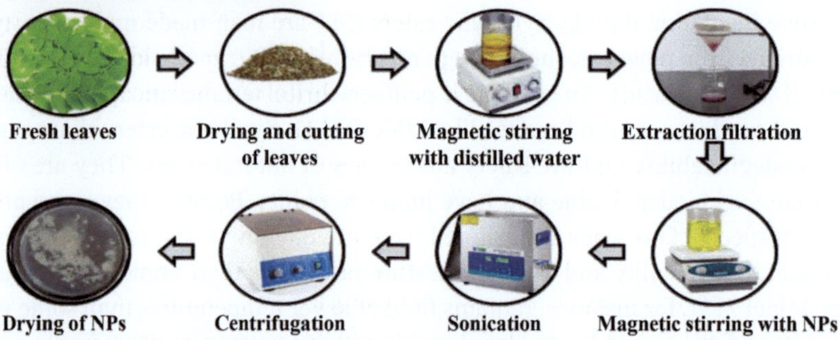

Figure 11.4 (a) Experimental setup for evaluation of relative permittivity and values. (b) Green synthesis method for preparing nanoparticles. *Source:* [33] / MDPI / CC BY 4.0.

conventional performance. Moreover, they introduced a green synthesis method for preparing the nanofluid, aligning the production process with the eco-friendly theme.

Green liquid insulating materials have transitioned from experimental options to proven solutions for a range of transformer and electrical equipment applications. Natural esters, in particular, stand out as a successful innovation marrying technical performance with sustainability. With continued improvements in formulation and accumulating service experience, it is conceivable that by the mid-21st century, biodegradable insulating fluids could become the default choice, relegating mineral oil to niche or legacy status. This would greatly mitigate the environmental risks associated with the vast oil-filled infrastructure of the electric power industry.

11.4 Gaseous Insulating Materials: Toward SF_6-Free Solutions

Gaseous dielectrics are employed in power systems predominantly in gas-insulated switchgear (GIS), high-voltage circuit breakers, gas-insulated transmission lines, and similar equipment. The function of the gas is to provide electrical insulation (and in breakers, arc quenching) in a compact apparatus. For the last several decades, sulfur hexafluoride (SF_6) has been the unrivaled choice for these applications due to its exceptional dielectric strength and arc-quenching properties. However, SF_6 has a dark side. It is the most potent greenhouse gas known, with a GWP ~23,500 over 100 years and an atmospheric lifetime of ~3200 years. Even small leakage rates accumulate to a significant climate impact. Therefore, it is necessary to develop and promote new types of green insulating gases. This section explores the state-of-the-art in SF_6 alternatives, including new fluorinated low-GWP gases, gas mixtures, vacuum technologies, and solid insulation approaches.

11.4.1 The Challenge of Replacing SF_6

Replacing SF_6 is nontrivial because few substances match its unique combination of properties. An ideal GIS insulating gas needs high dielectric strength, thermal stability, chemical inertness, nonflammability, low toxicity to humans, a suitable boiling point, and environmental acceptability. SF_6 excelled in all but the last. The same chemical inertness that makes SF_6 ideal inside equipment is what makes it a persistent greenhouse gas (GHG), it does not break down in the atmosphere.

Any alternative gas will likely have to compromise on one or more attributes. The approach has been to use gas mixtures. By mixing a new insulating gas with a buffer gas (typically air, nitrogen, or CO_2), one can tailor the overall properties. The bulk buffer gas can manage aspects like pressure/boiling behavior while reducing costs, while the minority component provides dielectric strength.

11.4.2 New Fluorinated Gases: Fluoronitriles and Fluoroketones (Air Mixes)

The most prominent SF_6 alternatives today are the proprietary 3 M™ Novec™ insulating gases, notably Novec 4710 (a fluoronitrile, chemical name perfluoro (2-methyl-3-pentanone nitrile), C4F7N) and Novec 5110 (a fluoroketone, chemical name perfluoro-2-butyl-3-oxazolidinone, often simplified as C5F10O) [34]. These are developed specifically to replace SF_6 in GIS applications. Both of these compounds have significantly lower GWP than SF_6. They also have much shorter atmospheric lifetimes because they are not completely inert. These gases are used not in pure form, but in mixtures. The performance of these mixtures has been found to be satisfactory in medium-voltage and even high-voltage applications. Owens et al. reviewed that gas mixtures with Novec gases can match the dielectric performance of SF_6 systems at similar ratings [34]. For instance, 170 kV GIS and 72.5 kV circuit breakers have been designed and put into service with fluoronitrile/CO_2 mixtures or fluoroketone/air mixtures, delivering insulation and interruption capabilities on par with SF_6 designs.

These new gases are currently more expensive than SF_6. However, the volume of gas needed in equipment is small, and since regulations are forcing the elimination of SF_6, economies of scale will likely bring costs down. The cost factor is not expected to be a barrier in the long run, especially when weighed against the external cost of SF_6 emissions.

11.4.3 Vacuum Interruption and Solid Insulation – Beyond Gas

While new gases are one route, another is to avoid gases altogether where possible. Medium-voltage switchgear is a domain where SF_6 has been used for insulation and breaking, but an alternative is vacuum and solid insulation. Vacuum interrupters are already universally used for medium voltage (MV) circuit breakers (up to ~52 kV class) to break currents. They extinguish arcs without any gas. These vacuum interrupters would still be housed in SF_6 gas for insulation external to the interrupter. Now, designs are moving to "dry air" or solid dielectric insulation around vacuum interrupters. For example, some MV switchgears use epoxy or silicone encapsulation around the interrupter, along with normal air or nitrogen at atmospheric pressure in the tank (or slight pressurization). These arrangements eliminate SF_6 entirely. For insulation performance, one might use a combination of cast resin and air. At transmission voltages (>100 kV), vacuum technology is also being considered for breakers.

There is also research into entirely new gas chemistries beyond the 3 M Novec ones. One direction is hydrofluoroolefins (HFOs). These are short-chain fluorinated alkenes, which are currently used as refrigerants. Some have dielectric properties that might be useful, though generally their dielectric strength is lower than SF_6 and they are often flammable. Another approach is to use nitrogen or carbon dioxide with electronegative additives that are not fluorine-based, for example, a small amount of a rare gas or an organic inhibitor. Xenon actually has dielectric strength comparable to SF_6, is inert and GWP-free, but it's exceedingly expensive and scarce.

CO_2 and N_2 by themselves can be used for insulation if one accepts a performance reduction. They have no GWP issues. Some equipment manufacturers offer CO_2-based GIS for certain classes. These typically require much higher pressure and possibly lower ratings. The advantage is simplicity and environmental safety.

11.4.4 Implementation Status and Future Outlook

The key to gaining widespread acceptance is to demonstrate that its reliability and safety are equivalent to SF_6. SF_6 is a well-known, highly reliable technology. Over time, new alternatives must demonstrate a similar low failure rate. Standardization is another aspect. For example, how to name and classify gas mixtures, how to test the purity or moisture content of fluorocarbon/carbon dioxide mixtures. It is worth highlighting that the elimination of SF_6 greatly reduces the carbon footprint of GIS equipment. A life-cycle assessment cited by Owens et al. shows that using fluoronitrile or fluoroketone mixtures can cut greenhouse gas emissions by more than 99% over the equipment life [34]. Figure 11.5 qualitatively illustrates dielectric strength comparisons for a few gases and gas mixtures relative to SF_6. Pure Novec 4710 and 5110 actually have higher dielectric breakdown voltage than SF_6 at 1 atm, but since they must be diluted for practical use, the mixtures achieve somewhat lower strength.

The prospects for green gaseous insulation are very positive. Through a combination of new gases, mixtures, and alternate technologies like vacuum, the power industry is on the cusp of a major shift away from SF_6. This will drastically reduce the GHG emissions associated with electrical equipment. Over the next decade, we can expect SF_6 usage to peak and then rapidly decline as new installations favor sustainable options.

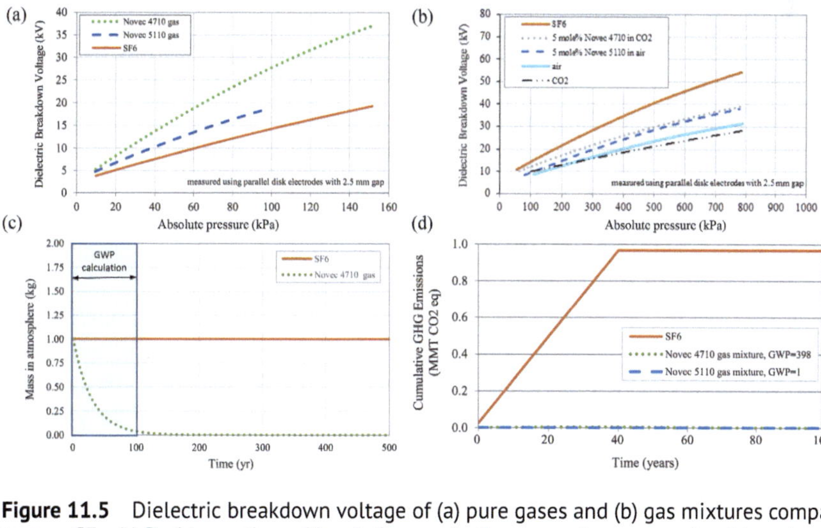

Figure 11.5 Dielectric breakdown voltage of (a) pure gases and (b) gas mixtures compared to pure SF_6. (c) Residence time of insulating gas in the atmosphere, assuming 1 kg release of each compound at time zero. (d) Cumulative greenhouse gas emissions, assuming emission equivalent to 1 T/yr. of SF_6 for 40 years of operation. *Source:* [34] / MDPI / CC BY 4.0.

11.5 Innovation Pathways

Discuss by emphasizing cross-disciplinary innovation approaches for green insulation technologies that will shape the future. These include the use of bio-based materials, dynamic covalent (recyclable) networks, recyclable thermosets, smart/self-healing systems, and AI-designed materials. Each of these pathways offers a toolkit of techniques and concepts that can be applied to multiple types of insulators [35–49]. For example, bio-based approaches are relevant to solid polymers and liquids. Dynamic covalent chemistry can make both solid components and even gels recyclable. AI design can optimize anything from a polymer's molecular structure to a fluid's composition. It is also necessary to consider how to integrate these innovations into actual electrical and electronic applications, and to discuss global research and development trends indicating how insulation technology will develop in the future.

11.5.1 Bio-Based and Renewable Raw Materials

The drive to incorporate bio-based raw materials into insulating media is motivated by the desire to reduce reliance on fossil resources and improve end-of-life outcomes. Natural ester transformer oils are replacing mineral oil, cellulose (paper) continues to serve as a renewable solid insulator, and emerging biopolymers are being used in electronics. Moving forward, this trend will deepen in several ways:

1) Research will yield bio-based polymers that match or exceed the electrical and thermal performance of petrol-polymers. For instance, bio-derived thermoplastic polyesters or polycarbonates could be tailored for dielectric uses. Lignin-based epoxy resins are being explored. Lignin is rich in aromatic rings and can potentially impart high dielectric strength and thermal stability when converted into resin components. Early experiments show that incorporating lignin derivatives can also reduce the flammability of the resin.
2) Instead of traditional inorganic fillers, we may see more natural-origin fillers. Cellulose nanocrystals (CNC) or nano-clays from biomass could reinforce polymers. These can improve mechanical strength and possibly act as dielectrics themselves. For example, a composite of polylactic acid (PLA) with cellulose fibers might yield a biodegradable insulation with decent mechanical strength for use in low-voltage devices or as cable sheathing.
3) We can expect industry standards to begin recognizing biocontent or biodegradability as a desirable attribute. Possibly, a future IEC standard might specify a class of insulating material as "Bio-based Class 1" if more than 50% of the carbon is renewable, among the criteria. Government procurement could encourage this.

The convergence of performance and sustainability is key. For widespread adoption, bio-based insulators must not ask engineers to trade off reliability or efficiency for greenness. Perhaps a new bio-derived thermoset will emerge that offers superior tracking resistance and a higher dielectric strength than a classic epoxy, while also being recyclable or bio-sourced, making it a win-win.

11.5.2 Dynamic Covalent Networks and Recyclable Thermosets

Dynamic covalent chemistry (vitrimers) is discussed as a means to create recyclable thermoset insulators. This is an innovation pathway with huge implications for the life cycle of electrical equipment. If all retired insulation materials from transformers, cables, and capacitors were not disposed of as landfill waste but were systematically recycled or reprocessed into new equipment or other products, the circular economy could truly take root in the power industry.

Vitrimer-based cable insulation, where after 30 years in service, the insulation can be removed, melted, filtered, and re-extruded onto new conductors for another life. Already, research on polypropylene blends (not exactly vitrimer but recyclable) for high-voltage direct current (HVDC) cables indicates the feasibility of recycling cable insulation multiple times with minimal performance loss. Extending this to cross-linked systems via vitrimers is plausible. One could incorporate a small amount of dynamic cross-linker in polyethylene to create a "vitrimer-like" thermoplastic that is easier to recycle. For transformer solid insulation, using a vitrimer epoxy means those parts can be reformed or repaired. If a bushing develops a crack, one could imagine heating the area to weld the crack shut. On a manufacturing level, scraps from casting processes could be melted and reused, improving factory efficiency and reducing waste. If materials are recyclable, components might be designed to be more easily disassembled. People can use vitrimer adhesives instead of permanently bonding and curing everything. These adhesives can be decoupled at high temperatures. This allows metal parts, plastic parts, and other materials to be separated at the end of their service life. The polymers can be reprocessed. The combination of dynamic covalent networks with self-healing also stands out. A material that can rearrange bonds can often also heal if conditions allow. This dual function could yield insulators that are not only recyclable at the end of their life but can also self-repair during their life. That might justify slightly higher material costs because of the extended lifetime and recoverability.

Dynamic covalent network technology is expected to transition from lab to industry within this decade. We may soon find that new classes of insulating components are advertised as "fully recyclable." The chemistry is maturing, and as manufacturing techniques adapt, we'll see uptake in niche applications first, then expansion into high voltage (HV) once thoroughly proven.

11.5.3 Smart and Self-Healing Insulation Systems

The concept of smart insulation includes materials that can adapt, respond, or communicate their state. Self-healing is one aspect. Self-healing polymers may change color when a crack occurs. Research is also being conducted on "self-sensing" insulators. By adding a small amount of carbon nanotubes to polymers, the impedance of the polymer changes when microcracks form, acting as a distributed sensor. The insulator can self-heal while alerting the system to the occurrence of damage, enabling maintenance teams to decide whether the component needs to be replaced or can be retained. Artificial intelligence (AI) can also play a role in operations. AI algorithms in monitoring systems can interpret data from smart insulators to predict faults or optimize performance. This is more about using AI in operations than in material design, but it completes the picture of an active smart grid. In terms of materials, self-healing may first find specific high-value use cases. One possibility is high-voltage DC cable terminals. These are stress insulation points where failure is costly. Self-healing silicone or epoxy can repair small surface discharges before they propagate, thereby preventing costly downtime. Another possibility is large rotating machines (generators, motors), where stator coil insulation can benefit from self-healing of partial discharge pits. Also, coatings for outdoor insulation (like hydrophobic, self-healing coatings on ceramic insulators or polymer) are being developed. A coating that heals erosion from ultraviolet (UV) or electrical activity can significantly prolong the life of outdoor insulation.

11.5.4 AI-Driven Materials Design and Discovery

Artificial Intelligence and Machine Learning are becoming invaluable in materials discovery, especially given the complex optimization problems presented by insulating materials [50, 51], as shown in Figure 11.6. Designing an insulator means balancing electrical, mechanical, chemical, thermal, and economic factors. Traditional trial-and-error or even classical computational approaches can be slow when the design space is huge. AI shines in finding patterns in large data and suggesting candidates that meet multiple criteria. People can envision using deep learning models to predict a polymer's dielectric constant, breakdown strength, and even aging rate from its molecular structure. There are efforts like the Polymer

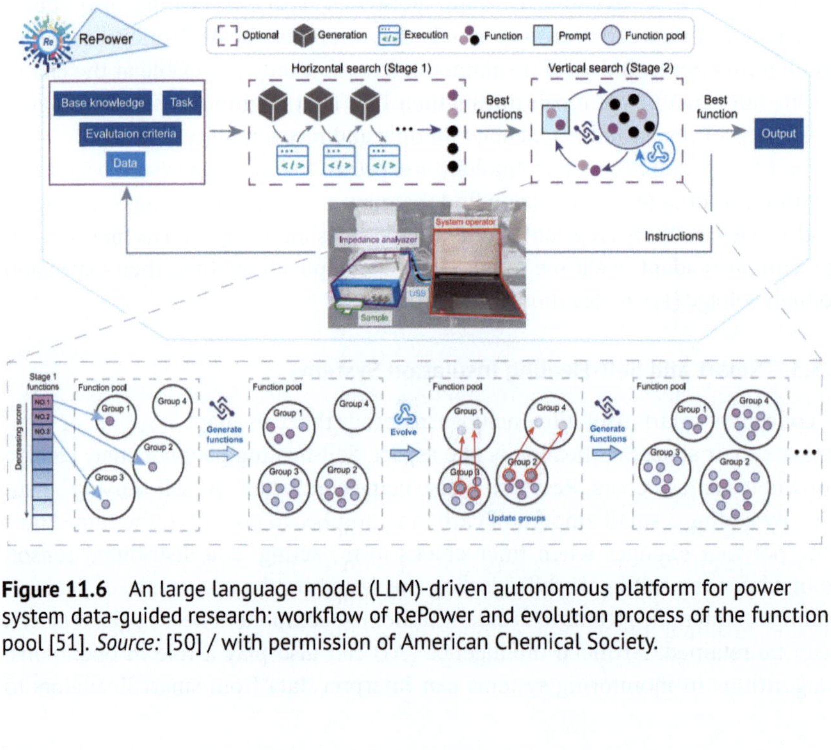

Figure 11.6 An large language model (LLM)-driven autonomous platform for power system data-guided research: workflow of RePower and evolution process of the function pool [51]. *Source:* [50] / with permission of American Chemical Society.

Genome Project that aim to use machine learning (ML) on big polymer property datasets to allow inverse design.

For gaseous insulation, AI could help identify new molecules or mixtures. In liquids, a similar approach can be used, employing AI to propose antioxidant additives or nanofiller types and concentrations to optimize fluid performance while maintaining biodegradability. Bayesian optimization algorithms could be employed in lab experiments to efficiently converge on optimal formulations. Beyond designing the material itself, AI can help in processing optimization. For example, predicting the curing profile for a vitrimer to achieve the best network structure, or forecasting how a certain insulation system will age under different stresses.

The infusion of AI into material science accelerates innovation and opens up vast combinatorial spaces that were previously impractical to explore. This means the next decade could bring forth completely new insulating compounds that a lone chemist might not have thought of, but an algorithm did, subsequently verified by experiments. Such materials might break the traditional performance limits.

11.6 Conclusion and Perspectives

The electrical power and electronics industries are undergoing a paradigm shift in how insulating materials are conceived, used, and managed. Traditionally, insulation is chosen purely for technical and economic reasons, with little regard for environmental impact. Today, with urgent climate and sustainability challenges, insulating materials are being reimagined through a green and sustainable lens. In this chapter, we explored the broad prospects for next-generation insulating materials. This covers solid, liquid, and gas media, with a focus on innovations that promise to make insulating materials low-carbon, environmentally friendly, and sustainable. At the same time, they enhance or at least maintain the high performance expected in electrical applications.

Solid insulators are increasingly utilizing bio-based content to reduce their environmental footprint. Breakthroughs in dynamic covalent chemistry have made it possible to create recyclable thermosets, which can address the long-standing issue of cross-linked insulator disposal. These materials can be repaired and reprocessed, marking a shift toward the circular use of materials in high-voltage equipment and electronics. The introduction of the concept of self-healing insulation, which allows materials to autonomously recover from electrical or mechanical damage, has the potential to revolutionize maintenance. Although still in the experimental stage, self-healing dielectrics have demonstrated the potential to extend equipment lifespan and prevent catastrophic failures.

In liquid insulation, replacing mineral oil with natural ester fluids is a prime example of green innovation already in commercial practice. This enhances safety and sustainability without compromising functionality. The development of high-performance plant-based nanofluids further demonstrates that the dielectric properties of environmentally friendly fluids can even surpass those of traditional oils. Synthetic esters and new formulations offer flexible options, all of which are biodegradable and fire-resistant. The trend in liquid insulation is clearly moving toward environmentally harmless solutions. This represents a significant shift for industries where oil spills and fires were once considered major risks.

For gas insulation, SF_6, once considered indispensable, is gradually being phased out. Thanks to intensive research and development, viable alternatives such as fluorocarbon-carbon dioxide mixtures and fluoroketone-air mixtures have been identified. These are now in operation in actual power grid equipment, reducing greenhouse gas emissions by over 99%. Additionally, the adoption of vacuum and solid insulation technologies as alternatives to gas, where feasible, reflects a trend toward simplification and the elimination of reliance on any greenhouse gases. In summary, the field of gas insulation is at the forefront of a major environmental victory, achieving these advancements without compromising the reliability or compactness of electrical equipment.

Throughout these developments, bio-based materials help reduce carbon footprints and are often biodegradable. Dynamic covalent networks enable recyclability and longevity. Artificial intelligence is accelerating the pace of discovery and optimization, ensuring that new materials can meet complex multi-physics electrical insulation requirements in a fraction of the time required by traditional methods. As these avenues converge, we expect insulation systems to become not only more environmentally friendly, but also smarter and more robust.

Looking ahead, we can expect standardization and policy to further drive these innovations. Technical standards will evolve to include testing and certification of environmentally friendly insulation materials. Utilities and electronics manufacturers may adopt procurement policies that favor sustainable materials, creating market appeal. Governments may implement regulations to encourage or mandate the use of more environmentally friendly materials.

There are still many opportunities and challenges for researchers and engineers. Developing high-voltage biodegradable solid insulators with tracking resistance. Ensuring the long-term stability of natural ester fluids under extreme conditions is crucial. Economically scaling up the manufacture of glass polymers. Integrating self-healing materials into large systems. Continuing to push the boundaries of material performance through sustainable choices. Interdisciplinary collaboration between electrical engineers, chemists, materials scientists, and data scientists is critical to solving these complex problems.

The future of green insulation materials is very bright. The trajectory of innovation shows that sustainable insulation does not come at the expense of performance. On the contrary, it will enhance the resilience and safety of power systems while fulfilling our collective responsibility to the environment. Insulation materials are typically invisible components that play an important role in the sustainability narrative of modern technology. Through creativity and collective effort, sustainable insulation materials will become a shining example of the next generation of low-carbon and eco-friendly electrical engineering.

References

1 Berardi, U. and Nosrati, R.H. (2018). Long-term thermal conductivity of aerogel-enhanced insulating materials under different laboratory aging conditions. *Energy* 147: 1188–1202.

2 Pernebayeva, D., Irmanova, A., Sadykova, D. et al. (2019). High voltage outdoor insulator surface condition evaluation using aerial insulator images. *High Voltage* 4 (3): 178–185.

3 Wan, B., Dong, X., Yang, X. et al. (2023). Rising of dynamic polyimide materials: a versatile dielectric for electrical and electronic applications. *Advanced Materials* 35 (39): 2301185.

4 Yang, X., Huang, W., Dong, H., and Zha, J.W. (2025). Smart polydimethylsiloxane materials: versatility for electrical and electronic devices applications. *Advanced Materials* 37 (17): e2500472.

5 Aditya, L., Mahlia, T.M.I., Rismanchi, B. et al. (2017). A review on insulation materials for energy conservation in buildings. *Renewable and Sustainable Energy Reviews* 73: 1352–1365.

6 Sevindir, M.K., Demir, H., Ağra, Ö. et al. (2017). Modelling the optimum distribution of insulation material. *Renewable Energy* 113: 74–84.

7 Wan, B. and Zha, J.W. (2023). Dynamic polyimide dielectric film: a concept and application perspective. *Advanced Functional Materials* 34 (13): 2312829.

8 Kulandaivel, S., Keng, N.W., Samykano, M. et al. (2025). Mxene nanofluids in advanced applications: an in-depth review of thermophysical characteristics and technological innovations. *Materials Today Sustainability* 29: 101084.

9 Khan, B., Saleem, J., Khan, F. et al. (2019). Analysis of the dielectric properties of R410a gas as an alternative to SF_6 for high-voltage applications. *High Voltage* 4 (1): 41–48.

10 Wan, B., Yang, X., Dong, X. et al. (2022). Dynamic sustainable polyimide film combining hardness with softness via a "mimosa-like" bionic strategy. *Advanced Materials* 35 (2): 2207451.

11 Liu, R.-L., Wang, J., Shen, Z.-H., and Shen, Y. (2024). AI for dielectric capacitors. *Energy Storage Materials* 71: 103612.

12 Rozga, P. and Beroual, A. (2021). High voltage insulating materials—current state and prospects. *Energies* 14 (13): 3799.

13 Zhang, B., Xue, J., Chen, X. et al. (2021). Review of surface transient charge measurement on solid insulating materials via the pockels technique. *High Voltage* 6 (4): 608–624.

14 Wan, B., Xiao, M., Dong, X. et al. (2023). Dynamic covalent adaptable polyimide hybrid dielectric films with superior recyclability. *Advanced Materials* 36 (52): 2304175.

15 Zhu, L., Islam, A., Alinejad, F. et al. (2024). Highly silanized cellulose biocomposites for sustainable insulation materials. *Materials Today* 81: 95–103.

16 Khoukhi, M., Dar Saleh, A., Mohammad, A.F. et al. (2022). Thermal performance and statistical analysis of a new bio-based insulation material produced using grain puffing technique. *Construction and Building Materials* 345: 128311.

17 Jiang, Y., Li, J., Li, D. et al. (2024). Bio-based hyperbranched epoxy resins: synthesis and recycling. *Chemical Society Reviews* 53 (2): 624–655.
18 Yang, Y., Dang, Z.M., Li, Q., and He, J. (2020). Self-healing of electrical damage in polymers. *Advanced Science* 7 (21): 2002131.
19 Gallos, A., Paës, G., Allais, F., and Beaugrand, J. (2017). Lignocellulosic fibers: a critical review of the extrusion process for enhancement of the properties of natural fiber composites. *RSC Advances* 7 (55): 34638–34654.
20 Raza, M., Farhan, A., and Abu-Jdayil, B. (2024). Lignocellulose–based insulation materials: a review of sustainable and biodegradable solutions for energy efficiency. *International Journal of Thermofluids* 24: 100844.
21 Santoso, L.L., Prakoso, S.P., Bui, H.K. et al. (2024). A green high-k dielectric from modified carboxymethyl cellulose-based with dextrin. *Macromolecular Rapid Communications* 45 (12): e2400059.
22 Lee, C., Bae, J., Noh, Y. et al. (2022). Automatic pressure gelation analysis for insulating spacer of gas insulated switchgear manufactured by bio-based epoxy composite. *Applied Sciences* 12 (20): 10202.
23 Ranefjard, O., Strandberg-de Bruijn, P.B., and Wadso, L. (2024). Hygrothermal properties and performance of bio-based insulation materials locally sourced in sweden. *Materials* 17 (9): 2021.
24 Orrillo, A.G. and Furlan, R.L.E. (2022). Sulfur in dynamic covalent chemistry. *Angewandte Chemie International Edition* 61 (26): e202201168.
25 Luo, C., Wang, W., Yang, W. et al. (2023). High-strength and multi-recyclable epoxy vitrimer containing dual-dynamic covalent bonds based on the disulfide and imine bond metathesis. *ACS Sustainable Chemistry & Engineering* 11 (39): 14591–14600.
26 Gablier, A., Saed, M.O., and Terentjev, E.M. (2020). Transesterification in epoxy–thiol exchangeable liquid crystalline elastomers. *Macromolecules* 53 (19): 8642–8649.
27 Wang, S. and Urban, M.W. (2020). Self-healing polymers. *Nature Reviews Materials* 5 (8): 562–583.
28 Zhou, Y., Li, L., Han, Z. et al. (2022). Self-healing polymers for electronics and energy devices. *Chemical Reviews* 123 (2): 558–612.
29 Gao, L., Yang, Y., Xie, J. et al. (2020). Autonomous self-healing of electrical degradation in dielectric polymers using in situ electroluminescence. *Matter* 2 (2): 451–463.
30 Xiang, P., Wan, B., Huang, W. et al. (2025). Synergistic dual effect dynamic network regulation of self-healing silicone rubber composites for outdoor insulation. *Composites Communications* 56: 102418.
31 Raeisian, L., Niazmand, H., Ebrahimnia-Bajestan, E., and Werle, P. (2019). Feasibility study of waste vegetable oil as an alternative cooling medium in transformers. *Applied Thermal Engineering* 151: 308–317.
32 Siddique, A., Yaqoob, M., Aslam, W. et al. (2024). A systematic review on promising development of cost-effective, biodegradable, and environment friendly vegetable based nanofluids as a future resource for green transformer insulation oil. *Journal of Molecular Liquids* 403: 124836.

33 Siddique, A., Shahid, M.U., Aslam, W. et al. (2025). Sustainable insulating materials for high-voltage equipment: dielectric properties of green synthesis-based nanofluids from vegetable oils. *Sustainability* 17 (4): 1740.

34 Owens, J., Xiao, A., Bonk, J. et al. (2021). Recent development of two alternative gases to SF_6 for high voltage electrical power applications. *Energies* 14 (16): 5051.

35 Lap, T., Goujon, N., Mantione, D. et al. (2023). Bio-based polyhydroxyanthraquinones as high-voltage organic electrode materials for batteries. *ACS Applied Polymer Materials* 5 (11): 9128–9137.

36 Zhou, X., Fu, F., Shen, M. et al. (2023). Research progress of bio-based polybenzoxazine materials. *Advanced Sustainable Systems* 8 (2): 2300372.

37 Deng, Y., Zhang, Q., Qu, D.H. et al. (2022). A chemically recyclable crosslinked polymer network enabled by orthogonal dynamic covalent chemistry. *Angewandte Chemie International Edition* 61 (39): e202209100.

38 Lin, Y., Chen, Y., Yu, Z. et al. (2022). Reprocessable and recyclable polymer network electrolytes via incorporation of dynamic covalent bonds. *Chemistry of Materials* 34 (5): 2393–2399.

39 Wan, B., Liu, H., Wang, J. et al. (2025). Self-healing and recyclable polyimide hybrid film with linear and crosslinked bridging structure. *Composites Part B: Engineering* 306: 112769.

40 Wan, B., Zheng, M.S., Yang, X. et al. (2022). Recyclability and self-healing of dynamic cross-linked polyimide with mechanical/electrical damage. *Energy & Environmental Materials* 6 (1): e12427.

41 Faye, I., Decostanzi, M., Ecochard, Y., and Caillol, S. (2017). Eugenol bio-based epoxy thermosets: from cloves to applied materials. *Green Chemistry* 19 (21): 5236–5242.

42 Shirazi, E. and Jadid, S. (2019). Autonomous self-healing in smart distribution grids using agent systems. *IEEE Transactions on Industrial Informatics* 15 (12): 6291–6301.

43 Liu, H., Wan, B., Wang, J. et al. (2024). Dual regulation of flexible chain segments and dynamic covalent bonds for a self-healable copolyimide. *Journal of Materials Chemistry A* 12 (32): 21190–21199.

44 Ma, S., Wei, J., Jia, Z. et al. (2019). Readily recyclable, high-performance thermosetting materials based on a lignin-derived spiro diacetal trigger. *Journal of Materials Chemistry A* 7 (3): 1233–1243.

45 Wan, B., Dong, X., Yang, X. et al. (2022). High strength, stable and self-healing copolyimide for defects induced by mechanical and electrical damages. *Journal of Materials Chemistry C* 10 (31): 11307–11315.

46 Chen, S., Ding, J., Zhao, Y. et al. (2024). Design of robust and recyclable covalent elastomeric networks through dynamic enamine–urea bonds. *Macromolecules* 57 (8): 8813–8825.

47 Yin, H., Chen, Y., and Wang, X. (2022). Deep learning-based density functionals empower AI for materials. *Matter* 5 (8): 2452–2455.

48 Baylis, C. and Cruickshank, C.A. (2023). Review of bio-based phase change materials as passive thermal storage in buildings. *Renewable and Sustainable Energy Reviews* 186: 113690.

49 Zhou, Y., Yuan, C., Li, C. et al. (2019). Temperature dependent electrical properties of thermoplastic polypropylene nanocomposites for HVDC cable insulation. *IEEE Transactions on Dielectrics and Electrical Insulation* 26 (5): 1596–1604.

50 Liu, D.-F., Dong, W.-Z., Wang, X.-J. et al. (2025). Generating high-temperature polymer dielectrics with a graph convolution and reinforcement learning framework. *The Journal of Physical Chemistry C.* 129 (22): 10020–10026.

51 Liu, Y.X., Jia, M., Zhang, Y.X. et al. (2025). Repower: an LLM-driven autonomous platform for power system data-guided research. *Patterns* 6 (4): 101211.

Index

Next-Generation Electrical Insulating Materials: Low-Carbon, Eco-Friendly, and Sustainable Solutions, First Edition. Edited by Jun-Wei Zha.

e

g

h

i

j

k

l

m